*When Geologists Were Historians*

When Geologists Were Historians, 1665–1750

RHODA RAPPAPORT

Cornell University Press

ITHACA AND LONDON

Cornell University Press gratefully acknowledges a subvention from Vassar College, which aided in publication of this book.

First published 1997 by Cornell University Press.
Printed in the United States of America.

Library of Congress Cataloging-in-Publication Data

Rappaport, Rhoda.
When geologists were historians, 1665–1750 / Rhoda Rappaport.
p.   cm.
Includes bibliographical references and index.
ISBN 0-8014-3386-X (alk. paper)
1. Geology—Europe—History—17th century.   2. Geology—Europe—
History—18th century.   3. Communication in geology.   I. Title.
QE13.E85R36   1997
551'.094'09032—dc21   97-24040

 IN MEMORY OF HENRY GUERLAC

Contents

Acknowledgments   ix

Introduction   1

1   The Republic of Letters   7

2   Certainty and Probability   41

3   Natural and Civil History   83

4   Fossil Questions   105

5   Diluvialism, For and Against   136

6   Alternatives to Diluvialism: Some Ingredients   173

7   Alternatives to Diluvialism: Some Syntheses   200

8   Buffon and the Rejection of History   235

Abbreviations   263

Bibliography   265

Index   302

Acknowledgments

In those optimistic days when I thought that one book, on the subject as I conceived it, could deal with the whole of the eighteenth century (through 1815), I was the happy recipient of a grant from the National Science Foundation. More recently, Vassar College's generous policy of faculty leaves has facilitated the completion of this book. To the knowledgeable and accommodating staff of the College Library, especially the Reference Department and the Interlibrary Loan Office, I owe more than I can say. Colleagues who have provided intellectual and moral support include Evalyn A. Clark, Thomas J. McGlinchey, Donald J. Olsen, and Margaret R. Wright.

At Cornell University, the staffs of the Olin and Kroch Libraries have welcomed me with unfeigned pleasure, friendliness, and help. I only once glimpsed the riches of the History of Science Collections at the University of Oklahoma, but I am grateful to the late Duane Roller for trotting out the relevant treasures during my brief visit . In Europe, I have virtually lived at the Bibliothèque nationale, but I have also found indispensable the resources of two other Paris institutions: the Archives of the Academy of Sciences and the Bibliothèque centrale of the Museum of Natural History. To the *conservateurs* of these collections, Christiane Demeulenaere-Douyère and Yves Laissus, I am indebted for their aid, as I am also to Maryse Schmidt-Surdez at the Bibliothèque publique et universitaire in Neuchâtel.

Encouragement as well as substantive help have been forthcoming during years of intermittent conversations with Roger Hahn, the late Jacques Roger, Martin Rudwick, Alice Stroup, and Kenneth Taylor. Irwin Primer has been an unfailing source of comments, questions, and bibliographical suggestions; and Blossom Primer has struggled helpfully with some of the bastard Latin of the decades around 1700.

To Jeanne Casañas I owe the debt of a computer-moron to a computer-expert: she prepared the disk, working her way through a combination of typescript and scribbles. In addition to doing so with skill and intelligence, her calm competence made bearable the frantic life of this author. At a more recent stage in the proceedings, two *fonctionnaires* of the Cornell University Press, Roger Haydon and Teresa Jesionowski, have more than earned my gratitude.

RHODA RAPPAPORT

Poughkeepsie, New York

 *When Geologists Were Historians*

Introduction

Intellectuals living in the decades around 1700 often considered themselves as belonging to an international "Republic of Letters," characterized by the search for truth and the frank communication with colleagues across the boundaries of religious, social, political, and linguistic differences. Peculiar to that Republic, although not a requisite for membership, was the willingness and ability of members to deal with a range of subjects—from ancient history to mathematics, from epistemology to anatomy—that inevitably startles the modern specialist.

In respecting the absence of boundaries of the Republic, this book seeks to put early geology into a number of contexts that scholars usually investigate separately. One such context is the practice of that communication considered essential to the Republic and facilitated by the establishment in 1665 of two pioneering periodicals: the *Journal des savants* (Paris) and the *Philosophical Transactions* of the Royal Society of London. The role of these learned journals and the many others that followed is examined in Chapter 1, as is another vehicle of communication: the learned society or academy. The several scientific societies that proliferated in the later seventeenth and early eighteenth centuries have been subjected to fine-grained analyses by many scholars, so that structures, functions, and local conditions have become increasingly familiar to modern readers. Here these groups are treated primarily as centers of correspondence, the members striving to keep abreast of work by foreign colleagues and by nonmembers sending their writings to the formal organizations. After this preliminary discussion, it should come as no surprise in later chapters to find Swiss naturalist Johann Jakob Scheuchzer transmitting news of one of his discoveries to four learned journals or to encounter Luigi Ferdinando Marsili requesting judgment of his projects from the Royal Society of London and the Royal Academy of Sciences in Paris.

Other contexts especially pertinent to early geology are epistemology in general and the study of history in particular. As a way of illuminating these connections, it is instructive to consider the modern tension between the mathematical sciences on the one hand and the experimental and historical sciences on the other. On the whole, the former are thought to provide the model to which the others should aspire; if not quite in rebuttal, but certainly in defense, various writers have eloquently pointed out the value and legitimacy of the historical sciences in particular (historical geology, paleontology, evolutionary biology).[1] In the period discussed here, 1665–1750, such tensions came into existence as part of a larger concern about what constituted "knowledge." The mathematical sciences were being accorded unprecedented admiration because, it was argued, only mathematics could achieve certainty. As an inevitable corollary, other disciplines were judged to be merely factual and descriptive, subject to the vagaries and biases of human interpreters, and thus only probabilistic in their conclusions. Indeed, peculiar to the seventeenth century were assaults on the study of history, considered to be hopelessly unreliable, and historians during those decades had to find ways to defend the value of such scholarly pursuits. Under these circumstances, geologists faced acute problems in seeking to reconstruct the earth's past. If we are now accustomed to the existence of historical sciences, early philosophers were not, so that even to contemplate *a science of the past* seemed to be a contradiction in terms.

Chapter 2 deals with the epistemological hierarchy as this was developed in the last decades of the seventeenth century. Because the Bible was one of the premier sources for knowledge of ancient history, part of this chapter discusses the degree to which methods of historical scholarship were deemed applicable to the sacred text. That early geologists considered human (or "civil") history to be a crucial adjunct to the study of nature forms the subject of Chapter 3; included here is an examination of the kinds of history thought to be most relevant and reliable, and the uses of history in supplying proofs of the way nature works.

---

[1] For example, Simpson 1963, chap. 7, and Gould 1989, 277–81. See also Mott Greene 1985, 97, 115; Guntau 1989; Laudan 1982.

Readers impatient with Chapters 1 and 2, where early geologists are mentioned but not discussed in any detail, will find the start of such analysis in Chapter 3, where epistemological issues in particular are shown to have entered into specific geological texts. Chapters 4–7 deal with subjects familiar in traditional histories of geology: fossils, diluvialism, volcanoes and the earth's heat, the timescale, and so on. For these issues, however, traditional approaches require reconsideration. Diluvial geology, for example, is here redefined, its historical basis analyzed, and the arguments it engendered recounted in some detail. Similarly, the timescale requires analysis not primarily as a problem of measurement, but as one of perception and perspective: Why did so many naturalists feel comfortable—the usual interpretation being that they felt constrained—with what any modern would judge to be a very short timescale?

In treating these and other topics as matters of international debate, it seems advisable to comment on how I have judged particular writings, issues, and individuals to be "international" in stature. One obvious recourse is to notice in any given text whom and what its author cites; another is the examination of correspondence; and a third is consultation of the learned journals to see which publications were reviewed. In the latter instance, the sheer number of journals defeats any effort to be exhaustive, and I cannot claim to have used more than a fraction of this huge literature. At times, nonetheless, I have chosen to discuss or omit particular authors depending on whether, in conjunction with other evidence, their writings received attention in some of the more prominent journals. ("Prominent" here generally means francophone, the French language increasingly having replaced Latin, as one may see from the number of French-language journals published in Germany, the Netherlands, and other places beyond the borders of France.) This accounts, for example, for the attention I have given to seemingly obscure persons such as Johann Jacob Baier and the neglect of others, chiefly German and Swedish, who had a limited press outside their own countries before 1750. In other words, internationalism as an ideal of the Republic of Letters had some curious limits at that time, and those limits are reflected in this book.

In ending with Buffon's theory of the earth (1749) and the immediate aftermath of its publication, I have been influenced by the nature of the theory itself and by changes that began to occur in the 1750s. My emphasis on fossils and diluvialism signals the chief geological problem of earlier decades: the dynamics of sedimentation. In this area, Buffon's theory marks the culmination and selective synthesis of earlier studies. At the same time, significantly missing from his theory was that concern for human history so common among his predecessors. Whether Buffon persuaded his readers to follow in his own path is a major desideratum for future research; here he is regarded as consciously trying to bring to an end one tradition in geology, and the reactions of readers are only briefly sketched.

To come to a halt shortly after 1750 also involves an interpretation of the next decades and the preceding ones. As readers will notice, my international cast of characters consists chiefly of writers British, French, and Italian, with the occasional Swiss, German, and Swede. This seemingly lopsided focus was dictated by two considerations. For one, I found remarkably little and oddly selective reference in major book-review journals to German and Swedish publications, and only a little more to Swiss. If this phenomenon might be dismissed as journalistic bias, my impression was confirmed by *the virtual lack of chemistry* in geological publications emanating from Britain, France, and Italy before the 1750s. This factor, taken in conjunction with the relative silences of the book-review journals, suggested to me that two cultural zones existed in Europe: one was the British-French-Italian complex, the other the Swedish-German, with the Swiss perhaps belonging more to the former than to the latter. To put this cultural division more clearly as it is related to the history of geology, I may say that I was initially baffled by Rachel Laudan's fine book *From Mineralogy to Geology* (1987), because the early chapters lack so many of the familiar British, French, and Italian names. Laudan took her point of departure from the German and Swedish chemists whose writings received little attention in the other cultural zone. In Laudan's study, chemistry became an integral part of reconstruction of the earth's past; in my own research, I have found that chemistry played at best a minimal role until after 1750.

The absence of chemistry from Buffon's theory of the earth may be taken to represent the end of this geological tradition in one European cultural zone. Or, to put it more precisely, Buffon's readers eventually began to see that his emphasis on mechanical processes needed to be supplemented by attention to chemical analysis. At least, as I have argued elsewhere, this was true in France, and Ezio Vaccari's study of Giovanni Arduino offers evidence that the same may be true for Italy.[2] Eventually, during the latter half of the eighteenth century, the mechanical and chemical traditions existed in combination in the writings of most European geologists.[3]

A word must be said about the vocabulary used throughout this book. In the decades around 1700, terms such as geology, fossil, science, and history did not possess their modern meanings. Even "naturalist" could signify anyone who studied nature—from the structure of the cosmos to

[2] Rappaport 1994; Vaccari 1993, pt. 2.
[3] Rudwick 1990, 388–89.

the taxonomy of plants—or anyone devoted to wholly naturalistic explanations of the mysteries of the physical world. Rather than strangle on circumlocutions, I use all these terms in their modern meanings, unless the text explicitly indicates otherwise. As an exception, "philosopher" will now and then be used in its traditional meaning, namely, to designate a person engaged in the study of subjects amenable to human reason; the term thus applies equally to Robert Boyle and to a medievalist such as Jean Mabillon.

I have avoided terms all too familiar to modern readers: catastrophism, uniformitarianism, actualism, neptunism, vulcanism, and plutonism. These labels have their own history, and they now possess connotations that impede historical analysis. If, for example, one calls Anton Lazzaro Moro a uniformitarian in principle but a catastrophist in practice, the words do not tell us that he assumed that nature works in uniform ways (a commonplace) and that a main natural mechanism is the volcanic eruption (a most uncommon assumption). Further confusion results if both Moro and Thomas Burnet are dubbed catastrophists, since Burnet used a single, worldwide cataclysm, the Flood, whereas Moro's eruptions were all local events occurring at various times. Catastrophism has also come to signify the use of inexplicable and even miraculous causes. Both Burnet and Moro, however, were resolutely naturalistic writers, opposed to the very method and viewpoint sometimes said to be typical of catastrophists. These limited examples should suggest why I have chosen to abandon misleading "-isms." The sole exception is diluvialism; as used here (and clarified in my text), the word does not signify all theories incorporating the Flood, but only those in which the Flood played the most important role in shaping the earth's crust. Burnet was a diluvialist; Nicolaus Steno was not.[4]

Other aspects of my vocabulary require explanation of a different kind. Hélène Metzger once delivered a lecture on whether "the historian of science should make himself the contemporary" of those scientists he studies. In defending the affirmative, Metzger argued that, unless we can read older works the way their first audience did, we run the risk of misunderstanding both the works themselves and the evolution of the sciences.[5] To be sure, no historian—of science or any other subject—can

[4] Unrecognized confusion about -isms can be found in Bynum 1981, in articles such as "catastrophism" and "diluvialism." A. V. Carozzi 1971 examines whether Cuvier was a "catastrophist" in the usual sense of that term. Modern geologists' reactions to the use of "catastrophes" are indicated in Baker 1978. See also Rappaport 1982, esp. 31–38, on "revolutions," and Ellenberger 1989.

[5] "L'historien des sciences doit-il se faire le contemporain des savants dont il parle?" (1933), in Metzger 1987, 9–21. Especially thoughtful about the use of modern terminology in another field of history: Cobban 1964.

discuss the past wholly in the terms used in the past, but nor does modern terminology convey adequately the levels of knowledge possessed by our ancestors. It may help the modern reader to know that mammoth and mastodon had not been distinguished from one another or from the elephant in the 1720s, but more important to Sir Hans Sloane and his contemporaries was whether these fossil remains belonged to human giants, or elephants, or a living creature that superstitious Siberians called the "mammuth." Similarly, Réaumur would have found incomprehensible the label "Tertiary" applied to the faluns of Touraine, for his problem in 1720 was to envisage the physical process that had produced such deposits; to use "Tertiary" implies a classification and historical sequence that did not then exist. On the whole, then, I have opted for a non-modern vocabulary, with minimal clarification when necessary, and with references in the notes to studies where the interested reader can find further information.

CHAPTER 1

The Republic of Letters

W hen Denis de Sallo launched the *Journal des savants* early in 1665, he addressed himself to an international community that he called "the republic of letters." If he invented neither the phrase nor the concept, Sallo did recognize that intellectuals needed an organized way of exchanging information and keeping abreast of "the chief works published in Europe." The title of the journal indicated that this would be no mere "gazette," appealing to the fashionable world; rather, it would be a sober compendium dealing with and addressed to the world of learning. In the event, Sallo's journal and its many successors and imitators paid especial attention to two large areas of learning: scholarship of all kinds and the sciences. On the whole, such categories virtually excluded both imaginative literature and works of piety.[1]

Earlier in the century, the extensive correspondence of Marin Mersenne (1588–1648) had served to alert philosophers to the work of distant col-

---

[1] Quotations are from "L'Imprimeur au lecteur," preceding the first number of *JS*, 2 January 1665. (All references to *JS* are to the original Paris edition, unless Amsterdam 12mo is specified in the notes.) The standard study by Morgan 1928 must now be supplemented by the research of Vittu, in Bots 1988, 167–75, Moureau 1988, 303–31, and Sgard 1991, 645–54. Nonlearned journalism has received less attention, but see Mattauch 1968. Literary scholars tend to regard the shortlived learned journals in England as a prelude to the work of Addison and Steele; the pioneering study of this kind is by Graham 1926.

leagues. At a later date, the even more massive correspondence of Antonio Magliabechi (1633–1714), librarian to the Grand Duke of Tuscany, served a comparable function, as Magliabechi both distributed books sent to him in multiple copies and acquired books for correspondents who requested them.[2] Indeed, as historians have often noted, the private correspondence of Henry Oldenburg (ca. 1618–77) would eventually provide the basis for material published in the *Philosophical Transactions* of the Royal Society of London.

After 1665, private correspondence continued to play a vital role in the dissemination of news, as the editors of the new journals cultivated relations with publishers, booksellers, and assorted individuals capable of supplying news about books published in major European cities. One such supplier of news, Pierre Des Maizeaux (1673–1745), a Huguenot who settled in London in 1699, became a correspondent for no fewer than eight journals published in France and Holland. In Florence, Magliabechi's role expanded, as he continued to provide books for individuals and for three learned journals in Leipzig, Parma, and Rome.[3] Furthermore, while journals alerted readers to new books, correspondence remained the necessary means of acquiring them. Thus, Giuseppe Valletta in Naples subscribed to several journals and cultivated epistolary friendships in order to acquire the books for his rich library.[4]

Denis de Sallo did not limit his journal to reviews of new books, but promised also to include eulogies of significant persons recently deceased, the decisions of "tribunals" ranging from courts to universities, and information about scientific experiments, observations, and inventions. Although original articles of these kinds did appear from time to time, the *Journal des savants* soon narrowed its focus to concentrate on book reviews. From Henry Oldenburg's perspective, the *Journal,* however admirable, remained too general in its coverage, and his *Philosophical Transactions* would emphasize the sciences. Sallo himself, already eager to receive English news, welcomed the first number of the new journal, but it was

[2] Magliabechi is now being studied and some of his letters published. For an example of his activities before the existence of book-review journals, see letters from Borelli, 1660–64, in Quondam and Rak 1978. Other valuable materials and studies include Doni Garfagnini 1977, 1981; Mirto 1984; and Waquet 1982.

[3] For Des Maizeaux, see Almagor 1989. On Magliabechi, see also Dooley 1991, 19.

[4] The basic study is Comparato 1970, to be supplemented by Mirto 1984, 77–79, and Dibon 1984. For editors and their correspondence, good examples are provided by the LeClerc letters 1991; Bots and Lieshout 1984; and Labrousse in Dibon 1959. For the relationship between letters and journals, see Waquet 1983 and esp. Goldgar 1995, chaps. 1–2.

Sallo's formula, not Oldenburg's, that would prove more attractive to subsequent editors and publishers.[5]

Of the several ways one might measure the success of Sallo's journal, private correspondence supplies some clues, as future journalist Pierre Bayle was soon receiving laudatory reports from his brother, and Gottfried Wilhelm Leibniz soon proposed that a comparable German journal be established to report on the novelties available at the great Frankfurt book-fairs.[6] Enterprising publishers began, as early as 1665, to issue unlicensed reprints of Sallo's journal, and imitators in the next decades would use the equivalent of "savants" in their titles: the *Giornale de' letterati* (Rome, 1668, and others of the same title in various Italian cities), the *Acta eruditorum* (Leipzig, 1682), the *Histoire des ouvrages des savans* (Rotterdam, 1087), and *The History of the Works of the Learned* (London, 1699). Eventually, the Paris original, issued in handsome quartos, was so often pirated in smaller, cheaper formats that the Paris edition began to appear in two formats: the quartos and a duodecimo. Sallo's periodical also became a regular source of articles translated for republication in other learned journals.[7]

The few journals just mentioned scarcely begin to suggest the magnitude of this explosion in learned journalism. In 1684 Pierre Bayle summarized what had been occurring:

> The plan to inform the public, using a kind of Journal, about interesting happenings in the Republic of letters was found so convenient and pleasant that, as soon as Monsieur Sallo . . . began to do so . . . several countries showed their enthusiasm either by translating the journal he was having printed each week or by publishing something similar. This imitation has grown more and more since that time.

Bayle added that, because the Netherlands had not yet produced such a journal, he was launching his own *Nouvelles de la république des lettres*

[5] Moray to Huygens, 3 February 1665 (O. S.), in Huygens 1888, 5:234. *JS* (30 March 1665), 156.

[6] Letters of 1667 and 1668, respectively, quoted by Labrousse in Dibon 1959, 99–100, and by Laeven in Bots 1988, 239–40.

[7] See the several works by Vittu, cited above (n. 1). One of the studies that attempts to identify the sources of articles in a particular journal is Gardair 1984. The shortlived *Weekly Memorials for the Ingenious* (London, 1682) also relied regularly on the *Journal des savants;* there were two London journals of this name, and the one I have examined is described in Crane and Kaye 1927, no. 935. The importance of francophone journals is stressed by Eisenstein 1992, chap. 1; the indispensable work is now Sgard 1991, but Hatin 1866 is still useful and Goldgar's analysis superb. For the first decades of the *Acta eruditorum,* see Laeven 1986.

(1684).[8] In the next years there would appear in Amsterdam alone a succession of journals with the titles *Bibliothèque universelle et historique* (1686), *Bibliothèque choisie* (1703), *Bibliothèque ancienne et moderne* (1714), and *Bibliothèque raisonnée* (1728)—the first three edited by Jean LeClerc. Eventually, editors began to specialize, devoting journals to national literatures that seemed of particular interest or that the "universal" journals allegedly neglected. After 1700, one thus encounters such titles as *Nova literaria helvetica* (1703), *Acta literaria sueciae* (1720), *Giornale de' letterati d'Italia* (1710), and another series of *Bibliothèques: angloise* (1717), *britannique* (1733), *germanique* (1720), and *italique* (1728). By 1718, the editors of *Europe savante* could estimate that some fifty journals had come and gone in various languages, and this figure did not include those still surviving.[9]

Thanks to the lively controversies aroused by Sallo, editors knew that the business of reviewing books could undermine public decorum. Authors might well be antagonized by reviews or by the prospect of having their works publicly judged, and other readers might expect certain books to be either censured or praised. Sallo's Paris successor and many other editors therefore proclaimed their impartiality or neutrality, sometimes warning readers that material found offensive was not by the editors themselves but was to be found in the books under review. Even neutrality, however, had its stated limits, as the editors of the *Acta eruditorum*, *Works of the Learned*, and the Jesuit *Mémoires de Trévoux* (1701) all declared that any books detrimental to established government and true religion would necessarily be criticized.[10] In fact, such limits were often breached, as reviewers rarely could or would refrain from passing judgment. The most explicit precedent had been set by Pierre Bayle, who insisted that he owed his readers a degree of intelligent guidance.

> We will maintain a reasonable medium between servile flattery and bold condemnation. If we sometimes judge a work, it will be without bias and spite, and in such fashion that we hope not to antagonize those concerned in our judgment; for . . . we do not claim to establish any presumption for

---

[8] Preface to *NRL*, March 1684, in Bayle 1727, 1:1.

[9] *Europe savante* (January 1718), v. A long discussion of existing journals can be found in Scipione Maffei's preface to *GLI* 1 (1710), 13–67. S. Barnes 1934, 257, has counted 330 journals in seven countries during the period 1665–1730.

[10] *JS*, "L'Imprimeur au lecteur," prefacing the first number for 1666. *Acta*, 1684, quoted in Prutz 1845, 280; also, Hensing in H. D. Fischer 1973, 36–37. *HWL*, preface to the first number, January 1699. *Trévoux* in Dumas 1936, 33; also, Rétat 1976, esp. 176–78. Goldgar 1995, 185, errs in saying *Trévoux* was alone in this policy.

or against authors: one would have to have a foolish vanity to claim such authority.[11]

Bayle's critical intelligence offered severe competition to the colorless and mediocre *Journal des savants* of the 1680s, and later journalists may have learned from this example. Or they learned by experience: the *Mémoires de Trévoux*, for example, after a decade of announced neutrality, found it necessary to inform readers that it would henceforth mingle some criticisms with its summaries, so as to offer readers a more effective guide to new books. In fact, as will be evident in a later chapter, *Trévoux* had been offering judgments from its earliest years.[12]

Common editorial professions of impartiality—usually expressed as the aim of providing "extracts" from new books—have given rise to the curious notion that these journals merely printed "abstracts," with little or no evaluation.[13] On the contrary, judgments ranged from the highly complimentary to the severely critical, with the middle ground occupied by the noncommittal or the single phrase or sentence that hints at evaluation. Furthermore, the choice of books to review occasionally suggests predilections by the editors, as when, in the long controversy over the origin of springs, the *Acta eruditorum* carried a not very favorable account of Antonio Vallisneri's treatise and ignored his chief opponent, whereas the *Journal des savants* paid attention to that opponent and ignored Vallisneri. In this instance, the bias of the Paris journal may perhaps be explained by the presence on the editorial board of Nicolas Andry who was conducting a fierce debate with Vallisneri on quite a different subject. As contemporaries knew, learned journals might also be influenced in a number of ways—by respected scholars, by members of their own staffs, by the State, and by editorial perceptions of public opinion. Just as "there is no innocent scholarship," there was scarcely any innocent (neutral) journalism.[14]

---

[11] Quoted in Hatin 1866, 33. For a defense of the functions of learned journals, see *BR* 1 (July–September 1728), v–xvii.

[12] Dumas 1936, 36–37; also, 33, 43–44, 49 for policies in matters of religious controversy. The vacillations of *Trévoux* on the subject of Noah's Flood will be discussed below, Chapter 5.

[13] Kronick 1976, chaps. 7–8 for taxonomy of journals. Comparable views were espoused by Graham 1926, 5, and are repeated by Forster 1990, 4–6.

[14] The quotation is from Rétat 1976, 187. Review of Vallisneri in *Acta* (November 1726), 488–95; review of Gualtieri in *JS* (June 1725), 376–87. The Andry-Vallisneri dispute concerned biological generation; see Roger 1993. For examples of editorial caution because of perceptions of public opinion, see Almagor 1989, 25 and n. 27, and 27–28. For an appeal to a scholar to influence a review, see Andreoli 1922, and for a surprising decision by a State *fonctionnaire*, see the comments in Woodbridge 1976, 338–39, on Richard Simon, abbé Bignon, and the *JS*.

From the evidence presented thus far, it is clear that the book-review journal created what some historians have called a "new cultural space," that is to say, a new realm of international communication.[15] This evidence, however, stems mainly from the nature and number of the journals themselves. When one looks for collateral information about circulation and readership, little has been discovered. Daniel Mornet's classic study of five hundred private libraries in France does reveal that one-fifth contained journals edited by Pierre Bayle and Jean LeClerc, with the *Journal des savants* running a close third. (The figure for Bayle is noteworthy, his journal having been officially banned from entering France. In practice, this meant that subscribers had to pay more to receive it through the mails.) Subscription figures have been found for one Italian journal, the *Novelle letterarie* (Florence, 1740), as has a list of cities to which another Italian journal was to be sent.[16] Valuable as such studies are, they still cannot tell us about who read the journals, about whether such readers were prompted to buy books, about the lending of books, or about the private libraries that welcomed visitors.

These questions may be unanswerable, but nor can one yet judge how adequately the journals covered the international world of learning. Some periodicals have been examined in order to determine the attention given to various fields of study: law, theology, travel, science, and so on. In a model of this kind, Jean Ehrard and Jacques Roger sampled the contents of the *Journal des savants* and the *Mémoires de Trévoux* during two periods, 1715–19 and 1750–54. Comparable analyses have been produced for the *Acta eruditorum*, Bayle's journal, and a few others. On the whole, most such studies are disappointing, chiefly because they employ the undiscriminating categories (like "theology and philosophy") used by the journals themselves. Even the best, by Ehrard and Roger, gives results that could have been predicted: both of the journals they examined remained remarkably cosmopolitan, and both eventually reduced the proportion of reviews of theological works and increased the number devoted to the sciences. Elsewhere, too, efforts to quantify generally confirm conclusions long known— for example, that Bayle concentrated on publications in French and Latin, with little attention to the sciences, whereas Jean LeClerc gave much space to English philosophy. If journals with editorial committees might be expected to show considerable range and balance in subject matter, this

---

[15] Ricuperati in Capra 1986, 79, and Gardair 1984, 11.

[16] Mornet 1910; Dooley 1982; Waquet and Waquet 1979. There exist worthwhile studies of Giovanni Lami, founder and editor of the Florentine journal, by Pellegrini 1940; Vaussard 1954; Rosa 1956.

actually depended on the composition of the committees. Thus, two Venetian journals carried quite a few articles on the sciences, thanks in both cases to the active participation of Antonio Vallisneri; on the other hand, the *Bibliothèque germanique,* edited by Protestant pastor-historians, emphasized theology and history.[17]

The extent of any journal's coverage depended also on its editor's access to new books. Even the fame of Bayle's journal apparently did not bring him a great influx of foreign books, and most of the works he reviewed were published in the Netherlands. (Dutch publishers, to be sure, had a vast clientele of foreign authors.) More fortunate were the *Acta eruditorum* and the *Giornale* of Benedetto Bacchini, as both received aid from the indefatigable Magliabechi in Florence. But the fragility of such arrangements may be illustrated in Bacchini's case. Despite the aid of Magliabechi, Bacchini's chief supplier was actually Gaudenzio Roberti, librarian to the Duke of Parma; when Roberti died in 1695, Bacchini's journal did not long survive.[18] Under such circumstances, it is hardly surprising that the *Mémoires de Trévoux,* able to rely on the international membership of the Society of Jesus, should have been in a good position to maintain its universal aims; even so, *Trévoux* suffered from a relative lack of English news, and its editors seem to have been erratic in their German and Italian coverage.

Another considerable difficulty hampered the internationally minded, namely, the need to find reviewers able to read English. (One could always find readers of Italian, and no one before about 1750 saw any necessity to read German.) Denis de Sallo himself faced that problem in announcing the first number of the *Philosophical Transactions,* but assured his readers that he had found "an English translator." His successor, abbé Jean Gallois, was a linguist, and it is possible that he himself produced the long and appreciative review of Robert Hooke's *Micrographia* published in the *Journal* in 1666.[19] In later years, the *Journal* was one of several periodicals to receive

---

[17] Ehrard and Roger in Bollème 1965, 1:33–59. The urge to quantify can also be misplaced, as in the study of a dozen libraries by F. Piva in Postigliola 1988; contrast H.-J. Martin 1969 and Viglio in Anelli 1986. The Venetian journals were *La Galleria di Minerva* (1696) and *Giornale de' letterati d'Italia* (1710). My judgment of the German journal is based on my scanning of the content for the first few years of its existence; the different impression conveyed by Kämmerer in Sgard 1991, 188–89, may well reflect changes in the journal after the first dozen or so years.

[18] Other pressures on Bacchini are indicated by Momigliano in *DBI,* s.v. Bacchini.

[19] Sallo's welcome of *PT* is in *JS* (30 March 1665) 156, the last number of the *JS* before its license was suspended. The enthusiastic review of Hooke is in *JS* (20 December 1666), 491–501, and it even includes reproductions of some of Hooke's illustrations. The sale catalogue, Seneuze 1710, indicates that Gallois owned many volumes in English. Valuable studies of Anglo-French relations include Ascoli 1930; Bonno 1943, 1948; Frantz 1934; Hall and Hall 1976; Jacquot 1953, 1954; LaHarpe 1941.

English news from Pierre Des Maizeaux in London. If Jean LeClerc in Amsterdam could read English, Francesco Nazari in Rome could not, but he decided to learn the language so that he might publish in his *Giornale* some "extracts" from the *Philosophical Transactions*. By 1718, the editors of *Europe savante* allowed that more Continentals knew English than had been true in past years, but that this was far from enough, because the English, living "in the bosom of liberty," could conduct their search for truth in unhampered fashion. The editors thus welcomed the new *Bibliothèque angloise* as performing an essential function.[20]

For the communication of scientific news in particular, no book-review journal gave this area of learning undue prominence. (The *Philosophical Transactions* did, of course, but it was not primarily a book-review journal.) But it would be an error to conclude, as David Kronick has, that scientific reviews and the occasional original article were merely buried in such periodicals. Before about 1750, this error is taxonomic, based on the assumption that scientists were specialists addressing other specialists. Henry Oldenburg disagreed. As he explained to Robert Boyle, if there were more journals like Nazari's, modeled on the *Journal des savants*, one would have "a good generall Intelligence of all ye Learned Trade, and its progress." To scientists like J. J. Scheuchzer in Zurich and Vallisneri in Padua, promoting the learned trade induced both to take active roles as editors of more than one learned journal. It has even been said of the *Mémoires de Trévoux* that this long-lived Jesuit publication did more to educate the public in erudite subjects, including the sciences, than the few more specialized periodicals of the eighteenth century.[21]

Some journals reported regularly on the publications of scientific societies and academies. The *Bibliothèque germanique*, for example, gave accounts of the volumes issuing from the Berlin Academy, and the Paris Academy was covered by both the *Journal des savants* and *Works of the Learned*. In fact, Sallo's successors at the *Journal* made heroic efforts to deal with an astonishing number of academies, with uneven results. Such reports were perforce selective, the reviewers being unable to do justice to all the contents of substantial academic volumes; here, be it said, is another

---

[20] *Europe savante* (January 1718), x–xi, and review of the *Bibliothèque angloise* in February 1718, esp. 317–20. For Jean LeClerc and John Locke, see the lucid analysis by Colie in Bromley 1960, 111–29.

[21] Oldenburg to Boyle, 24 March 1667/68, in Oldenburg 1965, 4:275. The judgment on *Trévoux* is by Daumas 1957, 107. Contrast Kronick 1976, 254–55, with the sounder views of Dooley 1991, 111–27. George 1952, 305–6, analyzes the proportion of original articles to book-reviews in *PT*, from a ratio of about 3:1 before 1700 to about 21:1 in the next decades.

area where reviewers implicitly passed judgment, as they chose those articles they thought worthy of detailed summary. But not all academies published such tomes, or did so only erratically rather than annually. Enterprising journalists could on occasion supply this lack, as when the *Journal des savants* carried reports on activities in the Paris Academy for part of that period (1666–99) before the Academy itself began to produce its annual *Histoire et Mémoires*. Although the Société royale de Montpellier did not manage until 1766 to publish a volume of the writings of its members dating from the period 1706–30, accounts of some of its public meetings could be found in the *Journal des savants* (then translated in *Works of the Learned*) and the *Mémoires de Trévoux*. And for the short-lived academies founded in Bologna by Anton Felice Marsili, virtually the only record of activities survives in the *Giornale* of Benedetto Bacchini.[22]

Regrettably, we know little about how individual readers of these journals used them. In a few cases, library catalogues—ranging from the riches of Giuseppe Valletta in Naples and Etienne-François Geoffroy in Paris to the more limited collections of some North American colonists—reveal sometimes remarkably large assortments of book-review journals.[23] But did the owners of such journals then buy books, or did they content themselves with merely reading reviews? The latter is exactly what some writers of books feared, foreseeing that the reader of reviews would consider himself to be well informed when he really had acquired only a smattering of knowledge. However understandable these fears, it must be said that journalism in the decades around 1700 was of a caliber that inspires our retrospective admiration. The editors achieved a degree of "neutrality" not by omitting criticism, but by summarizing with care the contents of books under review. Although it would be impossible to analyze a substantial number of the hundreds of journals, comment on the quality of a few will suggest their value not only to past readers but also to modern historians.

Chief among them was the *Journal des savants:* a model for many others, pillaged by its competitors, and reprinted by enterprising publishers

[22] For the strenuous efforts of the *JS*, one need only look at the detailed *Table générale* 1753, 1:181–236, for references to French provincial and non-French scientific societies. The notable absence of Swedish academies in the *JS* will be pertinent below, in Chapter 7. Accounts of the Montpellier meetings paid attention to a memoir by Jean Astruc that will surface again in later chapters of this study. For what is known about A. F. Marsili's two academies, see Cavazza 1981, 895–98.

[23] For Valletta, see above, n. 4. Geoffroy 1731, beginning with item no. 1824 of this inventory. For colonial libraries, see Fiering 1976. A recent study, by Carozzi and Bouvier 1994, indicates that Horace-Bénédict de Saussure would later use journals systematically to build his rich library.

beyond French borders. A journal of astonishing longevity, ensured by State protection, its quality varied especially in its early decades, with a low point reached perhaps in the years 1674–86. On the whole, however, the standard of reporting remained high, as editors had developed a pattern of reviewing that conveyed to readers the organization, content, and thesis (if any) of the books discussed—sometimes with detailed analysis of a chapter considered of importance. One could certainly come away from the *Journal* knowing if one wished to pursue further a particular book. After Sallo's initial boldness in roundly condemning an occasional book or praising one deemed not quite orthodox, criticism in the *Journal* was usually restrained, even confined to a mere sentence or phrase, so that evaluation did not become a substitute for careful reporting. Compliments required less restraint, but the journalists did not engage in mere flattery as they conscientiously reported the grounds for their approval. In 1763, Edward Gibbon offered his judgment of the *Journal:* "The father of the rest, it is still their superior."[24]

With fewer resources and more handicaps, Benedetto Bacchini succeeded in producing one of the most intelligent journals of this period. His *Giornale de' letterati*—published for a time in Parma (1686–90), then in Modena (1692–97)—had by then a good many precedents it might have followed, and it seems to have had no serious rival in Italy; the two Roman journals of the same title had expired before 1686. In small cities like Parma and Modena, Bacchini had the advantage of being able to acquire patronage and the disadvantage of finding it hard to hide from his critics. Above all, being removed from an Italian center of the book trade and the publishing industry—Venice led in both respects—he needed access to new books, and this he found by cultivating two ducal librarians, Antonio Magliabechi and Gaudenzio Roberti.[25]

A Benedictine, Bacchini was a disciple of the French Maurists and in that sense an admirer of the new critical scholarship of his day. When Jean Mabillon published his *Traité des études monastiques* in 1691, Bacchini reviewed it at length, recommending that it be read by all members of religious orders.[26] In addition to reviewing new works on such controversial matters as biblical chronology, Bacchini's appreciation of nontraditional

---

[24] Gibbon is quoted in Graham 1926, 2. For the ups and downs of the *JS,* see Morgan 1928, chap. 3. Sallo really was bold, and not merely a target of offended readers; his review of Charles Patin's little book on numismatics (discussed below, Chapter 2) damned the volume with only the faintest of praise.

[25] The best studies are by Ricuperati in Capra 1986, 89–106, and Mamiani in Cremante and Tega 1984, 373–79.

[26] Bacchini 1692, 319–32.

books appears also in his range, as he discussed Cartesian and anti-Cartesian publications, the microscopic observations of Leeuwenhoek, Tentzel's influential pamphlet on fossil bones, and a new Italian book on earthquakes. Typically, his reviews begin with a helpful paragraph, as informative today as it must then have been, setting the particular volume into the context of contemporary research or debates, and they continue with a systematic indication of main topics and evidence. If one compares his review of Bernardino Ramazzini's *De fontium mutinensium* (1691) with the analysis of the same work in the *Journal des savants*, Bacchini provides the more even-handed analysis of the whole book, whereas the Paris reviewer concentrates on one chapter he considered important.[27]

To appreciate Bacchini's courage in his presentations of "modern" learning, it pays to recall that one of his contemporaries, Marcello Malpighi, encountered much opposition to his own modernity from colleagues at the University of Bologna. Among Mabillon's several battles in France, one was waged against the head of the Trappist order who denied the value of books in the monastic life. Bacchini's courage bore unusual fruit in that one of his Modenese disciples was Ludovico Antonio Muratori, the great medievalist; in Muratori's generation, moderns such as Vallisneri, Scipione Maffei, and Apostolo Zeno considered themselves Bacchini's disciples not so much in the realm of historical scholarship as in modernity. This relationship helps explain the quality of the *Giornale de' letterati d'Italia* (1710), founded and edited by Vallisneri, Maffei, and Zeno. In the next generation, the *Raccolta* (1728) of Angelo Calogerà inherited, so to speak, this fine tradition. The editor received not only encouragement from Vallisneri but also a surprising amount of material that appeared in the pages of this long-lived journal during the last years of Vallisneri's life and for a year or so thereafter.[28]

As a third and last example of the quality of the book-review journal, *Europe savante*, published in The Hague, merits some attention not for its range or popularity but for its bold espousal of editorial criticism. Pierre Bayle having explicitly charted this course, some of his fellow Huguenots thought him too impartial in religious matters. More generally, so many of the periodicals emanating from the Netherlands were edited by Huguenots that one might suspect them of bias. Similarly, the Jesuit *Mémoires de Trévoux* could be viewed askance because of its source, even if *Trévoux*, like other

---

[27] Bacchini 1692, 31–42; *JS* (29 March 1694), 150–54.

[28] In addition to works cited above, n. 25, see the long eulogy of Bacchini in *GLI* 35 (1724), 340–73. The contributions of Vallisneri to Calogerà's *Raccolta* are visible in the first volumes; the chief historian of Calogerà, DeMichelis, has written the excellent article on him in *DBI*.

periodicals, vacillated in a way that modern historians have detected more readily than did suspicious eighteenth-century readers. And then there were the seemingly timid editors, willing to engage in a word or two of evaluation now and then. *Europe savante* claimed to be cutting through to the heart of criticism: it would evaluate evidence. Even in religious matters, it would not espouse any cause, but would deal with the logical structure of argument, the proper citation of sources, and "the spirit of moderation" in any book.[29]

Europe savante lasted only two years (1718–20), and there is no evidence that anyone considered it an important journal. Its editors included three brothers, two of whom went on to impress the scholarly world as critical historians: Lévesque de Burigny in his *Histoire de la philosophie payenne* (1724) and Lévesque de Pouilly for his several controversial memoirs presented in the 1720s to the Royal Academy of Inscriptions in Paris. In its first number, the journal described the virtues of having assembled an editorial committee such that each new book could be handled by an expert; this would ensure impartiality as well as informed judgment, and anyone who objected to the judgments rendered was invited to communicate with the journal.[30]

For its grand opening, *Europe savante* devoted more than a hundred pages (small format) to an uncritical account of an edition of the famous academic eulogies pronounced by Fontenelle, about which the editors remarked that it was hard to know "where one should applaud most."[31] But the sciences did not much interest the editors, for subsequent numbers are remarkable chiefly for discussions of historical scholarship where at least two of the editors did have expert knowledge. In reviewing the first volume of papers produced by the Academy of Inscriptions in Paris, the journal reported disagreement among scholars about the date at which Pharamond, a legendary Frankish chief, began his reign; in trenchant paragraphs, the reviewer identified questions the academicians had failed to ask, notably whether there was sufficient evidence that such a person as Pharamond had ever existed.[32] Bold, too, were the journal's comments on an effort to prove "geometrically" the truth of Christianity and on volumes

---

[29] *Europe savante* (January 1718), esp. xi–xiv. Studies of this journal are by Varloot in *La Régence* 1970, 131–41, and Belozubov 1968.

[30] *Europe savante*, as cited above, n. 29. Lévesque de Pouilly's confrontation with the Academy of Inscriptions will be discussed below, Chapter 2.

[31] *Europe savante* (January 1718), 105. The review of Fontenelle occupied pages 1–116. A faint aroma of criticism would later be detected by Camusat 1734, 2:171–72; Camusat's defense of Fontenelle will be discussed below at n. 65.

[32] *Europe savante* (July 1718), 20–22.

purporting to offer rules of historical scholarship. In the former case, author Jean Denyse had dealt reasonably well with his proofs of "matters of fact," but he had had too much confidence in the testimony of early Christians who, like persons in all times and places, might well have preserved some "false facts" that happened not to conflict with true ones. Orthodoxy in belief was no shield against credulity.[33] As for Honoré de Sainte-Marie, the journal declared that he had little true understanding of scholarly methods.

> The [critical] Rules he sets up in his first Volume are either familiar to everyone, or dubious, or false. Using these same Rules, in his second Volume he considers true some Facts which are so obviously false that one cannot accept them without being reputed a scholar of little judgment.

One of Honoré's aims had been to defend the validity of traditions transmitted orally and much later committed to writing. As the reviewer remarked, however, "it is almost impossible for an unwritten Fact to be preserved, without alteration, in human memory during several centuries." The same point, later presented by Lévesque de Pouilly to a meeting of the Academy of Inscriptions, would shock members of that scholarly audience.[34]

From this brief and limited survey, it should be clear that learned journalism often could and did attain high levels of intelligent reporting and criticism. Study of these journals has become a modern scholarly industry, and one may expect future benefits from this research. Meanwhile, one gets an impressionistic picture of erratic and unsystematic coverage of a flood of new books, a problem that many journals tried to address by publishing lists of books that the editors had not seen or could not, for whatever reason, review. Nonetheless, one may say that a reader living in 1700 could have obtained a reasonably good acquaintance with Oldenburg's "learned trade" by subscribing to the *Journal des savants* and two or three other periodicals. A somewhat clearer picture of communication among scientists in particular emerges if one turns to the activities of societies and academies and to the careers of a few individuals.

Late seventeenth-century societies, academies, informal meetings, and *conférences* all display a family likeness far from coincidental. So much is this the case that Marta Cavazza, immersed in the peculiar conditions affecting

---

[33] Ibid. (April 1718), 257–74, esp. 258, 264–65.
[34] Ibid. (June 1718), 215–73, esp. 227, 272.

the several academies of Bologna, has noted the "striking concord" in aims between the Bolognese and the Royal Society of London.[35] Many groups were avowedly "modern," sharing a dissatisfaction with the verbiage of Scholasticism, and promising to approach all subjects in a spirit of free and critical inquiry. Furthermore, recognizing that the study of nature in particular—more than history, literature, and language—was an international enterprise, the several groups and some individual members often made conscious efforts to communicate with like-minded persons and institutions in various cities and countries.

About many of these groups, little information survives except in the most fragmentary form. This is true, for example, of the meetings antedating the formation of the Royal Society of London, the several Paris groups both preceding and sometimes surviving the foundation of the Royal Academy of Sciences, and the numerous Italian academies in cities large and small. In many instances, formal organization followed some years of informal gatherings, and members of the newer institution might then try to reconstruct the history of the earlier meetings. This often proved difficult because records had been lost and members dispersed or deceased.[36] From such fragments as exist, from the records of some societies and academies, and from the correspondence of particular individuals, an effort will here be made to examine the common interests and aims and the concrete filiations that made these institutions part of the international Republic of Letters. Although discussion is here organized around three countries, links with institutions and individuals in other regions will also be indicated, briefly and selectively.

BRITAIN

With the foundation of the Royal Society of London, chartered in 1662, the metropolis became not so much a center of scientific activity as one of

---

[35] Cavazza 1981, 895.

[36] McClellan 1985 offers an excellent guide to institutions, structures, and publications. Other comparative analyses are in Boehm 1981; Hartmann 1977; and Voss 1980. Maylender 1926 provides an inventory of Italian academies; valuable studies of the cultural life of particular cities are in *Accademie* 1979. The difficulties of reconstructing histories of informal meetings are illustrated by Andreoli 1964 for Bologna and Gosseaume 1814 for Rouen. In Rouen, the careless keeping of early records was turned into disaster when fire destroyed such manuscripts as had been kept by the anatomist LeCat, prime mover of both the informal meetings and the chartered academy.

scientific communication. Fellows who spent little or no time in London—such prominent men as Edward Lhwyd, Martin Lister, Robert Plot, John Ray, and (for part of his career) Robert Boyle—considered the *Philosophical Transactions* to be an important vehicle of communication, even while they maintained a lively personal correspondence on matters not yet ripe for publication. If elements of an ideology promoted by the Society can be found in that superb apologia, Thomas Sprat's *History* (1667), more reliable indications of how the Society functioned and how it came to be perceived are in the letters of Henry Oldenburg and the content of the *Philosophical Transactions.*[37]

In encouraging British and foreign correspondents to communicate with London, Oldenburg at times found it necessary to explain that the Society was not interested in metaphysical or theological speculation. Nor did the Society much care for the building of "systems" in the fashion of Descartes, but preferred to receive reports of observations, experiments, and things "useful." In 1663 Oldenburg explained to astronomer Johannes Hevelius in Danzig that communication was essential "among those whose minds are unfettered and above partisan zeal, because of their devotion to truth and human welfare." The Society would thus welcome correspondence "with those who philosophize truly, that is, (in my opinion) with those who read the Book of the World, and make their thoughts on phenomena rigorous."[38] Precisely how to judge "rigorous" thinking seemed problematical to the framers of the Society's statutes in 1663, because here we read that Fellows were to report "matter of fact" and to separate such information from any "conjecture, concerning the causes of the *phaenomena.*" To visitor Balthasar de Monconys, who attended more than one meeting in 1663, the Fellows did indeed seem to be reporting observations and performing experiments without engaging in undue "reasoning."[39] This impression does not wholly convey what happened at meetings, the records showing that "much discourse" also took place. On the whole, however, Robert Boyle eventually became a symbol of the Society, as contemporaries in

[37] For analyses of Sprat's *History*, see Wood 1980 and Hunter 1989, chap. 2. These and other historians ask if the Society had an ideology, and they point out that disagreements existed among the Fellows on a range of methodological and philosophical issues. My own concern here is rather different: how the Society presented itself and was perceived by the learned world. M. B. Hall 1991, esp. chap. 9, is valuable for the Society's relations with foreign countries and individuals.

[38] Oldenburg 1965, 2:27–28.

[39] The statutes of 1663 are in *Record* 1912, 119. Monconys 1695, 2:45, 69. Monconys admits to some incompetence in English, but he got Oldenburg to clarify the proceedings in French.

Britain and on the Continent praised his experimental ingenuity and his interpretative restraint.[40]

That the Society's chief focus would be the study of nature was conveyed not only by Oldenburg, Boyle, and Sprat, but also in the early numbers of the *Philosophical Transactions,* which included Lawrence Rooke's "Directions for Sea-men, bound for far Voyages" and Boyle's "General Heads for a Natural History of a Countrey, Great or small." Boyle tucked into his instructions, however, an inquiry about local traditions, and, in fact, the Society had already in 1664 admitted to membership at least one scholarly antiquarian, Isaac Vossius. Erudition of a historical kind did interest many Fellows, Hooke and Halley among them, but publication on such topics would not be a regular feature of the *Philosophical Transactions* until about 1680.[41]

In addition to welcoming a great range of information, Oldenburg and the *Philosophical Transactions* made clear that the Society considered verification to be one of its tasks. In a review of Athanasius Kircher's *China Illustrata,* for example, the *Philosophical Transactions* remarked on one observation reported by the author:

> Of the *Serpent,* that breeds the *Antidotal* stone; whereof he relates many experiments, to verifie the relations of its vertue: Which may invite the *Curator* of the *Royal Society,* to make the like tryal, there being such a stone in their *Repository,* sent them from the *East Indies.*[42]

This critical attitude applied not only to the credulous Jesuit, as the Society also attempted to repeat anatomical experiments reported by Nicolaus Steno.[43]

---

[40] "Much discourse" was sometimes not recorded, presumably because this was opinion, not fact. For an example of recorded discourse, see Birch 1756, 1:247–48; for the omission of such discussion, see Rappaport 1986, 142 and n. 60. For a small sample of praise of Boyle, see Borelli 1663 in Adelmann 1966, 1:216, and Henry Power 1664 in Dear 1985, 156. That rationalists like Spinoza and Leibniz considered Boyle to be *too* cautious is well known (below, Chapter 2).

[41] The articles by Rooke and Boyle are in *PT* 1 (1666), 140–43, 186–89, 315–16, 330–43, Boyle's comment on human traditions being on page 189. Schneer 1954 and Hunter 1971 discuss archaeological topics investigated by Fellows. That articles of this kind did not make any regular appearance in *PT* until about 1680 suggests that Oldenburg (d. 1677) may not have valued such material as new knowledge.

[42] *PT* 2 (1667), 486. This stone would be examined by Francesco Redi in 1671; produced by the serpent itself, the stone was supposed to cure snakebite.

[43] On Steno's experiments, see Oldenburg 1965, 4:206, 224, 234–35. Also, Waller 1937, 55–56.

The repute of the Society on the Continent may be illustrated by examples of scientists who felt isolated or unappreciated in their own surroundings, and who perceived the Royal Society as offering a welcome for their researches. From Delft, Antoni van Leeuwenhoek in 1674 began to communicate to the Society lengthy reports of his microscopic observations; these would be published for many years in the *Philosophical Transactions*—and later summarized in the *Journal des savants* and the *Acta eruditorum*. In Bologna, Marcello Malpighi found it hard to obtain approval from his university colleagues, and Oldenburg began in 1667 to encourage him to seek a more receptive audience at the Royal Society; from soon after that date, Malpighi's writings were regularly published (in Latin) in London. Among Malpighi's disciples, Antonio Vallisneri had achieved considerable success as a physician and in being named to a professorial chair at Padua in 1700, but, as he informed Antonio Magliabechi, he wished to be associated with groups of experimental philosophers, for this would sharpen one's wits "as one knife sharpens another." When he began to correspond with Hans Sloane in 1703, Vallisneri identified himself not only as a student of Malpighi and a friend of Francesco Redi but also as a man dedicated to precise observation and experiment.[44] Already in correspondence with Sloane, Swabian naturalist Balthasar Ehrhart asked in 1735 if he might not be elected a Fellow of the Royal Society. This honor, he explained, would increase his stature locally and help him acquire patronage in pursuing his studies of natural history. The next year, in 1736, K. N. Lang (Langius) in Lucerne sent to Sloane a published tract describing a very peculiar fossil; although, noted Lang, his little pamphlet had been well received locally, he considered Sloane, then President of the Royal Society, to be the most "competent judge" of the curious powers of nature that had produced his unique fossil.[45]

In addition to Oldenburg's solicitations and the repute that attracted such unsolicited contributions as Lang's, the Society's luster was increased

[44] Vallisneri 1991, 1:195, 232. That scholars recognized the welcoming attitude of the Royal Society has been pointed out by M. B. Hall 1975, 184. For Malpighi's relations with London, see Adelmann 1966, vol. 1, chap. 21. More generally, for Italy and the Royal Society, see M. B. Hall in Cremante and Tega 1984, 47–64, and Cavazza 1980.

[45] Ehrhart to Sloane, 22 January 1735, and Langius to Sloane, 14 February 1736. Accompanying the letter of Langius was a copy of the brochure describing his fossil: *Appendix ad Historiam Lapidum figuratorum* . . . (1735). The letters and brochure are in Royal Society, "Early Letters and Classified Papers, 1660–1740," which I have consulted in the microfilmed version of 1990, produced by University Publications of America. The microfilming preserves the texts but does not allow identification of volume and page (or folio) numbers in the manuscript collection. Materials in the collection are more or less alphabetically arranged, so that the letter from Ehrhart can be found under the letter *E* and that from Langius under *L*.

by the activities of individual Fellows, not only by their publications but also by their visits abroad and by their correspondence. Sloane here was exemplary in his close contact with French savants and with abbé Jean-Paul Bignon, patron and administrator of the several French academies. In what may be a unique instance, the *Mémoires* of the Academy of Sciences even carried the full text of two articles by Sloane—as a "foreign associate" of the Academy, Sloane was not eligible to publish in the *Mémoires* but only to have his work summarized in the annual *Histoire*.[46]

Other Fellows of international repute included John Woodward, whose researches in natural history and antiquarianism entailed an enormous correspondence with philosophers living anywhere from the Baltic ports to the North American colonies. Whether Woodward ever became a symbol of the Royal Society may be questioned, and this role, in the first generation of the Society's existence, was doubtless played by Robert Boyle. Viewed from the Continent, Robert Hooke, so crucial a figure in the Society's meetings, was a relative unknown, but Boyle's writings appeared regularly in Latin translations and sometimes in vernacular languages as well. If modern scholars are increasingly fascinated by the moral, theological, and metaphysical components of Boyle's writings, his contemporaries for the most part cited his works for their observations and experiments. Even his short and seemingly insignificant tracts—such as "Relations about the bottom of the sea" (1671)—received attention well into the eighteenth century, for Boyle was the "typical" English philosopher: infinitely curious, ingenious, cautious, and reliable. Some, to be sure, complained that Boyle built no synthesis on his enormous range of observations and experiments, but the experiments themselves were often replicated, and Boyle's laboratory became something of a tourist attraction.[47]

Beginning in Oldenburg's day, the Royal Society attempted to establish regular communication with Continental institutions, but results were erratic at best. This was true not only in the well-known case of the Paris Academy but also for the Florentine Accademia del Cimento and for one of the Roman academies of the 1680s. In a period when the free commu-

[46] One of Sloane's memoirs, on the fossil bones of "elephants," will be discussed below, Chapter 4. For the fame of his collections, see Gersaint 1736, 43–44. His intimate relations with France, evident in his unpublished letters, are examined by Bonno 1943 and Jacquot 1954. For Bignon's career, see Bléchet 1979.

[47] The complex inner life of Boyle is being reexamined now, with some emphasis on surviving manuscripts, as in Hunter 1994. For a good analysis of how Boyle was viewed on the basis of his publications, see Pighetti 1988. His tract on "the bottom of the sea" will be discussed below, Chapter 6. Woodward's repute is discussed in Chapter 5; the standard study is by Levine 1977, but also worthwhile is an essay by Porter 1979 which highlights the personality and mannerisms that antagonized Woodward's contemporaries.

nication of new knowledge was thought to be a duty among citizens of the Republic of Letters, such failures are nonetheless understandable, because scientists jealously guarded their own projects until they reached a publishable state. Oldenburg himself, even while he urged correspondents to report on work in progress, did glimpse the problem of priorities, as he occasionally wrote to defend Fellows such as Boyle and Hooke for research they had been conducting but had not yet published—some competitor had "scooped" them.[48] Typically, then, news came from individual reporters, not from institutions. Just as Francis Vernon and Henri Justel for a time told London about activities in Paris, Marcello Malpighi and Sir Thomas Dereham were major conduits for Italian news.

The main vehicle of communication between London and the Continent, in matters of natural philosophy, should surely have been the *Philosophical Transactions*, but Continental philosophers and journalists regularly noted and complained that English writers had a tendency to write in English. As the use of Latin declined—and was sometimes explicitly repudiated as the language of Scholastic learning—philosophers had to become linguists, and knowledge of English was sufficiently rare on the Continent as to distinguish Jean LeClerc's several *bibliothèques* from the journals of his competitors. To serious readers, wishing for more detail than the book-review journals provided, translating the *Philosophical Transactions* became a project of some importance. Partial translations were eventually published in France and Italy. On the whole, however, journals like LeClerc's and Nazari's, as well as the *Acta eruditorum* and the *Journal des savants*, performed vital service in transmitting to the Continent at least some fragments of English philosophy. More than fragments were conveyed by the appearance in Latin of Robert Boyle's writings; by way of contrast with Boyle, Hooke's works, never translated, achieved less fame in Europe, and John Woodward's theory of the earth became known abroad mainly after J. J. Scheuchzer produced a Latin version.[49]

If both the Society and the *Philosophical Transactions* achieved international renown, the Society as an institution had rather less success as a model of organization. Imitated in Dublin and Oxford, the London group nonetheless had no visible structure for Continentals to emulate. Moreover, it soon became evident to onlookers that the Society's royal charter did

[48] For Oldenburg's defensiveness, see, for example, *PT* 2 (1667/68), 628, and Steno 1671, translator's preface.

[49] A Latin translation of Hooke's *Micrographia* was actually planned but never materialized; see Oldenburg 1965, 3:31 and 33 n. 2. For an extraordinary effort to summarize the *Micrographia* orally, see Adelman 1966, 1:366–67. For editions, abridgments, and translations of *PT*, see Kronick 1976, 137–38, 148.

not entail royal patronage of the kind established in Paris, and some Fellows eventually expressed envy of the Paris Academy in this respect. Furthermore, it is likely that the structure of the "renewed" Academy of 1699—specifying both ranks of members and large fields of specialization—appealed to a sense of decorum as well as to a practical desire that various scientific disciplines be properly represented. In other words, informal gatherings may have followed the example of London, but chartered institutions on the Continent more often imitated Paris.[50]

FRANCE

Much has traditionally been made of Continental rationalism, in contrast to British empiricism, but so stark a contrast will not stand up to investigation. In France in particular, an indigenous experimentalism was associated with the writings of Pierre Gassendi and Marin Mersenne, and the philosophy of Descartes also stimulated inquiry into the workings of nature. By the 1650s, Paris could boast some half-dozen discussion groups, varying among themselves in the attention given to natural problems. These groups may be contrasted with the earlier assemblies sponsored in the 1630s by Théophraste Renaudot, where meetings were chiefly devoted to speech-making, each topic addressed by more than one speaker. In addition to tackling such subjects as respiration, volcanoes, the origin of mountains, and the certainty attainable in natural knowledge, participants in the Renaudot meetings discoursed about whether one should marry, the nature of "panic" fear, and why familiarity breeds contempt. In style and argument, these discourses had little in common with the later, "modern" approach to nature, but they were reprinted from time to time, and selections were even translated into English with the misleading indication that these texts were by "the virtuosi of France."[51]

Speechmaking continued in certain groups of the 1650s and thereafter, but with increased awareness of the need for experiment. Two of the Paris assemblies, conducted by Nicolas Lemery and Jacques Rohault, were

---

[50] In general, see McClellan 1985 who offers a much broader range of societies and academies than I do in this chapter. The distinctive and contrasting models were clearly London and Paris. London envy of French patronage is discussed by Hunter and Wood 1986, 62–63.

[51] In the basic study by Solomon 1972, see esp. 65–66. Renaudot 1666, 3:305–23; 4:134–44; 5:387–96. An English translation appeared in two volumes, the first of them entitled: *A Collection of Discourses of the Virtuosi of France* (1664). The second volume (1665) had "French Virtuosi" in its title.

unusual in being science courses, the experiments performed by these skilled teachers. Other meetings—sponsored or conducted by Bourdelot, Denis, Justel, Montmor, and Thévenot—featured a mixture of speeches and experiments, with some of the speeches actually being reports on observations or experiments.[52]

Although it is hard to know precisely what happened at these gatherings, a letter from Samuel Sorbière to Thomas Hobbes informs us of the ideals adopted in 1657 by the group meeting at Montmor's home. The Montmor assemblies would be devoted to "the clearer knowledge of the works of God, and the improvement of the conveniences of life." At each meeting, a "question" would be posed for the next meeting, and two persons would be designated "to report their opinion" on the subject. These reports, written "in concise and reasoned terms, without amplification or citation of authorities," would be read to the assembly and then discussed at large.[53] If these statutes do not mention experiments, some were conducted from time to time, as when Rohault brought to a meeting "all his apparatus of lodestones" or when the anatomist Pecquet performed dissections.[54] On another occasion, Sorbière himself delivered a speech intended to entertain the gathering until more people should arrive to hear Rohault and others.[55]

The Montmor gatherings did not impress such visitors as Christiaan Huygens and Henry Oldenburg, perhaps because attention to experiments did not yet figure prominently at the time of their visits.[56] In the 1660s, however, experiments seem to have been a fairly regular feature of the Bourdelot, Justel, and Thévenot groups. Rohault and Nicolaus Steno participated at Bourdelot's, as did itinerant Italian botanist Paolo Boccone. At Justel's home on at least one occasion, the topics included discussion of an experiment performed earlier that day at the new Academy of Sciences. Thévenot's assemblies were for a time enlivened by the presence of Steno, whose dissections aroused the great enthusiasm of both the *Journal des savants* and a visiting provincial, André Graindorge, who reported these

[52] For all of these groups the basic study is Brown 1934. The content of Rohault's meetings can be inferred from his *Traité de physique* (1671) and Lemery's from his *Cours de chymie* (1675). Both works summarize lectures given in earlier years, and both were reprinted many times, Lemery's as late as 1756. Additional information is in the eulogy of Lemery by Fontenelle 1709, 1720, 2:174–76; Perkins 1958, 407; and Niderst 1972, 112–16. For Rohault, see Clair 1978 and McClaughlin 1977.

[53] Quoted in Brown 1934, 75–76. Also in Huygens 1888, 4:514, and Sorbière 1660, 631–36.

[54] Brown 1934, 127.

[55] Sorbière 1660, 194.

[56] Oldenburg 1965, 1:278, 287. Hahn in Bos 1980, 58–59.

events to the little academy in Caen (Normandy). In all these instances, it seems clear that experiments were performed mainly when a capable person happened to be present and when the facilities or specimens happened to be available. Thus, Rohault brought his own "apparatus" and Boccone arrived in Paris with his own collection of fossils and dried plants. Even the "conferences" of Jean-Baptiste Denis, himself an anatomist, combined actual performances with speechmaking and reports on new books.[57]

By the time the Academy of Sciences was established (1666), there existed in Paris a receptive audience for "modern" challenges to traditional learning. In the new institution, the order of the day resembled that of London: new inquiries, new observations and experiments, and, in short, new knowledge. As in London, the work of individuals had to be verified by colleagues—a requirement that ultimately entered into the statutes of 1699—but the Parisians also adopted a plan for collaborative research reminiscent of both Francis Bacon's philosophy and the practices of the Florentine Accademia del Cimento. Such collaboration had collapsed in Paris by 1699.[58]

Because the Paris Academy published so little for more than thirty years, and because Oldenburg failed to establish regular communication with it, the Academy has sometimes been viewed as a "closed" institution, resistant to outside influences and secretive about its proceedings. Certainly, Leeuwenhoek never deluged Paris with the observations he regularly sent to London, so that news of his research was eventually brought to Paris by Huygens.[59] But in contrast to this closed image, abbé Gallois, academician and journalist, often reported academic activities in the pages of the *Journal des savants* while he was its editor (1666–74), and such news was probably transmitted by the academicians themselves to gatherings like Henri Justel's. Scattered through the records of early meetings are also examples

[57] In addition to Brown 1934, passim, see the following: Magalotti 1968, 217–18, and Brown 1933 on Justel, McClaughlin 1975 on Thévenot, and Lux 1989 on Graindorge. The ecstatic comment on Steno is in *JS* (23 March 1665) 141. Proceedings at the Denis meetings are in Denis 1672, at Bourdelot's in LeGallois 1674. For the Royal Society, M. B. Hall 1991 attempts to determine the frequency with which experiments were performed at meetings.

[58] The standard treatment is Hahn 1971, 25–33 for issues treated here, but indispensable now for the day-to-day functioning of the Academy before 1699 is Stroup 1990; also 1987. The brilliant essay by Salomon-Bayet 1978 lacks the documentation that historians require. The statutes of 1699 (Art. XXV on verification) were printed in *HARS* [1699] 1702, and reprinted in Fontenelle 1709, 1720, vol. 1.

[59] AAS, P-V, t. 7, fols. 176r, 185r–v (16 July and 30 July 1678); the text conveys the astonishment of members at the tiny size of the lens and the incredible numbers of organisms it revealed.

of correspondence with such outsiders as Paolo Boccone; and when the anatomists in 1681 gave their attention to the dissection of an elephant, Edme Mariotte—unique among members for his knowledge of both English and German—introduced his colleagues to Caspar Horn's compendium of tales about and descriptions of the animal.[60] When the Academy did begin to publish its *Histoire et Mémoires,* the *Histoire* summarized communications from nonmembers who had volunteered information to the Paris institution. Outsiders occasionally appeared at meetings, as when Cassini I invited Edmond Halley to report his observations of the transit of Mercury or when Jesuit missionaries conferred with the Academy about observations they might make in China. In these ways, the Academy was less "closed" than has been alleged, although appearances by outsiders were rare. When Boccone was in London, he could visit the Royal Society; in Paris, he lectured to the assemblies of Bourdelot and Denis.[61]

To assess the repute of the Academy, in France and abroad, is difficult for the early decades. Certainly, the glories of royal patronage soon were visible in the building of the Paris Observatory, but those who looked forward to detailed accounts of academic research must have been disappointed when the great folios assembled by the anatomists in 1671 were distributed as gifts and thus not easily available for consultation.[62] To inquirers, however, private correspondence supplemented the articles in the *Journal des savants* and eventually became more reliable than the *Journal.* Leibniz, for example, had a number of informants in Paris, some of them members of the Academy, and Cassini I maintained friendships in Bologna. With the publication in 1698 of J.-B. DuHamel's history of the Academy, book-review journals welcomed this systematic account of three decades of academic research.[63]

The annual publication of the *Histoire et Mémoires* provided material not only for journalists but also for enterprising printers. As early as 1708, a

[60] As an example of Boccone's communication with the Academy, AAS, P-V, t. 10, fols. 8v–9r (31 January 1680). Mariotte's presentation of Horn 1629 is in AAS, P-V, t. 10, fol. 72v (18 June 1681); what he said about the book is not recorded.

[61] Birch 1756, 3:87. Brown 1934, 238–40. Boccone went first to Paris and then to London in search of patronage, and he apparently failed in both places. Details of his career are in *DBI.* For Halley, AAS, P-V, t. 9, fol. 99v (10 May 1681). Because the Academy examined inventions, the inventors sometimes attended meetings to present their work and answer questions.

[62] This disappointment is recorded in the English translation of Perrault 1688, "The Publisher to the Reader."

[63] DuHamel's history was reviewed at length in *JS* (19 January 1699), 32–36, continued 26 January, 37–42; *HWL* (April 1699), 216–20, and (May 1702), 283–88; and *Trévoux* (November–December 1701), 292–307. The post-1700 dates are reviews of the second edition of DuHamel (1701).

Paris firm reprinted in a single volume Fontenelle's superb "preface" of 1699, the new statutes of the Academy, and a collection of those academic eulogies for which Fontenelle became so famous. As we have seen, *Europe savante* began its short career in 1718 by reviewing at length another reprint of this kind. Editions of the eulogies in particular continued to appear. A few of the more important ones, notably of Isaac Newton and Peter the Great, were also issued separately and in translations.[64] In these elegant and eloquent biographies, Fontenelle conveyed his version of the work and ideals of the Academy and its members. To one observer, these eulogies were virtually beyond criticism because Fontenelle "has so much accustomed us to believe and think as he does." The same writer noted, too, that Fontenelle's annual *Histoire* succeeded in giving academic researches "a clarity they could not in themselves have attained."[65]

The Academy's new statutes and publications inspired considerable emulation in France and elsewhere. To be sure, local conditions required some adjustments in the Paris model. In smaller cities, it was not always possible to find specialists in some of the disciplines—especially the mathematical sciences—as defined by Paris. Nor did it seem feasible or desirable to separate the sciences from other areas of erudition, for, as some French provincial academicians noted, the most useful role for local societies was to examine regional natural and civil history. (Not until the reorganization of 1785 was "natural history" accepted as a specialty in the Paris Academy.) Furthermore, the various French and European academies usually could not publish annually, or even regularly, accounts of the works of their members.[66]

With these serious modifications in mind, one may still find the Paris model expressed in the ordering of ranks and listing of specialties so notably absent in the Royal Society of London. As an especially striking example, the Accademia degli Inquieti in Bologna abandoned in 1704 the formlessness of earlier Bolognese groups and adopted a structure in explicit imitation of Paris. Further transformed a few years later into the "Academy of the Institute of Sciences," the Bolognese in 1731 published a

---

[64] I have not tried to compile a systematic list of editions of the Fontenelle eulogies. The catalogues of the Bibliothèque nationale and the British Library amply demonstrate their popularity. See also Charles Paul 1980. On the taste for such biographies, see Goldgar 1995, 147–54, which puts no emphasis on Fontenelle.

[65] The quotations are from Camusat 1734, 2:169, 172. That even a critical mind like Adam Smith's could be wholly seduced by Fontenelle's version of what science and scientists were all about appears in Smith's *Theory of the Moral Sentiments* 1759; see Smith 1976, pt. 3, chap. 2 (in this edition, pp. 125–26).

[66] McClellan 1985 and other comparative works, above, n. 36. For local history in provincial French academies, see Roche 1988, chap. 6; see also Barrière 1951, 92–96, 308–9.

first volume of *Commentarii,* divided into the equivalent of "histoire" and "mémoires." The permanent secretary, Francesco Maria Zanotti, tried to emulate Fontenelle not only in the elegance of his prose but also in introducing comments on and evaluations of the academic memoirs he was ostensibly summarizing.[67]

Comparable structures of membership and publication can be found in academies from St. Petersburg and Berlin to the small groups assembled in such French cities as Dijon, Montpellier, and Rouen. In fact, when members of the informal assembly held in Rouen decided to constitute a more official body, they appealed for advice in drawing up their regulations to Fontenelle, a *rouennais* by birth.[68] Just as the Paris Academy before 1699 had had difficulties in finding monies to support an annual publication, so, too, did its imitators. Even the Montpellier "Société royale," formally affiliated with Paris, did not until 1766 find a way to produce a volume of papers dating from 1706 to 1730, and the Society never did manage to compile the natural history of Languedoc envisaged by its early members. A comparable project, announced with some fanfare by the academy of Bordeaux early in the century, also failed to materialize.[69]

In ways other than structure and publications, the Paris Academy achieved a renown somewhat different from that of the Royal Society of London or the Florentine Cimento. Like the other groups, it avoided taking a corporate stance on matters of interpretation or theory. At the same time, it was repeatedly called on by the State to judge the value of new inventions, and it did not hesitate to reject the work of individuals who claimed to have squared the circle or to have invented a perpetual motion machine. When, however, such a reputable author as L. F. Marsili submitted for judgment a sketch of his oceanographic studies, the Academy responded simply by complimenting Marsili on his experimental ingenuity.[70] In this respect, the Academy and the Royal Society of London were in full accord, for the Londoners also resisted efforts to appeal to their cor-

---

[67] Cavazza in Cremante and Tega 1984, 109–32, reprinted with some modifications in Cavazza 1990, chap. 4. Also, Ambri Berselli 1955.

[68] Gosseaume 1814, 1:8.

[69] Barrière 1951, 208–9. Montpellier, *Mémoires,* vol. 1 (1766), preface, and correspondence of one of the early secretaries of the Society with abbé Bignon in Paris, 1726–27, at BN, MS fr 22229, fols. 37–38, 41–42, 57–58. For patronage problems of the Paris Academy before 1699, see Stroup 1987. Kronick 1976, 155, explains the slender publication record of French provincial academies by saying they had "open to them" the volumes published in Paris. In reality, only the Montpellier society, being affiliated with Paris, was permitted to send one memoir per year for publication in the Paris *Mémoires.*

[70] Marsili 1711, 22–23.

porate judgment. And yet it seems that the Paris Academy did acquire the reputation of being an authoritative tribunal.

Within France, at least, submitting to the judgment of Paris was inevitable, given the pyramidal organization of intellectual life and the elite membership of the Academy. More generally, French intellectual eminence was acknowledged in various ways, as when an Italian resident of Vienna congratulated astronomer J.-N. Delisle on his return from St. Petersburg to Paris, "the Capital of the learned world."[71] Commonplaces of this kind are less significant in this study than the remark of Louis Bourguet in Neuchâtel that there existed in Paris an academic "school" of naturalists— the "school" consisted only of Réaumur and Antoine de Jussieu, with Fontenelle interpreting and applauding their writings.[72] Bourguet here paid tribute to the authoritativeness of publications emanating from the Paris Academy, but he also, unwittingly, paid tribute to Fontenelle. Formally and officially, the Academy did not pass judgment, but Fontenelle did. As will be argued in a later chapter, the Academy, thanks to Fontenelle, played a role in natural history that no other scientific society could or would have assumed.

ITALY

Among the more obvious difficulties inherent in attempting to describe Italian academies is the astonishing number of such groups and the little information that survives about so many of them. In no other country was there so long and lively an academic tradition, dating back to the fifteenth and sixteenth centuries. This tradition underwent some "modernization" in the seventeenth century, stemming in part from the influence of Galileo and his disciples and in part from the awareness in some Italian cities of developments north of the Alps.

Historians of post-Galilean Italian intellectual life point, with varying degrees of emphasis, to the vigor of the Galilean heritage.[73] To Galileo and such disciples as Benedetto Castelli, Evangelista Torricelli, and Vincenzo

---

[71] Marinoni to Delisle, Vienna, 24 April 1748, in AOP, MS B. 1. 4, fol. 110.

[72] Bourguet 1729, 177–80. Antonio Vallisneri, a passionate francophobe, accused the Academy of being dictatorial not only in science but also in taste and culture; discussed in Rappaport 1991b, 80–81 and n. 30. For the Academy's power and prestige, and its function as judge, see Hahn 1971, esp. chap. 3.

[73] The basic work is U. Baldini 1980. The best English treatment of Italian cultural and intellectual life, from the later seventeenth century until about 1750, is in Carpanetto 1987, chaps. 6–9.

Viviani are commonly attributed both the experimentalism of the Florentine Cimento and the adoption by so many intellectuals of mechanical philosophies. Even the dialogue form, used so brilliantly by Galileo, has attracted attention as a distinctively Galilean style imitated in the next generation or two by writers such as Marcello Malpighi and Antonio Vallisneri. (Dialogue, of course, was an ancient form, and it could be—and was—used by anti-Galileans, such as Filippo Buonanni, S.J.[74]) When experimentalists such as Francesco Redi and Giovanni Alfonso Borelli developed conflicting styles and viewpoints, they and their supporters nonetheless either claimed a Galilean parentage or inadvertently revealed one in their writings. To this powerful indigenous allegiance, scholars in recent years have added evidence of Italian awareness of foreign philosophers (Bacon, Boyle, Descartes, Gassendi) and of such foreign institutions as the Royal Society of London and the Academy of Sciences in Paris.[75]

The Florentine Cimento, during its brief existence (1657–67), maintained a degree of secrecy about its proceedings, some of its members being cautious about revealing their projects before publication. But information about this academy was known in London and Paris by way of private correspondence addressed to such people as Oldenburg, Montmor, and Thévenot. When the Cimento's *Saggi* finally appeared in 1667, its publication had been anticipated for at least two years by Fellows of the Royal Society. The luxurious volume, however, proved disappointing, for, as Oldenburg remarked to Boyle, the experiments on air pressure could no longer be considered new.[76]

Characteristics of the *Saggi*—the dispassionate focus on experiment, not interpretation, and the exclusion of all personal names from the volume— have generated different assessments by modern scholars. On the one hand, it has been said that the "peculiar empiricism of the Accademia del Cimento can be understood *only* in [the] context" of fears spawned by the condemnation of Galileo in 1633.[77] In a more subtle vein, Paolo Galluzzi has presented evidence that theoretical disagreements among academicians—"moderns" versus Aristotelians in the small membership—naturally dictated that the *Saggi* confine itself to "fact." Others have suggested that the prudent avoidance of metaphysics and theology merely resulted in an

---

[74] For dialogue in particular, see Altieri Biagi 1980, introduction; Torrini 1977, 102–3; Adelmann 1966, 1:155–56, 166–67. For Vallisneri, Maugain 1923, 567, points to a newer model: Fontenelle's *Dialogues des morts* (1683). Basile 1987, 148, indicates that in Buonanni's dialogue of 1691, Francesco Redi was satirized as Simplicio.

[75] For example, Cavazza 1980, 1985; Pighetti 1988.

[76] Oldenburg 1965, 4:186, and Waller 1937, 50. For communication of the Cimento with Paris, see Cochrane 1973, 244–45.

[77] Middleton 1971, 333, italics added.

epistemology shared by the Cimento with other academies not subject to such pressures.[78]

These several assessments ignore the fact that there are some theoretical allegiances visible in the *Saggi*, such as the series of "experiments to prove that there is no positive levity." Other Aristotelian concepts were also overtly challenged: Are heat and cold "qualities" or substances? The Florentine motto, "testing and retesting," applied not only to new experiments but also to the detection of errors in received texts—the latter reminiscent of the *nullius in verba* of the Royal Society of London. If omitting all personal names from the *Saggi* suggests fear to some historians, quite a different motive emerges from Lorenzo Magalotti's preface, especially when compared with the similar rationale stated by Claude Perrault on behalf of the anatomists of the Paris Academy. Both Magalotti and Perrault declared that they were publishing only what their respective groups could agree on, such agreement signaling the discovery of true matters of fact; to attach a *personal* name to any observation or experiment meant, by contrast, the publication of an unverified report or opinion.[79]

Individual members of the Cimento, especially Borelli and Redi, produced treatises that circulated rapidly north of the Alps. Redi in particular aroused great interest on several occasions, as when his published reports of experiments undermining belief in spontaneous generation (1668) prompted debate in London, Paris, and Geneva. In fact, he had earlier stirred up a long controversy about venomous snakes (1664), with repercussions in London and Paris. When he tested the so-called serpent stone, a reputed cure for snake-bite, his results were presented in Paris at one of the "conferences" of J.-B. Denis; when Denis, in turn, claimed to have discovered a "styptic water" that would stop bleeding, a supply sent to the Grand Duke of Tuscany was turned over to Redi for examination.[80]

Among successors to the Cimento—and Italian academies regularly invoked the Cimento as their predecessor—the several Bolognese institutions have been the subject of much analysis in recent years, there being an unusual amount of material surviving about these groups. One such gathering, to which Malpighi belonged in the 1660s, devoted itself to anatomical dissection, to supplement the bookish learning of the Univer-

---

[78] Cochrane 1973, bk. 4, chaps. 1–2; Galluzzi 1981, esp. 800–815; U. Baldini 1980, 405–20.

[79] In Middleton 1971, 221–25, 339–44. For Magalotti's preface, I have used the translation by Richard Waller (1684). Cf. Perrault 1671, preface. The epistemological issues will be analyzed in Chapter 2.

[80] Adelmann 1966, 1:392, 393; M. Baldini 1975, chap. 2; Basile 1987, chaps. 2, 4; Denis 1672, 193–94; Heyd 1982, 89; Lux 1989, 49, 66–67.

sity of Bologna. More broadly experimental were the Accademia della Traccia ("tracks" or "footprints") and the Accademia degli Inquieti ("the restless"), as well as the private meetings that took place for a short time at the home of Anton Felice Marsili. Most vigorous were the Inquieti, for a while led by the young Eustachio Manfredi and later, in 1704, reorganized by Giambattista Morgagni. By that date, the Academy had accepted the hospitality of L. F. Marsili, whose generous endowment in 1711 helped transform the Inquieti into an academy attached to the university; in 1714, the reformed structure was officially inaugurated.[81]

Modern historians have detected in these Bolognese groups interests, attitudes, and activities revealing their place in larger European contexts. As in the Cimento, experimentalism had become the foundation of inquiry into nature; but the Bolognese, unlike the Florentines, identified such modernity with Francis Bacon, Robert Boyle, and the Royal Society—they also, to be sure, admired Galileo and the Cimento. In addition to their commitment to experimentation, individual members sought to test such modern ideas as Gassendist atomism, and some gave explicit allegiance to mechanical philosophies as more comprehensible than the "qualities" associated with Scholasticism. If the several historians who emphasize the dangers of the "new learning" in post-Galilean Italy are correct in their analyses, one may conclude that the Bolognese had found a solution to this problem: they had linked their new academy to a bastion of the old learning, the University of Bologna.[82]

As with the Cimento, historians have tended to see the Bolognese delay in producing the *Commentarii* (1731) as a post-Galilean problem of satisfying the censors. Worth noticing, however, in the correspondence of Zanotti, the Bolognese secretary, are such problems as getting the academicians to submit articles in publishable form and judging which contributions actually deserved publication. Comparable difficulties afflicted Fontenelle, as a committee of Paris academicians wrangled about whether the work of their colleagues should be included in the *Mémoires*.[83] One may suppose that Zanotti, like Fontenelle, also found it difficult to become familiar with and to summarize writings in a considerable range of scientific fields. More readily documented, Zanotti certainly did face problems

[81] Cavazza 1990. See also Limiers 1723 and Marsili 1728, as well as Adelmann 1966, 1:103–5, 126–27, and Stoye 1994, chap. 10.

[82] In addition to studies of Bologna, cited above, n. 75, much of value about Italian university life, with emphasis on Padua, can be found in Dooley 1984, 1988.

[83] Correspondence of Zanotti in Morgagni 1875, esp. 112–38 and passim. For Fontenelle's problems, see Delisle to Louville, Paris, 8 July 1718, in AOP, MS B. 1. 1, fol. 99. De Zan puts some emphasis on Zanotti and the censors, in Predaval Magrini 1990, 230–33.

of reconstructing the Academy's history, a task he considered essential as a preface to the *Commentarii*.[84]

The several Bolognese academies, but especially the formal one that opened its doors in 1714, attracted much attention in the north. The first of the Cassinis, settled in Paris, advised the astronomers of the Inquieti on matters ranging from the need for an observatory to the best books to buy. Relations with London were cordial, as the Bolognese expressed their admiraton of two icons, Bacon and Boyle. Recognizing the importance of Dutch publishers, the Bolognese also campaigned for attention in the Netherlands as a way of ensuring the further diffusion of their writings. Not only did books by Malpighi and Guglielmini appear in Dutch editions, but the new Bolognese Institute received publicity in at least one of the Dutch francophone journals and in an excellent book-length account of its founding and structure, for which the chief founder, L. F. Marsili, provided the requisite information.[85]

Almost as well known north of the Alps was the Neapolitan Accademia degli Investiganti, formally organized in 1663, but in existence for some time before that date. The precise dates of the Investiganti are far from clear: 1663–70 can be agreed upon; some decline in activity followed 1670; and the trial of the supposed "atheists" in 1688–93 seems to have shifted the attention of academicians from natural philosophy to matters political and social.[86] Some Neapolitans regarded the famous trial as an attack on that *libertas philosophandi* so much discussed in Italy and northern Europe in this period, but others focused on the trial as a political and social crisis, pitting the Roman Inquisition against Neapolitan autonomy and the local aristocracy against the citizenry of Naples. The Investiganti seem to have survived the crisis, in some informal way, but the next known academy in Naples, the Medinacoeli, apparently had little interest in natural philosophy.

Readers of English received a glimpse of the Investiganti in John Ray's account of the "grand tour" on which he accompanied Philip Skippon.[87] In 1664 the travelers attended one of the weekly meetings, noting the large

[84] Andreoli 1964. See the Fontenelle letters of 1701, 1702, and 1707, in Rappaport 1991a, 286, and *Fontenelle* 1989, 22.

[85] Cavazza 1990, chaps. 3, 4, 8; M. B. Hall in Cremante and Tega 1984, 47–64. Notices in journals, Dutch (francophone) and Italian, are indicated in Cavazza, chap. 8. See also Limiers 1723.

[86] The basic studies are by Fisch in Underwood 1953, 1:521–63, and Torrini 1981. For Neapolitan intellectual life, Badaloni 1961 must now be supplemented by Dibon 1984; Osbat 1974; Ricuperati 1970, 1972; Suppa 1971; and especially Rossi 1969, 1984. Ferrone 1982 also has much to say about Naples, the city attracting so much scholarly attention as background to the writings of Pietro Giannone and Giambattista Vico.

[87] What follows is from Ray 1673, 271–72. Skippon's account is in Churchill 1732, 6:607.

audience of about sixty when the academy itself numbered only about fifteen members. The proceedings began with a Torricellian experiment, followed by discussion. Other members then "recited . . . discourses composed" for the occasion, each followed by discussion. In an oft-quoted judgment, Ray marveled to find "such a knot of ingenious persons," displaying such "freedom of judgment," in a Catholic country. And he noted, too, that academicians were familiar with such philosophical writings as those of Bacon, Gassendi, and Descartes, as well as the more recent publications emanating from England. At the same time, Ray and Skippon were told that the local clergy disliked all this novelty; indeed, after Ray left Naples, a rival Accademia degli Discordanti was formed to defend ancient learning.[88] If the Investiganti as a group published nothing, individual members such as Tommaso Cornelio, Leonardo di Capua, and G. A. Borelli (who belonged to more than one Italian academy) produced treatises well known in the north. Famous, too, was the library of Giuseppe Valletta, open to scholars and notable for the owner's systematic acquisition of foreign publications.

Among the innumerable Italian learned societies, the "physico-mathematical academy" founded in 1677 by Giovanni Ciampini in Rome deserves at least brief mention because accounts of its activities would be transmitted to Cassini in Paris and to the Royal Society in London. At a later date, when Countess Clelia Borromeo established in her home in Milan an academy devoted to science and erudition, Sir Thomas Dereham, a diplomat vitally concerned with matters of natural philosophy, thought that news of the new academy merited a report to the Royal Society. Some half-dozen years later, when Montesquieu visited Milan in 1728, this Accademia de' Vigilanti had shifted its focus to political topics.[89]

This brief sketch of a few Italian academies and their international filiations hardly does justice to the place of Italy in northern intellectual life. Northerners regularly traveled to Italy, either as part of the gentlemanly "grand tour" or for more scholarly purposes. To the list of several visitors of great fame—the young Boyle, Ray, Leibniz, Mabillon, Montesquieu, and others—one may add names especially pertinent to this study, such as the Scheuchzer brothers of Zurich and Louis Bourguet from Neuchâtel. Johann Scheuchzer served for a time as secretary to L. F. Marsili, and his older brother, Johann Jakob, became an intimate correspondent of

---

[88] Fisch in Underwood 1953, 1:537.

[89] For the account of Ciampini sent to Paris, see Gardair 1984, 152; for London, Birch 1756, 4:55. See also Middleton 1975 and Akerman 1991, chap. 15. The Milanese academy is mentioned by Dereham in 1722 (Royal Society, "Early Letters," s.v. Dereham) and in letters of the same period from Vallisneri to Bourguet (Neuchâtel, MS 1282). Montesquieu 1950, 3:914. The *JS* caught up with the Milan group belatedly, September 1731, 540–44.

Antonio Vallisneri. The brothers would also become members of the Inquieti in Bologna. Bourguet visited Italy often and even lived there for a time; his many Italian correspondents came to include Vallisneri and Scipione Maffei, and in 1728 he would become one of the founders and editors of the *Bibliothèque italique*.[90]

If Malpighi never crossed the Alps, other learned Italians did. The secretary of the Cimento, Lorenzo Magalotti, carried with him to London and Paris copies of the luxuriously printed *Saggi* of his academy. Not a formal emissary like Magalotti, Antonio Conti came to enjoy life in London and Paris, and he reported to friends such as Vallisneri in Padua what he thought would interest them.[91] By the time L. F. Marsili went north—to France in 1705, to England and Holland in 1721—he had long been known by correspondence with and some few writings sent to scientists in both London and Paris. While still a soldier on campaign in eastern Europe, he had begun to communicate with the Paris Academy (1683) and the Royal Society (1691), and in 1700 he would dedicate to the latter the *Prodromus* of his study of the Danube basin. When his military career ended, he settled for a time in southern France to pursue research that he eventually dedicated to the Paris Academy.[92]

Such examples hardly convey the attractions of Italy and Italians to northerners. This was, after all, the country of the Renaissance, the land of Galileo, and the birthplace of that ancient literature most familiar to educated Europeans. The condemnation of Galileo lingered in memory as a sign that all was not well in Italy—and such Italians as Muratori and Vallisneri agreed that a cultural revival was necessary—but some of the journalists thought they knew better. Louis Bourguet and his collaborators argued that Italian books already merited more attention than they had been receiving north of the Alps; other journalists welcomed the erudition of Muratori; and Dutch publishers could be persuaded to produce editions of the works of the Bolognese. Even Fontenelle came close to admitting that he had been wrong. In one of those famous eulogies, he had con-

---

[90] For Swiss-Italian relations, there are valuable studies by Cavadini-Canonica 1970; Crucitti Ullrich 1974; and Kurmann 1976. Less useful is Hans Fischer 1973. On Franco-Italian scholarly relations, see Waquet 1989a. Graf's old study 1911, chap. 4, is still useful for Anglo-Italian relations. For Leibniz, see Robinet 1988.

[91] In general, Badaloni 1968. Letters to Vallisneri are in Conti 1972, 386–99, and one reply, 434–35.

[92] AAS, P-V, t. 11, fol. 4r (28 July 1683); letters of 1691 and 1725 in Royal Society, "Early Letters," s.v. Marsili; *PT* 20 (1698), 306; and above, at n. 70. See also Stoye 1994, 111–12, 162–63, 263–70, and the journal of his 1721 voyage published by McConnell 1986. As indicated earlier, nn. 81, 85, his endowment of the Bologna Academy of the Institute also received attention in northern publications.

demned Italian learning as too much tied to things "old," even though he had already acknowledged that at least one Italian, Domenico Guglielmini, had exhibited signs of the most admirable "modernity." Without admitting earlier lapses in judgment, Fontenelle eventually pronounced the new academy of Bologna, endowed by Marsili, as possibly even "the [new] Atlantis of Bacon."[93]

The several means of communication explored in this chapter should no doubt be judged as erratic and inefficient, especially when compared with the ideals of the Republic of Letters. As Françoise Waquet once pointed out, journals in particular had their shortcomings: many were short-lived, their international circulation cannot be assumed, wars could disrupt the mails—and readers sometimes found journals offensive rather than informative.[94] Nor did learned societies readily communicate to the public or to other institutions the projects their members were engaged in; personal correspondence might be less restrained, but philosophers on the whole preferred to let the learned world wait for the finished printed product. In fact, to speak of savants as united by two institutions, the book-review journal and the learned society, is to tell only a partial tale, because correspondence continued to be a crucial link among members of the Republic of Letters.

To admit the inadequacies of journals and learned societies should be taken not as a counsel of despair but rather one of prudence. If, for example, it took Giovanni Bianchi in Rimini almost two years to track down a copy of the *Mémoires de Trévoux* in which he had been harshly criticized, this does not mean that one should cease to use *Trévoux* as an international journal. Caution merely dictates that one seek reviews of a single publication in more than one journal before one can say that that publication achieved international repute.[95] Similarly, one cannot assume that even famous academic journals such as the *Philosophical Transactions* and the *Histoire et Mémoires* of the Paris Academy were instantly available to and read by those members of the Republic of Letters most vitally concerned with the sciences. Again, this is a message of prudence and proof.

Communication in the twentieth century has been revolutionized, in the first instance, by Alexander Graham Bell, and more recently by the com-

[93] A balanced assessment is in *Fontenelle* 1989, 577–88. Eulogy of Marsili, in Fontenelle 1825, 2:277; of Guglielmini, in Fontenelle 1709, 1720, 2:62–88.
[94] Waquet 1983. The wholly erratic effects of war are indicated by Ultee in Neumeister 1987, 2:535–46.
[95] Problems exist even here, as Goldgar 1995, 95–97, offers examples of journalists not wishing to duplicate notices of books published in other journals.

puter, so that we associate communication with speed. In 1700, despite the later anxieties of Giovanni Bianchi, everyone awaited the mails, even though they were slow and had irregular rhythms, as Jean-Michel Gardair has noted in his perceptive comment that the distance from London to Rome was shorter than that from Rome to London.[96] Nonetheless, and however slow, erratic, unpredictable, and inefficient, communication among men of letters worked sufficiently well to ensure a common language and a common outlook, so that even those resistant to "modern" scholarship learned to manipulate the ideas and vocabulary of 1700.

[96] Gardair 1984, chap. 11, esp. 300.

CHAPTER 2

Certainty and Probability

D ebates about what constitutes truth and knowledge have a history long antedating the seventeenth century, but the issues acquired unusual depth, breadth, and urgency at that time. As scholars like René Pintard, Richard Popkin, and Henry Van Leeuwen have shown, one of the fruits of the sixteenth-century Reformation was prolonged dispute about the claims to certainty of the rival religious confessions. Received truths of a different kind were also shaken by the telescope's revelations of sunspots and lunar topography. At the same time, the revival of ancient skepticism posed a threat to the reliability of all knowledge. Philosophers and theologians responded in a variety of ways to such challenges, some, following the route of Michel de Montaigne and Pierre Charron, moving toward skepticism coupled with religious fideism. To Pierre Gassendi, the rejection of Aristotelian philosophizing entailed an emphasis on such probabilistic conclusions as could be discovered by experiment. For René Descartes, doubting became the first step toward the re-establishment of a route to certainty. And Francis Bacon's proposals were meant to lead to truth by way of what he called a new method of induction.[1]

Amid these diverse yet overlapping solutions to epistemological problems, the activities of natural philosophers in particular required scrutiny.

[1] Pintard 1983; Popkin 1968; Van Leeuwen 1963.

How solid were their claims to be establishing not only a new but a better philosophy? What confidence could one place in the truth of the discoveries celebrated by John Ray and others, and were there ways to strengthen that confidence? Since so much knowledge of the world depended on the testimony of others, how could one evaluate that testimony? For similar reasons (and for others to be explored later), historians, too, found themselves on the defensive, as they sought to put their own discipline on a more solid foundation.

These fundamental questions captured the attention of both scientists and historians to such an extent that Sergio Bertelli has described the later seventeenth century as a period of "crisis"—a crisis of confidence—in the scientific and erudite disciplines.[2] What had been occurring was no less than a re-examination of the nature of knowledge. Traditionally, nothing had qualified as "knowledge" except the deductive, the causal, the certain: Aristotelian physics met these criteria. Increasingly, however, certainty was being confined almost exclusively to mathematics, all other kinds of knowledge being probabilistic to some degree. Sometimes disconcerting, sometimes stimulating, this state of affairs led philosophers to try to enlarge the realm of certainty or at least to find ways in which the less probable might be elevated to the more probable. In this quest, the Scholastic argument from authority was consciously rejected: "authority" became merely "authors" whose views were to be evaluated.[3]

Although debate would continue far into the eighteenth century, European intellectuals had by about 1720 reached a marked consensus about fundamental matters of truth and probability. This is not to say, of course, that they agreed on every issue, but even disagreements were expressed in a language understood by all. By the 1720s, one observer judged that intellectuals had become so methodologically self-conscious that they—he addressed Italian philosophers in particular—should write autobiographies: by recounting their own discoveries and methods, the contributors to this project would serve as models for future generations.[4]

---

[2] Bertelli 1955.

[3] In addition to works cited in the preceding two notes, I have found Borghero 1983 exceptionally valuable; also useful are Dear 1988; Hacking 1975; Patey 1984; and Shapiro 1983. Shapiro ranges widely in British writings, and her passing remarks about the Continent are confined to contrasts between British empiricism and Continental rationalism (e.g., p. 270); for critical comments, see Aarsleff 1982, 9.

[4] Porcia discussed this project with friends as early as 1721, and public announcement appeared in 1728 in Calogerà's *Raccolta;* discussion is in DeMichelis 1979, chap. 3. The first work in the series was the famous autobiography of Giambattista Vico. Porcia himself gathered from Vallisneri the materials for a biography, published in Vallisneri 1733, vol. 1, and now available in a critical ed., Porcia 1986.

MATHEMATICAL AND DEDUCTIVE KNOWLEDGE

When Montesquieu recorded in his *Pensées* that mathematicians deal with truth and falsity, and not with the intermediate realm of probability ("in this respect, there is no more or less in mathematics"), he echoed what had already been a truism in Aristotle's day. For Aristotle, as for Galileo, mathematics offered the perfection of truths not open to qualification or argument. In 1632 Galileo's Salviati would declare that only in mathematics does the human intellect achieve "absolute certainty," or the certainty that comes of "understanding necessity."[5] A comparable assertion occurs in Descartes's *Discourse on Method* (1637) in the famous discussion of the author's own education. Admitting that he had found both pleasure and intellectual dissatisfaction in poetry, history, and other subjects, he continued: "Above all I delighted in mathematics, because of the certainty and self-evidence of its reasonings. But . . . I was surprised that nothing more exalted had been built upon such firm and solid foundations."[6]

In elevating mathematics as tool and model, Galileo and Descartes consciously attacked the Aristotelian position that mathematics applies to very little of the physical world and that each discipline has its own methods, proposing instead the "transfer" of mathematics to nature. Debates early in the century—in the Jesuit college in Rome and in the circle of Marin Mersenne in Paris—centered on some issues crucial to such a transfer. Could mathematics, being noncausal, provide true knowledge? Was it legitimate to isolate for study the mathematizable aspects of natural bodies? Should mathematical analysis be trusted if it conflicted with the established principles of philosophy?[7] To Galileo, nature was fundamentally mathematical in structure, and the proper method of study required that one "deduct the material hindrances"—that is, the planes that are not perfectly smooth, the spheres that are not perfectly spherical, the secondary qualities insofar as these cannot be reduced to particles in motion. Among his disciples, however, Evangelista Torricelli seems to have been unusual in pursuing further this kind of mathematical physics. More typical among Galileans were Torricelli's own barometric experiments and the activities

---

[5] Galileo 1962, 103. Also, *The Assayer* (*Il Saggiatore*, 1623), in Galileo 1957, 237–38. Aristotle, *Metaphysics*, XI, iv, and *On the Heavens*, III, vii. Montesquieu 1949, 1:1181.

[6] In Descartes 1984, 1:114. For somewhat comparable statements about the sterility of education and the appeal of mathematics to one of Descartes's contemporaries, see Pagel in Underwood 1953, vol. 1, esp. pp. 491–92.

[7] Crombie 1975; Dear 1988, chap. 4; Funkenstein 1986, 36–37, 296–97; Feldhay 1987.

of the Cimento: mathematics here meant the measurement of any quanti-
ties that happened to be measurable. When Leopold of Tuscany, patron of
the Cimento, said that a good philosopher must be "a good geometer," his
academicians had for some years been pursuing the program of precise
measurement.[8]

In his well-known assessment of Galileo's work, Descartes complained
that mathematics had been applied only to selected subjects. As a result,
Galileo had described discrete phenomena but had not sufficiently
explained these because he had neglected "first causes in nature," and thus
"had built without foundation."[9] For Descartes, then, as for many thinkers
who rejected larger or smaller elements of his philosophy, truth resided
not only in mathematics but also in those deductive procedures modeled
on geometry. In the epistemological language of the period, Descartes sub-
sumed mathematics under metaphysics. Robert Boyle thus analyzed
Descartes's procedure as beginning with God's existence, a "metaphysical
certainty," so that he would have a firm foundation before venturing
to deal with geometrical demonstrations.[10] Other writers did not always
pause to dissect Descartes's method in this fashion, but they tended
nonetheless to equate metaphysical and mathematical as the highest
degree of certainty. Nor did they necessarily start with God's existence, but
rather with definitions of matter and other physical "suppositions" considered indubitable. In this realm lay truths to which assent was "compelled,"
or, in Descartes's language, ideas so clear and distinct that their opposites
were impossible.[11]

Confidence in the geometric (deductive) mode became extraordinarily
pervasive. Precision and rigor had earlier typified treatises on theology,
mathematics, and astronomy, but by the end of the seventeenth century,
Fontenelle could write of a "geometric spirit" not confined to mathemati-

---

[8] Galileo 1962, 206–8; 1957, 273–78. For Torricelli, see the article by Mario
Gliozzi in the *DSB*. Leopold's letter to Huygens, 10 February 1668, quoted in Middleton
1971, 325.

[9] Descartes to Mersenne, 11 October 1638, in Mersenne 1945, 8:95. Also quoted in
Popkin 1968, 152–53, where "first causes" is changed to "the primary cause."

[10] Boyle, "The Excellency of Theology," published 1674 but written about a decade
earlier, in Boyle 1772, 4:42. The most wide-ranging account of Continental controversies
aroused by Descartes remains Bouillier 1854. Among the more important modern works
are Clarke 1989 and *Descartes* 1950, as well as the durable classic by Mouy 1934. For Britain,
see Lamprecht 1935 and Rogers in North and Roche 1985. A valuable analysis of certain
aspects of Cartesian historiography as well as early responses to his reductionism is in
Borghero 1983, chap. 1.

[11] Shapiro 1983, chap. 2, offers British examples that show some variations in how to
identify and classify what is most certain.

cal subjects but manifesting itself in "the order, clarity, precision, and exactitude that have been prevalent in good books for some time."[12]

At times, admiration for mathematics showed itself in loose form, as when in discussions of language—from the comments by Thomas Sprat to the treatises by John Wilkins and the Port-Royal grammarians—philosophers insisted that a mathematical model should be borne in mind if language were to achieve clarity and precision.[13] But the geometric mode proper can be found in Spinoza's *Ethics* (1677), organized into definitions, axioms, postulates, and propositions (with corollaries), and in Pierre-Daniel Huet's *Demonstratio evangelica* (1679), a reply to Spinoza. If Thomas Hobbes insisted that the deductive model was the only sure route to a proper political philosophy, John Locke thought that moral philosophy might attain certainty if one began with proper definitions and proceeded with "the same indifferency" typical of the mathematician's search for truth. John Craig in 1699 tried to assess in mathematical fashion the credibility of human witnesses, and William Whiston in 1702 marshaled axioms, hypotheses, and propositions in his examination of Old Testament chronology. From Frankfurt came a treatise on experimental chemistry, "ex principiis mathematicis demonstrata" (1681), and Leibniz provided a "demonstration" of the suitability of one candidate for the elective throne of Poland. Among unpublished efforts, Neapolitan Carlo Buragna produced a philosophical treatise "demonstrated in the geometrical manner," and Dubliner Samuel Foley wrote an "Essay attempting in the Geometrical Method to Demonstrate an Universal Standard whereby one may judge the real Value of Everything in the World."[14] Indeed, the language of demonstration and deduction turns up in some poetry, as John Dryden's Adam

[12] Fontenelle, "Préface sur l'utilité des mathématiques...," originally published in *HARS* [1699] 1702, the translation here quoted from Marsak 1961, 89. See also the comment by Niderst 1972, 49, on the importance of Cartesianism and mathematics in Fontenelle's eulogies of academicians. Contrasts between the sixteenth and seventeenth centuries may be extracted from Rossi 1970 and Febvre 1982, chap. 11 and passim. Febvre alludes not to mathematics but to the lack in the sixteenth century of a concept of "law" (of nature). "Chronology" was also meant to be a precise discipline, but the problems inherent in ancient texts used to calculate the age of the world meant that results were always open to question and modification.

[13] Of the many discussions of the reform of language, I have found most valuable Knowlson 1975; M. Cohen 1977; and Aarsleff 1982. Among alternatives to the mathematical model, some thought language might be made to correspond to either the "essence" of natural objects or the perceived order of the natural world. To find a "natural order" and a language that expressed it became a subject of prolonged dispute among botanists, beginning late in the seventeenth century.

[14] Huet will be discussed later in this chapter. For Hobbes, see Pocock 1973, 149, 153–54; Locke, *Essay*, IV, iii, 18–20; Craig [1699] 1964; Whiston 1702. I have not seen Johann Helfrich Jüngken's *Chymia experimentalis curiosa ex principiis mathematicis demonstrata*

uses the *cogito* to discover his own existence, and Sicilian poet Tommaso Campailla's Adam receives angelic aid to discover the Cartesian structure of the universe and the ascent to God.[15]

Throughout this period, the demonstrative ideal shows itself in the writings of experimentalists as something like a goal, a hope, or an unrealizable dream. The goal seemed clear enough to Nicolaus Steno, who in 1667 announced that progress in anatomy depended on the application of mathematics to the study of muscles. That Steno's treatise was geometrical in structure did not, however, disguise its experimental content. Similarly, Isaac Newton adopted "a kind of axiomatic presentation" in the heavily experimental *Opticks* (1704), but several years later added to the text an explicit avowal that the demonstrative ideal had not been and could not always be attained: "although the arguing from Experiments and Observations by Induction be *no Demonstration* of general Conclusions; yet it is the best way of arguing which the Nature of Things admits of."[16]

Demonstration being the ideal, both Newton and Boyle were sometimes criticized for failing to provide it, the critics unconsciously echoing the complaints of Descartes about Galileo. For Spinoza, Hobbes, and Leibniz, the mechanical philosophy could readily be illustrated and rendered probable by experiments, but it could also be demonstrated by the simple expedient of making a few "suppositions" from which one could deduce explanations of experimental phenomena. A remarkably similar complaint came from Robert Hooke, that talented experimenter, in his critique of Newton's theory of light and colors. According to Hooke, Newton's experiments, admirable in themselves, had no "necessary" connection with the theory Newton claimed to derive from them.[17]

(Frankfurt/M, 1681), listed in the catalogue of the British Library. The Leibniz "demonstration" is mentioned by Akerman 1991, 242, and the treatise of Buragna (d. 1679) by Fisch in Underwood 1953, 1:536. For Foley and other Dublin examples, see Hoppen 1970, 33, 81–82, 122–23. Jüngken's book was reviewed in *HWL* (July 1706), 387–92, without any comment on its "mathematical" aspirations; as will be indicated below, Chapter 3, the same silence on this method greeted Steno's *Prodromus* (1669).

[15] Dryden, *The State of Innocence* (1674), in Bredvold 1956, 67–68. Campailla, *L'Adamo ovvero il Mondo creato* (1709, 1728), in Garin 1966, 2:879–80.

[16] The remark on Newton's "axiomatic presentation" comes from Henry Guerlac, in *DHI*, 3:386. Newton's quoted statement (italics added) was added to the *Opticks* in the 1717–18 edition as part of the famous Query 31. Steno, [1667] 1969, 1994, and analysis in Bastholm 1950, 142–63.

[17] Hooke's critique of 1672 is in I. B. Cohen 1958, 110–15. The issues dividing Boyle and Spinoza, expressed in the latter's correspondence with Oldenburg, are analyzed by A. R. Hall and M. B. Hall in *Mélanges* 1964, 2:241–56, and by Colie 1963. For Boyle and Hobbes, see Shapin and Schaffer 1985, esp. chap. 4. For Boyle and Leibniz, see Loemker 1955.

As long as chains of reasoning were carefully constructed, the merits of the geometrical mode lay in both the certainty of the conclusions and the *nature* of those conclusions. That is to say, deduction yielded causal explanation and thus true knowledge. By contrast, Isaac Newton knew that the mathematical laws of the *Principia* (1687) were descriptive, not causal; he had not provided those chains of reasoning that would explain why gravity varied inversely with the square, rather than (say) the cube, of distance. Nor, of course, had he tried to explain what gravity was To his disciples, Newton had displayed admirable restraint in offering laws and admitting his ignorance of causes. To critics who admired Newton's mathematical brilliance, this was still only mathematics in a descriptive mode—not deductive certainty, not causal explanation, not what a French reviewer called genuine physics. This assessment in the *Journal des savants* states the case accurately for its time, and French Newtonians of a later generation eventually allowed what Newton himself could hardly bring himself to acknowledge—namely, that to seek the causes of gravitation was irrelevant to the acceptance of Newtonian physics.[18]

The limits of man's ability to arrive at a knowledge of causes were analyzed not by Newton but by Robert Boyle, who, for modern historians as for his contemporaries, was the quintessential empiricist. For Boyle, even seemingly undeniable axioms, such as the impossibility of creating something out of nothing, displayed the limits of human knowledge, because God had in fact created the world ex nihilo. To Edme Mariotte, the same axiom was merely a law based on experience, but Boyle found it impossible and undesirable to exclude the divine from natural philosophy. Furthermore, Boyle distrusted the geometric mode—mere measurement was a different matter—for a number of reasons: postulates and chains of reasoning may contain error, mathematical laws are noncausal, and many aspects of nature cannot be reduced to mathematics.[19] And yet, for all his important doubts, Boyle seems to have wished otherwise, as when he remarked: "it is not always necessary, *though it be always desirable,* that he, that propounds an hypothesis . . . be able *a priori,* to prove his hypothesis to be true, or demonstratively to show, that the other hypotheses proposed

[18] *JS* in I. B. Cohen 1958, 428–29. Newton's critics are also discussed by Koyré 1965, Appendices. For French Newtonians, see Guerlac in Wasserman 1965, 307–34.

[19] Among his many discussions of mathematics, see esp. Boyle 1772, 2:740–42. Shapin 1988 offers a different analysis. I cannot agree with Sargent, in Hunter 1994, esp. 65–66, that Boyle had no "probabilistic epistemology." To be sure, he did not develop such views with the kind of rigor that a post-Humean philosopher would require. Mariotte will be treated in the next pages.

about the same subject must be false."[20] Admitting that this would be very hard to do, Boyle here acknowledged as an ideal a method he himself found useless and even unusable, for there were, to him, no undeniable axioms from which an a priori demonstration could proceed.

Other empiricists, too, considered "demonstration" to be a desirable but not always attainable goal. Most, however, would not have followed Boyle in so closely integrating God's freedom into their examinations of the natural world or the limits of human knowledge. In this respect, Jacques Rohault expressed a typical point of view when he asserted that science deals not with what God could do, but what He has done. For Rohault, a Cartesian, certainty was at least sometimes within human grasp, although he also admitted that nature (not God) might well employ more causes than philosophers had discerned.[21] In comparable fashion, Bolognese physicist Domenico Guglielmini would preface his *Della natura de' fiumi* (1697) with comments on the certainty of mathematics as compared with the troubles of the experimenter. Among the several difficulties indicated in this often reprinted treatise, so many variables exist in nature that nature (not God) is thus likely to employ more causes than man can immediately detect. Not surprisingly in this period, Guglielmini's treatise was organized as a set of theorems.[22]

Some members of the Academy of Sciences in Paris—notably Huygens, Roberval, and Claude Perrault—agreed on the superior status of mathematics, acknowledging the lesser, probabilistic value of other researches. In this area of discussion, Edme Mariotte's *Essay de logique* (1678) may be taken as a representative view of what his colleagues thought.[23] Mariotte begins with "principles of knowledge, or first truths," defined as "propositions so certain and self-evident that, provided one properly understands their meaning, one cannot doubt their truth; and they are accepted as certain

[20] "Of the excellency and grounds of the corpuscular or mechanical philosophy," 1674, in Boyle 1772, 4:77. "A priori" is italicized by Boyle; other italics are added. In the same paragraph he refers to nature as "God's epistle written . . . in mathematical letters."

[21] I have consulted both Rohault 1671 and the excellent English version (1723; 2d ed. 1728–29). Passages cited are in pt 1, chap. 5, secs. 12–13.

[22] I have used the reprint of Guglielmini in *Raccolta* 1723, 2:227–30. He also cites here Boyle's early essays on the problems of replicating experiments. A Latin translation was also published in Guglielmini's *Opera omnia* (Geneva, 1719).

[23] In addition to the *Essay*, as reprinted in Mariotte 1717, 2:609–701, I have also benefited from the analyses by Brunet 1947; Rochot 1953; and especially Coumet in *Mariotte* 1986, 277–308. Alan Gabbey, in *Mariotte* 1986, 205–44, has argued for Mariotte's debt to Roberval at least for elements in the short part 1 of the *Essay*. The Gabbey textual evidence is persuasive but does not destroy Rochot's suggestion that the Academy engaged in discussions of the same sort as in the Mariotte *Essay*. See Salomon-Bayet 1978, esp. 99–100, as well as examples to be discusssed below.

and infallible, without the assumption of any prior knowledge." As examples, he offers mathematical axioms: the whole is greater than the constituent parts, two things equal to a third are equal to each other, and so on.[24] Mariotte then explains that in experimental philosophy the search for causes does not proceed from "first truths," but from inferences drawn from effects.

> To explain natural phenomena, one is often satisfied to seek a single cause; and yet, ordinarily, several causes work together and contribute in various ways to produce an effect; from which it follows that it is impossible properly to explain most effects, because one does not know most of their causes, and it is easy not to know causes because they are not sought.[25]

Inference, Mariotte adds, does not require that we construct regressive chains, but only that we discover causes "sufficient" to explain the effects. At times, inference can also yield what he calls "natural maxims" or "empirical principles," one of these being that nature cannot produce something out of nothing. While we do not know the causes behind such principles, the principles themselves become fruitful tools of further discovery and inference.[26]

What Mariotte has done is to introduce his *Essay* with a discussion of deduction, necessary to a treatise on logic, and then to ignore the matter. Although it has been said that he had no need to elaborate further on so familiar a topic, one of Mariotte's comments on mathematics suggests a different reason for his decision.[27] In remarking that one need not seek causes for "natural maxims," Mariotte adds that scientists should "in this respect imitate geometers, who do not try to prove the first principles they employ, but devote themselves to extracting from such principles ever new consequences." Mariotte must have known that his recently deceased colleague, the mathematician Roberval, had been attempting to reduce to a minimum the number of unproven first principles used by geometers. To acknowledge that even geometry relies on unproven axioms seems to have meant to Mariotte that geometry was not markedly different from experimental philosophy.[28]

---

[24] Mariotte [1678] 1717, 2:613–14.

[25] Ibid., 2:659.

[26] Ibid., 2:655–56.

[27] For the no-need suggestion, see Brunet 1947, 27, and the excellent article by Mahoney in the *DSB*.

[28] Mariotte [1678] 1717, 2:659.

If this reading of Mariotte is correct, then he was not alone in pointing out the vulnerability of mathematics and in seeking to elevate empiricism. Not only Roberval but also, in their diffcrent ways, Pascal and Leibniz had commented on the desirability of strengthening the foundations of geometry. In 1679, Pierre-Daniel Huet, having abandoned his earlier enthusiasm for Cartesianism, probably thought he was delivering the coup de grâce by exposing the weaknesses of geometry. Mathematicians, he declared, not only deal with things having no real existence (points, lines), but they also have disagreed on such fundamental matters as the provability of postulates, the validity of particular proofs, and the best logical order of propositions. One must therefore conclude that geometry is no more infallible than "moral demonstrations" based on the evidence of the senses and the experience of mankind.[29] In raising the issue of points and lines, Huet alluded to a debate that had made its appearance earlier in the century and that stems ultimately from Aristotelianism: Is mathematics applicable to the physical world? Pierre Gassendi, for example, had queried the concept of "infinite divisibility," arguing that it was irrelevant to the physical world of atoms. At a later date, Roberval took part in an impassioned debate at Montmor's Academy in Paris on whether points and lines have any real existence. Still later, these questions would be summarized in Pierre Bayle's *Dictionary* (1697), Bayle favoring the Gassendist position.[30]

Huet and company did not immediately succeed, as Huet's own work shows. His effort to weaken confidence in mathematics was embedded in his *Demonstratio evangelica,* which employs the deductive mode in order to show that the truths of Christianity can be proven with "all the rigor of geometry."[31] In the next decades, Malebranche's Cartesianism would reinforce confidence in deduction, and Newton's *Principia* would reveal the power of mathematics as applied to the physical world. On the other hand, extravagant efforts to apply the geometric mode to matters ranging from ethics to chemistry had run their course, fewer treatises of this kind appearing in the eighteenth century.[32] If Fontenelle expressed interest in efforts

---

[29] For Huet, see Tolmer 1949, 436–43; see also Huet 1810, 1:23–30, 201–2; 2:157, 162–64, and the English translator's summary and judgment, 2:470–72. For other critics of mathematics, see Salomon-Bayet 1978, 242–43, and Victor Cousin 1865, 3:232–33.

[30] Rochot 1957. For the debate chez Montmor, see the letter of Constantyn to Christiaan Huygens, 18 November 1660, in Huygens 1888, 3:178. For Bayle, Labrousse 1963, 2:226–28.

[31] Tolmer 1949, 539, and important discussion in Borghero 1983, 173–74.

[32] For example, the geometric mode can still be found in the John Mitchell essay on skin color, in *PT* 43 (1744), 102–50, in William Warburton (in Evans 1932, 52–53), and,

to quantify human affairs, this was a far cry from the geometric mode, as he selected for admiration works on statistics and probability: Jakob Bernoulli's *De arte conjectandi* (1713) and William Petty's *Political Arithmetic* (1690). Indeed, a number of writers late in the seventeenth century and through the middle years of the eighteenth, returned to the Aristotelian position, often citing Aristotle himself, that different disciplines have their own methods. Some adopted a view adumbrated early in the seventeenth century and later given pithy expression by Fontenelle: "there is in geometry *only what we have put there;* only the clearest ideas that the human mind can form about magnitude."[33] Recognizing that mathematics is a discipline unto itself, Nicolas Fréret, of the Royal Academy of Inscriptions, would defend historical study as having its own methods and validity, and Diderot and Buffon would produce arguments in behalf of natural history as an inherently and legitimately nonmathematical discipline.[34] If Diderot and Buffon are sometimes presented by historians as being opposed to mathematics, their position may be more accurately described as an attempt to define disciplinary boundaries. Both having started their careers as mathematicians, they came to emphasize that there is more in nature than can be analyzed by or expressed in mathematical terms. Buffon would eventually argue that one can achieve virtual certainty about nonmathematical aspects of nature.

THE EVIDENCE OF THE SENSES

The geometric mode being a method of proof, not a method of discovery, even its most dedicated proponents knew that some room must be allowed for the empirical study of nature. In the extreme case of Leibniz, that room

---

most notoriously, in Vico's *Scienza Nuova*. Needless to say, these writers employed "axioms" that were no more "indubitable " than Craig's in 1699 or Whiston's in 1702. G. Bianchi's *De conchis* (1739) was severely criticized in *Trévoux* (April 1740), esp. 608–19, for his reliance on "propositions."

[33] The Fontenelle quotation, 1696, is taken from Marsak 1961, 95, italics added. Fontenelle's interest in statistical studies of human behavior is signaled by Niderst 1972, 498–99, and his shift from human history as a record of caprice to history as revealing the regular laws of human nature is discussed by Dagen 1966, esp. 634–35. Aristotle, *Nicomachean Ethics*, I, iii, 4.

[34] Diderot's *pensées* (1754) are almost always seen as anti-mathematical, but to me his text indicates a shift in priorities: it was time to devote attention to the non-mathematical sciences. Buffon overtly challenged the supremacy of mathematical truth, as will be apparent below, Chapter 8. The most complete study of Fréret is by Renée Simon 1961, but more analytical is the discussion by Mercier in *La Régence* 1970, 294–306. There is no substitute for the trenchant prose of Fréret's articles published in *MARI*.

might be small, as when he predicted in a letter to Henry Oldenburg that philosophers would soon have "as certain knowledge of God and the mind as we now have of figures and numbers. . . . And when these studies have been completed . . . men will return to the investigation of nature alone, which will never be entirely completed." As further clarification, Leibniz added that, deductive certainties once established, future generations would "always philosophize in the manner of Boyle."[35] Since Leibniz himself spent many years studying such far-from-certain subjects as history and geology, his comments to Oldenburg—and comparable remarks, more moderately expressed, pervade his writings—must be viewed as judgments of the epistemological value of different methods of philosophizing.[36]

The rationalist ideal of certainty did not prevent natural philosophers from pursuing empirical inquiries, and indeed the closest rival to mathematics for certainty was the *fact*. Some writers even went so far as to say that facts were as certain as mathematics, but what seems to have been the prevailing judgment was expressed by a learned journalist: "one must not forget that the Demonstration of a fact, even when done in a geometrical way, is always a moral Demonstration, because it is based only on the judgment of probabilities."[37] Defenders of the fact liked to point out that some were surely indubitable, but trouble arose instantly when one tried to pass beyond the fact to its causes. The existence of Stonehenge was a fact; explanations of its origins were not. The behavior of mercury in the Torricellian tube was a fact; how to explain the phenomenon was not.

Establishing facts became an essential part of the programs of institutions like the Cimento, the Royal Society, and the Paris Academy. Authors both ancient and modern had to be tested for accuracy, and new facts about nature, discovered by academicians, had to be confirmed by their colleagues. As a stupefying example of the latter procedure, botanist Nicolas Marchant brought a single plant to the Paris Academy each week for more than a decade; he read a description of the specimen to the members, they compared his words with the object, and they pronounced the description accurate.[38] In a more thoughtful vein, Rohault's *Traité de physique*

---

[35] Leibniz to Oldenburg, 28 December 1675, in Leibniz 1969, 166. Also in Oldenburg 1965, 12:98.

[36] E.g., Leibniz 1969, 133, 187–89, and Leibniz 1981, bk. 4, chap. 12. See also his confession that to study "the nature of the universe" was more "solid" an enterprise than empirical work of all kinds, in Leibniz 1875, 3:182. Nonetheless, he actively engaged in promoting the establishment of learned academies; these activities occupy much of Leibniz 1859, vol. 7.

[37] *Europe savante* (April 1718), 258. Cf. Filleau 1672, 15–16; Wilkins [1693] 1969, xix, 8–9. See also Van Leeuwen 1963, 69–70; Shapiro 1969, chap. 8.

[38] Based on scrutiny of AAS, P-V, from 1666 until Marchant's death in 1678.

(1671) outlined three varieties of fact, all of which he called "experiments" (*expériences*).

> The first is, to speak properly, only the mere simple using of our Senses; as when accidentally and without Design, casting our Eyes upon the Things around us, we cannot help taking Notice of them. . . . The second Sort is, when we deliberately and designedly make Tryal of any Thing, without knowing or foreseeing what will come to pass. . . . The third Sort of Experiments are those which are made in Consequence of some *Reasoning* in order to discover whether *it* was just or not.[39]

Rohault judged "the third sort" to be most valuable in natural philosophy.

Fact-collecting—Rohault's "first sort"—often in the seventeenth century meant collecting the *objects,* not just the information. Now and then, the owners of museums tell us why they engaged in this activity. Apart from whatever prestige may have been attached to ownership, and apart from the urge to possess the curious and the rare, some collectors pursued the Renaissance tradition of seeking to illustrate and clarify ancient texts like Pliny's.[40] Later in the century, some writers came to view collecting as serving the more "modern" purposes of testing such texts and of enlarging knowledge of nature. As John Ray put it in 1690, moderns were engaged in "the Study of Things." The ancients had excelled in the language arts, but, "Words being but the Pictures of Things . . . to be wholly occupied about them, is to fall in love with a Picture, and neglect the Life." More explicitly than Ray, Robert Hooke and John Woodward explained that to collect "Natural Bodys," and not just rarities, would be essential to the progress of knowledge.[41]

In Britain, the collecting phenomenon is usually associated with the philosophy of Francis Bacon, but it should probably be pushed back to the inspiration of Camden's *Britannia* (1586), that much-reprinted exercise in the collection of miscellaneous information. If it is harder to detect the models or inspiraton behind the many Continental collections, some were certainly both famous and extensive, such as Ole Worm's museum in

---

[39] Rohault [1671] 1728, author's preface. *Expériences* has no English equivalent. In the *Regulae ad directionem ingenii,* not published until 1684 (in Dutch) and 1701 (in Latin), Descartes expressed disapproval of what Rohault called his first two sorts; see Descartes 1984, 1:15–16.

[40] Borel 1649, preface and p. 108; he also explicitly sees himself as continuing in the tradition of sixteenth-century physicians who were collectors. See Findlen 1994, chap. 2.

[41] Ray to Robinson, 15 December 1690, in Ray 1718, 241; also in Ray 1848, 229. Hooke 1705, 280 (text dating from ca. 1667–68). Woodward 1696.

Copenhagen and Athanasius Kircher's in Rome. One French collector, Pierre Borel, claimed in midcentury that he knew of 163 other collections, general and specialized, in France alone.[42] By the end of the century, the dictionaries of Antoine Furetière (1690) and the French Academy (1694) indicated that "curieux" had acquired a new meaning, now signifying a person wanting to know all sorts of miscellaneous things, especially about the novel and the rare.[43] Even Pierre Bayle's *Dictionary* may be described as the work of a lifelong "curieux," engaged in the collection (and evaluation) of historical facts. Seeking a coherent way to present such miscellany, Bayle selected a conventional (alphabetical) taxonomy.[44]

As Rohault remarked, random observations—which he coupled with more or less random experiments (his "second sort")—should not be condemned, because "Knowledge is continually enlarged by them." Noticeably uneasy, however, was Nathaniel Fairfax, who complained to Henry Oldenburg about Joshua Childrey's *Britannia Baconica* (1660).

> I have found [in Childrey] soe many storyes coming either to quite nothing, or changing into quite another thing, when dived into by a wary person, yt I dare scarce count any thing a phaenomenon of nature, yt is barely founded on ye relation of an Historian.[45]

Fairfax captured well the problem of mere miscellany: how could one evaluate such material, and did these "stories" mean anything? When Fairfax doubted the accuracy of Childrey's reports—"Historian" here meaning reporter or describer—he raised the issue of verification.

If Pascal, with vigor and polemical skill, insisted that no authority could change or condemn matters of fact, other writers pointed out that facts themselves must first be verified before one embarked on seeking causes.[46]

---

[42] In general, see Impey 1985. For French collections (and Venetian as well), see Pomian 1987 and the spirited study by Schnapper 1988. For Italy, Findlen 1994 is very informative, despite strained efforts to see continuity between Aldrovandi and the later seventeenth century. Among studies of Camden and his tradition, especially useful is Fox 1956. The sheer joy of collecting is evident in the above works. Schnapper 1986a evaluates the notion that royal power and prestige were enhanced in France by royal collecting.

[43] Contrast the earlier lexicon by Randle Cotgrave [1611] 1950, where "curieux" is defined as "quaint, nice, daintie, precise; doubtfull, scrupulous, heedfull, busie; too too diligent; more carefull than needs."

[44] Labrousse in *Religion* 1968, esp. 62–64.

[45] Fairfax to Oldenburg, 15 April 1667, in Oldenburg 1965, 3:401. An early (1666) essay by Hooke may also be read as the work of a confused *curioso*, fascinated by detail and struggling to find guides through the morass. Hooke 1705, 1–65.

[46] Pascal [1650s] 1941, 585–88, 614–15, where even popes are fallible in matters of fact. For verifying facts before seeking causes, see the discussion in Borghero 1983, 244–46. Borghero and other historians single out Fontenelle's *Histoire des oracles* (1686) for the cautionary tale of the gold tooth.

In some cases, one wonders how such verification could have proceeded, for the Repository of the Royal Society contained a witch's teat, Athanasius Kircher's museum exhibited a rib and the tail of a siren, and Lodovico Moscardo owned a "true basilisk" rather than one of the forgeries he knew were being fabricated.[47] To have a collection at all, however, was recognized as a first step on the road to verification, and illustrated books served a similar function.

One of the first illustrated guides intended in part for collectors was Charles Patin's *Introduction à l'histoire, par la connoissance des médailles* (1665), and the frequency with which this little book was reprinted suggests a healthy market among amateur numismatists or those general collectors who included coins among their enthusiasms.[48] The budding numismatist actually enjoyed advantages over other philosophers, for authentic coins usually existed in sufficient quantity so that a new find could be compared with a body of genuine material. Furthermore, the illustrated book was a convenient substitute for visits to public and private museums. Such options were less readily available to naturalists, there being few guides to whole classes of natural objects. When Filippo Buonanni's *Ricreatione dell'occhio e della mente* appeared in 1681, it was not the first to contain engravings of mollusks, but it was notably more accurate than its illustrated predecessors. More than seventy years later, despite the critics and despite the publication by then of comparable volumes, Linnaeus would still find Buonanni's illustrations useful.[49] But no compendium could exhaust the riches of nature, as Martin Lister discovered in 1698 when he gained entry into many Paris collections; on one occasion, finding specimens unfamiliar to him, he borrowed some mollusks in order to have drawings made. In fact, in an age when new botanical and zoological specimens or descriptions were arriving from non-European parts of the world, collections and illustrated books might be of little help, and "verification" sometimes consisted of fitting the novelty into an existing system of taxonomy.[50]

[47] These examples are drawn from Pomian 1987, 94–95, Findlen 1994, 92, and Skippon in Churchill 1732, 6:672–74. See also Copenhaver 1991; Hoppen 1976; Schnapper 1988; and the examination of a forged basilisk by Grondona 1969.

[48] Inferences about the popularity of Patin are gleaned from the catalogues of the British Library and the Bibliothèque nationale. The title used in 1665 was changed to *Histoire des médailles*, but the content remained the same. Although Denis de Sallo mercilessly attacked the first edition, the judgment of the *Journal des savants* apparently had no effect on publishers who knew a good thing when they saw one.

[49] For serious criticism of Buonanni's illustrations, see *Acta* (February 1686), 111. See also Impey 1985, fig. 71, for one of the "fantasy" animals depicted by Buonanni. The durability of Buonanni is indicated by Dance 1966, chap. 1 and p. 43, and by Gersaint 1736, v.

[50] Lister [1699] 1967, 59. See also Raven [1950] 1986, 193, 216, 231, 250–51, and Blunt 1986, chap. 19. On the recognized need for illustrated books, see Stroup 1990, 69–70, 80, and Dezallier 1727.

Corroborating events proved even more difficult than authenticating objects or confirming discrete observations. One could only gather reports—Fairfax's "relation of an Historian"—and try to assess their credibility. Some of the rules governing acceptance of human testimony were laid down in the influential Port-Royal *Logique* (1662), in which Antoine Arnauld and Pierre Nicole warned the reader that the possibility that an event can occur is never sufficient reason to believe that it actually has occurred. In other words, attention in this instance must shift from the report to the reporter.[51] Since Arnauld and others developed their criteria for evaluating testimony in connection with historical events, their arguments will be more pertinent later in this chapter. The criteria, to be sure, were similar for past and present testimony, but the past posed additional problems that made the historian's task even more difficult than that of the natural philosopher.

Rules governing the acceptance of testimony were well known when, in 1665, the Royal Society received conflicting reports from two reputable astronomers, Adrien Auzout and Johannes Hevelius, about the location of a comet observed by both men. Henry Oldenburg on this occasion informed Boyle that "ye difference depending principally upon matter of fact, 'tis ye authority, number and reputation of other Observers, yt must cast the Ballance."[52] Other astronomers actually had made the same observations, but John Wallis seems to have been unaware of this when he indicated that the "great controversy" had become in essence a matter of evaluating the characters of Auzout and Hevelius. "I have no reason to suspect," said Wallis, "that either of them would willingly falsify an Observation: And yet how both can be solved, without allowing two Phaenomena [two different comets], I cannot tell. That there should be two, seemes somewhat odde; yet it is not impossible."[53] The Wallis solution would have been perfectly understandable to Antoine Arnauld: two comets are possible; the two witnesses cannot be impeached; and it is therefore likely that there really were two comets.

Oldenburg's summary of criteria is misleading in one respect: less significant than the number of witnesses—providing that there was more than one—was their character and independence. The latter trait would be stressed by Boyle when he reported in one of his tracts that he had culled

[51] Arnauld 1964, pt. 4, chap. 13. As a sign of the importance of the *Logique,* the Wing *Short-title Catalogue* lists a number of Latin editions published in London, as well as English translations—all of these before 1700. See also, Labrousse 1963, 2:14–15.

[52] Oldenburg to Boyle, 30 December 1665, in Oldenburg 1965, 2:653.

[53] Wallis to Oldenburg, 19 January 1666/67, in Oldenburg 1965, 3:313. See Guerlac 1981, 35–36; Shapin 1987; and Hetherington 1972.

his information "from the credible Relations of severall Eye-witnesses differing in nation, and for the most part unacquainted with each other."[54] Laid down as an "axiom" by Arnauld and Nicole, this methodological rule would be stated clearly in Mariotte's *Essay de logique.*

> When several people, without consulting each other, individually testify to a natural phenomenon in the same way and with the same circumstances, one must accept the truth of that phenomenon as a basic sensory truth. For, since innumerable different ideas are possible, it is very difficult, although not absolutely impossible, that several men should have the same idea about the same phenomenon with all the same circumstances, unless they had actually observed that phenomenon.[55]

Some writers cited as clearly analogous the criminal justice system, in which the evidence of one witness was insufficient to prove guilt, but the agreement of two or more independent witnesses could not be plausibly dismissed as the result of coincidence or collusion.[56]

In addition to possessing independence, trustworthy witnesses were variously described as men of upright character, equipped with sufficient learning to understand what they reported, or (conversely) so ignorant and naive that their testimony was obviously "sincere." For John Wallis, character played an important role in his conclusions about Auzout and Hevelius: not only were both experienced astronomers, but neither had any reason to lie. When Boyle gathered information about the nature of the seabed, he carefully indicated that his informants had had long experience at sea and that many were known to be of upright character. On another occasion, Boyle would point out to Henry More that particular experiments had been witnessed by a number of gentlemen, two mathematicians, and one of More's friends. In this curious list, gentlemen could be deemed trustworthy, More himself could testify to the character of his friend, and mathematicians presumably had trained and perhaps even detached minds.[57]

[54] "Of the Temperature of the Subterraneal Regions, As to Heat and Cold," in Boyle 1671, p. 4 of this tract, each work in this collection being separately paginated.

[55] Mariotte [1678] 1717, 2:623. Cf. Arnauld 1964, IV, vii, axiom 11.

[56] Sprat [1667] 1958, 100, and other English examples in Shapiro 1983, chap. 5. See also Leibniz in Davillé 1909, 465–69.

[57] "Relations About the Bottom of the Sea," in Boyle 1671, 3–6 of this essay (see above, n. 54), and Boyle to Henry More in Shapin and Schaffer 1985, 218, where only the gentlemen are singled out for notice. Also, Findlen 1994, chap. 5; Frank 1980, 132, 145, 292; Sargent in Hunter 1994, 67. For examples of personal insult and loss of honor felt by scientists when their results were doubted, see Brown 1976, 186, and Shapin 1987, 423. The word "sincere," often used, retained its old meaning of "without wax," i.e., honest, unpolished, undecorated; I am indebted to Richard Rouse of U.C.L.A. for this point.

But what recourse did one have if one did not know the character of an author or, worse yet, if a report were anonymous? Tackling anonymity in particular, Francesco Maria Zanotti, secretary of the Bologna Academy, explained to readers why he had decided to include in the first volume of the Academy's *Commentarii* (1731) a text of unknown authorship and provenance. One could, Zanotti explained, verify observations recorded in this text; furthermore, the manuscript was written with "candor" and other signs of truthfulness.[58]

In many cases, one could assume that the phenomena reported or witnessed were sufficiently "neutral" to provide no self-interested motive for distortion. This clearly was an important element in the Auzout-Hevelius dispute, and Fontenelle would repeatedly use his famous *éloges* to argue that men who studied nature became as dispassionate and reliable as nature itself. But some writers pressed the issue further, asking whether one should believe reports of prodigies and of generally incredible events. For Lorenzo Magalotti, secretary of the Florentine Cimento, it was relatively easy to reject ancient tales of the phoenix and the bird of paradise, but there was too much testimony—some of it modern and seemingly unimpeachable— for him to display comparable skepticism about the unicorn.[59] (At a later date, that most unconventional of cosmogonies, *Telliamed* [1748], would marshal a truly astonishing array of witnesses testifying to the existence of mermen and mermaids.) When Joseph Glanvill sought to prove the existence of good and evil spirits, he could not claim that his investigation was entirely neutral, for he saw his studies as crucial for belief in the immortality of the soul. His own bias confessed, he went on to charge his readers with the bias of living in what he called "a searching incredulous Age." Glanvill then proceeded to compile a series of case histories, some based on his personal investigations, intended to establish what he called matters of fact.[60]

Glanvill's efforts to show that spirits do exist follow closely the methodological canons developed by Boyle and other Fellows of the Royal Society—canons less well understood by followers such as Richard Bovet and Increase Mather who displayed none of Glanvill's investigative skill. As contemporaries across the English Channel realized, however, to establish the "fact" of strange phenomena was not to establish their causes. If Glanvill

---

[58] Zanotti in *Commentarii* 1 (1731), 91–92. See also the anonymous translator's note prefacing *An Account of the late terrible earthquake in Sicily* (1693). This text was sent to the Royal Society and may have been translated by Richard Waller.

[59] Basile 1987, chap. 6. For the Fontenelle eulogies, see above Chapter 1, n. 65.

[60] Glanvill [1689] 1966, [58–59], 67–68, 321–38; see also Cope 1956, chap. 4; Prior 1932; and Jobe 1981.

thought demonic possession an obvious cause, French jurists, theologians, and doctors increasingly found natural causes to be more plausible. Few could be sure that the devil did not exist, but it seemed clear that delusions, hysteria, illness, and fraud were more common aspects of human behavior than provable sorcery.[61]

In seeking to establish facts, Glanvill shared with others—including Arnauld, Bayle, Fontenelle, and Locke—the conviction that this process must precede any search for causes. Not surprisingly, Fontenelle's "libertinism" took him a step further: the likelihood of the events themselves might be a criterion even more important than the testimony of witnesses. In the specific case of ancient oracles, belief should be withheld, the prophets being frauds and the witnesses being deluded.[62] As contemporaries knew, this was dangerous and delicate ground, for it challenged the traditional Christian view that pagan oracles had been silenced with the advent of Christianity; furthermore, to question oracles implied comparable doubt about prophecy and even about miracles. Such Spinozism was generally repudiated, Catholics and Protestants agreeing that God's power and freedom could not be thus limited. But in "a searching incredulous Age," it proved as difficult to recognize God's intervention as it did the devil's. As usually defined, miracles were supposed to be confirmations of "true doctrine," but the matter could seldom be pursued much further, thanks to confessional disputes on the one hand and uncertainties about the extent of nature's powers on the other.[63]

Unlike facts and events, experiments in principle posed no questions about witnesses and testimony, for they could be replicated at will. Nonetheless, as Filippo Buonanni pointed out, when we try to confirm the microscopic observations of Redi or Malpighi, do we actually see only what we want and expect to see? In analyzing this problem, G. A. Borelli confessed that he had difficulties reproducing Malpighi's techniques of sectioning and staining specimens, but that he trusted his friend's reports and would merely counsel Malpighi to explain his methods more clearly.[64] The

---

[61] Bovet [1684] 1951; Mather [1684] 1856; Shapiro 1983, chap. 6. For France, Mandrou 1968 is indispensable, but see also the elegant article by Labrousse in *Scienze* 1982, 249–75. See also Thomas 1971, chap. 18, and Webster 1982, chap. 4.

[62] The classic treatment of Fontenelle, Bekker, and others by Paul Hazard 1963, pt. 2, chap. 2, interprets this debunking of prodigies as an attack on miracles.

[63] See the pertinent remarks by Colie 1957, 71. Burns 1981 is useful but limited. As good examples of the problems: Calmet 1715, 1:163–64; LeClerc 1696, 325–52, 353–76; and Locke 1958, 85, 86. Cf. Platelle 1968, 29–37, for Tridentine rules about establishing miracles; the rest of the volume illustrates tensions between evidence and the will to believe.

[64] Buonanni in Basile 1987, 158; Borelli in Adelmann 1966, vol. 1, chap. 7; discussion by Roger 1993, 190–92.

problem was not new but became acute with new instruments and methods. As early as 1610, Galileo had described his telescope with some care, announcing that "Unless the instrument is of this kind it will be vain to attempt to observe all the things which I have seen in the heavens."[65] In 1665, Hooke, too, would describe his microscopes and suggest how best to obtain clear images. Other equipment, notably the air pump, proved even more difficult to produce and to operate. As in the Borelli-Malpighi affair, Newton would receive queries about experimental details not made sufficiently clear in his first optical paper, and a remarkable number of experimenters struggled to reproduce his results. And Robert Boyle, in two early essays, discussed his own efforts to replicate experiments when his chemicals might contain impurities different from those in the reagents used by earlier chemists.[66]

One solution lay in the collaborative performance of experiments, and another in the demonstration of experiments before an audience. In the Florentine Cimento, for example, even "the most esteemed authors" had to have their results tested in order to detect any "experimental errors" that might have crept into their works. The academy's *Saggi* thus presented what the membership as a whole could testify to. In the Paris Academy, as Claude Perrault explained it, the collection of anatomical dissections published in 1671 was "not the Work of one private Person, who may suffer himself to be prevail'd upon by his own Opinion." Instead, these memoirs "contain only Matters of Fact, that have been verified by a whole Society," or, rather, by all the participating anatomists, so that the illustrations were not engraved until all those present had agreed on their accuracy.[67] In London, as in the meetings conducted in Paris by Rohault and Lemery, the experimenter performed to an audience. If the audience did not possess the skill and knowledge of the performer, the willingness of the latter to subject himself to public scrutiny was a guarantee against private bias.[68]

Given the accepted epistemological hierarchy of this period, philosophers wanted to distinguish the relatively certain fact (including experi-

---

[65] Galileo [1610] 1957, 30–31, 51.

[66] Hooke [1665] 1961, author's preface. I. B. Cohen 1958, 156, 173–74, and Guerlac 1981, 78. Boyle [1661] 1772, 1:318–53, and Sargent in Hunter 1994, 69–73. For the air-pump, Shapin and Schaffer 1987, esp. chap. 6.

[67] Leopold to Huygens, and Magalotti's preface to the *Saggi*, both in Middleton 1971, 325, 92. Claude Perrault [1671] 1688, preface.

[68] Shapin and Schaffer 1987 give the audience a somewhat different role than I do. See the insistence by Charas 1673 that he performed his experiments on vipers before very *large* audiences. Findlen 1994, chap. 5, struggles with two interpretations of who constituted the important observers: men of status, or men of learning, the two groups overlapping but not identical. For the collaborative ideal, see Houghton 1942; Hahn 1971, esp. 24–28; and Stroup 1990, chap. 6 and passim.

mental facts) from the more probabilistic interpretation. Members of the Cimento thus insisted on the importance of "historical" (descriptive) reporting of experiments, for much the same reason that Henry Power could say about any experimenter:

> let his Opinions be never so false (his Experiments being true) I am not oblig'd to believe the former, and am left at my liberty to benefit my self by the latter: And though he have erroneously superstructed upon his Experiments, yet the Foundation being solid, a more wary Builder may be very much further'd by it, in the erection of a more judicious and consistent Fabrick.[69]

Similarly, the rules of the Montmor Academy and the early statutes of the Royal Society specified some distinction between a subject presented "with no amplification" and the freedom with which members might then debate interpretations. In a drastic effort to minimize bias, the Paris Academy excluded both Jesuits and doctrinaire Cartesians—these men had powerful allegiances and hence insufficiently open minds.[70]

For all their programmatic statements, philosophers found fact and interpretation hard to separate. After all, the whole purpose of experimentation was to imitate and thus to *understand* nature. Just as Glanvill had assumed an interpretation of the facts he tried to establish, when Moïse Charas in Paris attempted to reproduce Redi's experiments with venomous snakes, he claimed results so different from Redi's as to warrant a wholly different causal explanation. Indeed, as members of the Paris Academy eventually realized, the pointless empiricism of Claude Bourdelin—he endlessly presented chemical analyses of plants, duly "verified" by his colleagues—had to give way to some alternative: should one concentrate on expanding Bourdelin's limited methods, or focus on medicinal uses of plants, or supplement experiments with a more "philosophical" approach? The Academy tried all three.[71]

Thoughtful writers like Boyle, Malpighi, and Fontenelle all recognized

[69] Power [1664] 1966, preface; see also Webster 1967 for the eclectic philosophy of Power. For reporting "historically" (*storicamente*), see Middleton 1971, 68, 110, and Borelli's suggested revisions of a draft of the *Saggi* in Abetti 1942, 329, 330.

[70] Hahn 1971, 15, and the comment by Lister [1699] 1967, 97, on the Paris exclusion of regular clerics. As an important exception, some of the Jesuit missionaries to China were in 1684 named "correspondents" of the Academy in an effort to ensure regular communication; see *Index* 1954, s.v. Fontaney. For the Montmor rules and the Royal Society's statutes, above, Chapter 1, nn. 39, 53.

[71] For Redi and Charas, see Thorndike 1958, 8:20–33. Stroup 1990, chaps. 6–7; Salomon-Bayet 1978, 63–65, 68.

that the laboratory offered the opportunity to repeat and scrutinize natural processes. As Fontenelle put it, here one could select causal agents and observe their effects.[72] But how much could one understand in this fashion? As Spinoza and Leibniz asked, could one use experiments to prove the particulate structure of matter? Boyle himself tried for years to find experimental evidence for nonmechanical causes in nature, with unsatisfactory results.[73] On one occasion, Pierre Perrault, brother of the more famous Claude and Charles, discussed with Christiaan Huygens the difficulties of interpreting experiments. For Perrault, to explain the limited height of the mercury in a Torricellian tube by invoking the weight of the air was to exclude other causal possibilities, such as "attraction" and nature's abhorrence of a vacuum. He did not, he insisted, wish to personify nature or return to occult qualities, but only to suggest that other causes might act in conjunction with air pressure. In reply, Huygens could only say that the weight of air "satisfies so fully" the experimental phenomenon that no further causal suppositions are necessary.[74] In the famous terse language of Newton's first rule of reasoning in the *Principia*, "we are to admit no more causes of natural things than such as are both true and sufficient to explain their appearances."

These statements by Huygens and Newton do call for the philosopher to exercise his judgment of the "true and sufficient," and Huygens himself, having taken part in an inconclusive debate in the Paris Academy about the cause of weight, knew well that agreed-upon explanations of even a single phenomenon were hard to achieve. As Claude Perrrault remarked, the main aim of science was to explain effects as best one could—the French expression more clearly conveys pessimism: "le moins mal qu'il est possible." This striking phrase occurs in Perrault's preface to his *Essais de physique* (1680), a preface largely devoted to the same point earlier made by Henry Power: facts are reliable, but readers of facts will inevitably and legitimately differ about causal explanations.[75]

In the decades around 1700, there were two solutions available for the dilemmas of causal explanation. One could select a modified Cartesian

[72] Fontenelle in *HARS* [1700] 1703, 51–52, reported verbatim in *HWL* (August 1703), 453–54. See also Adelmann 1966, 1:570–71; Altieri Biagi 1980, 1082–85; Duchesneau in Righini Bonelli 1975, 113–14; J. J. Becher in Debus 1977, 2:459–60.

[73] Articles by Principe and Henry in Hunter 1994, 91–105, 119–38. For Spinoza and others, above, n. 17.

[74] Pierre Perrault to Huygens, May 1673, followed by the undated reply from Huygens, both in Huygens 1888, 7:287–301. This exchange would be published a year later in Perrault's *De l'origine des fontaines* (Paris, 1674).

[75] C. Perrault 1680, vol. 1, preface. For the debate in the Academy in 1669, see Huygens 1888, 19:628–45.

route, taking experiment as far as possible toward a likely explanation and then seeking to demonstrate that explanation in a deductive manner. Such a rationalist program, advocated by some of the Neapolitan Investiganti and clung to much later by Hermann Boerhaave and others, was not only losing favor, but even a Cartesian like Rohault admitted that certainty often remained beyond human grasp:

> we must content ourselves for the most part, to find out how Things may be; without pretending to come to a certain Knowledge and Determination of what they really are; for there may possibly be different Causes capable of producing the same Effect.[76]

Alternatively, many agreed with Boyle's careful conclusion about the mechanical philosophy. For Boyle, this philosophy was "sufficiently recommended" for its intelligibility, its consistency, and "its applicableness to so many phaenomena of nature."[77] This tentative statement had much in common with the Continental tradition stemming from Gassendi and Mersenne, in which "moral," not demonstrative, certainty might eventually be claimed for hypotheses that had shown their fruitfulness. Huygens would go further in this direction by suggesting that at times one could find a "perfect" correspondence between hypothesis and fact, notably when a prediction based on hypothesis is borne out by new facts; under these circumstances, one achieves "a degree of probability which very often is scarcely less than complete proof."[78]

On the whole, one may say that philosophers of this period were united in their desire to increase their stock of natural facts and to provide the best possible (or least bad) causal explanations. Just how to do this remained debatable, although by the early eighteenth century the deductive mode was no longer the universal panacea. Similarly, acceptable kinds of explanation had narrowed down to the mechanical, even while the newer mysteries of Newtonian gravitation and Leibnizian monadology were being discussed. If philosophers had provided no final solutions to methodological problems, they had developed an acute awareness of possible bias, the need for verification, and the probabilistic nature of their own conclusions.

---

[76] Rohault 1728, pt. 1, chap. 3, sec. 3. The "rationalist" option is discussed in Badaloni 1961, 44–48, 50–53, 65; Clarke 1989, chap. 5 and pp. 213–14, 220; L. S. King 1963; Roger 1993, 195–97, 204–5.

[77] "Of the excellency and grounds of the corpuscular or mechanical philosophy," 1674, in M. B. Hall 1966, 207.

[78] Huygens [1690] 1945, vi–vii.

KNOWLEDGE OF THE PAST

When seventeenth-century thinkers asserted that nature itself was unfalsifiable, they generally did not conclude that man's knowledge of nature was equally reliable. Nonetheless, one could hope to verify reported facts, replicate experiments, and achieve a degree of confidence about the resultant inferences. Facts about the past, however, could be neither replicated nor directly observed, so that knowledge of history depended on surviving testimony. In contrast to the sciences, where new discoveries abounded, historical evidence seemed more or less fixed in quantity and permanently fragmentary in nature. In some cases, pertinent testimony might be wholly lacking, as when Pierre Bayle remarked that we have Roman but no Carthaginian accounts of the Punic Wars. Nor could historians emulate scientists by assuming uniformity or patterns in human affairs. To some—a classic example being Bishop Bossuet—there did exist an overarching pattern, in that great epochs of history could be interpreted as revealing God's general plan for mankind. But in their daily tasks, historians could rely on no such framework, for history, in Fontenelle's words, "has for its object the irregular effects of the passions and caprices of men."[79]

Early in the century, three intellectual currents placed historians on the defensive, necessitating that they examine the levels of confidence attainable in study of the past. First, as in other areas of the epistemological debate, confessional antagonisms called traditional certainties into question, each side challenging the historical interpretations used by the other. For Catholics, this meant trying to undermine Protestant assurance that one might know the correct meaning of Scripture without resort to an authoritative interpreter, the Church. Protestant sects debated this matter among themselves, even as they maintained a united front against Catholic reliance on post-apostolic oral traditions, post-apostolic miracles, and the authority of Rome. Second, in the revived tradition of ancient skepticism, doubts about the reliability of the senses and the mind meant doubts also about the accuracy of all human testimony, past and present. Accurate reporting was not only inherently difficult, but in the realm of human affairs was further compromised by the human penchant to lie, to distort, and to be partisan. Finally, the opening pages of the *Discourse on Method* contain a wholesale condemnation of the study of history. As Descartes argued, no observer of even a single event could achieve accuracy and com-

---

[79] Fontenelle 1699, in Marsak 1961, 90. Bayle's remark on the Punic Wars is in Labrousse 1963, 2:20. For the writing of history in the seventeenth and eighteenth centuries, see the relevant chapters in Hay 1977; for England, Levine 1987.

pleteness of reporting, and defective accounts by past writers were further distorted by the modern historian's selection from such records. Even the traditional role of history in teaching moral lessons could be considered valueless, for inaccurate models were not worth emulating.[80]

Not all Cartesians agreed on so drastic a dismissal of history, as Arnauld and Nicole devoted part of their *Logique* to rules governing the acceptance of historical testimony. But Pierre-Daniel Huet turned to history only after abandoning his early enthusiasm for Cartesianism, whereas Nicolas Malebranche expressed a degree of contempt for mere pedantry that produced no certain knowledge.[81] For Adrien Baillet, such views occasioned serious reflection when he presented his biography of Descartes (1691) to the public. Baillet did a great deal of research, and he felt obliged not only to discuss his sources at length but also to insist that his inclusion of the smallest details was essential for the sake of truth in reporting. Furthermore, he tried to reassure the reader that, in addition to verifying facts, he had "paid especial attention to placing them in the order they held during the life of our Philosopher. We daily experience cases where disarranged truths deteriorate into untruths: and there are few Histories where the facts have as much need to be placed in proper order as in one on Descartes."[82]

For Baillet, then, the proper response to historical pyrrhonism was research, precision, detail, and order. More generally, historians on the defensive adopted an ideal of precision that some expressed by claiming to deal with "matters of fact." Thus, biblical scholar Richard Simon announced in his *Histoire critique du Nouveau Testament* (1689):

> Our Religion consisting principally in Matters of Fact, the Subtelties of [Scholastic] Divines, who are not acquainted with Antiquity, can never discover certainty of such matters of Fact: They rather serve to confound the Understanding, and form pernicious Difficulties against the Mysteries of our Religion.[83]

To be sure, most readers of Simon regarded his facts as highly controversial, but he was far from alone in his program. As a modern Toland scholar

---

[80] Descartes 1984, 1:113–14. In addition to works cited above, nn. 1, 3, see Bertelli 1973, chap. 13.

[81] For Cartesianism and the study of history, see Borghero 1983, chap. 1; Lévy-Bruhl in Klibansky 1936, 191–96; and Gouhier 1926, 37–48. Also, Labrousse's splendid discussion of how Bayle could be both Cartesian and *érudit*, in *Religion* 1968, 53–70.

[82] Baillet 1691, 1:ij–iij, vj–vij, xxj–xxvj. For all his care, Baillet did make errors, such as confusing Christiaan Huygens with his father; see Huygens 1888, 10:399–402.

[83] Simon 1689, preface. (The English translation was published the same year as the French.) Simon's addiction to "fact" is discussed by Cotoni 1984, 14–27.

has observed about this period, religion itself was being defined as a "matter of fact" and thus open to the same kind of investigation as experimental philosophy.[84]

For all kinds of history, sacred and profane, historians devised three strategies for transforming mere opinion into reliable fact: the reform of narrative, the critical evaluation of documents, and the singling out of "monuments" as especially unbiased sources. As a minimal response, authors of narrative histories tried to achieve greater fidelity to the available evidence. If they were not concerned to increase the stock of facts, they at least attempted to eliminate fiction, so that, for example, the ancient tradition of the invented speech, placed in the mouths of historical figures, gradually disappeared from narratives. To tell the truth, however, did not require telling the whole truth. Since narratives were meant to be readable and instructive, excessive detail would not do. These works might include prefatory discussions of critical methods, but, as the author of one such preface put it, he saw no need in a general history of France to ask if unpublished manuscripts could contribute anything of value.[85]

Other historians produced no narratives but engaged in the critical examination of documents, paying attention to the evaluation of witnesses, the dating and authentication of texts, and the study of ancient and medieval languages. A product of confessional disputes, the *Acta sanctorum* (1643) contains a methodological preface outlining the principles used by historians working in a peculiarly delicate field. The Bollandists acknowledged that saints's lives had often not been written down until centuries after their deaths, so that popular traditions, transmitted orally, had had time to accumulate. With no more immediate sources available in many cases, one could only advise caution in using this material. But even eyewitnesses, when such can be found, may disagree and may have motives for lying. (One clearly should not trust accounts of saints by heretics or schismatics.) Nevertheless, errors and distortions cannot wholly invalidate any account, or history could not be written at all—except under God's direction. Degrees of credibility must therefore be identified, the most impor-

---

[84] Iofrida 1983, esp. 58, also 46, 60. Pitassi 1987 emphasizes Jean LeClerc's concern for historical method and fact, his assumption being that the true message of the Bible will not be impugned by scholarship. On Christianity as "morally" (not "demonstratively") true, see Northeast 1991, 62.

[85] Daniel 1729, vol. 1, preface. According to Borghero 1983, 304 n. 26, the preface to Daniel's first edition (1696) emphasized style; discussion of method was added to the second edition (1713). See also the classic tale of abbé Vertot and the siege of Malta, in Hay 1977, 143.

tant authors being the eyewitnesses, then those who record the reports of eyewitnesses, and so on down to people who have collected and preserved reports and documents. All such persons deserve to be believed, with the added proviso that "they are upright men, of discretion, and . . . their writings are incorrupt and unfalsified."[86]

These rules, however hard to apply in particular cases, would be repeated, modified, amplified, and refined by Arnauld and Nicole, by John Locke, and by medievalists Leibniz, Mabillon, and Muratori. Understandably irrelevant to the Bollandists was the question of whether miracles had occurred: one judged the reliability of the witness, but not the likelihood of the event reported. The same principle was adopted by people like Arnauld, Bayle, and Locke, and was indeed used by John Wallis in evaluating reports from astronomers Auzout and Hevelius. Ideally, as in the law courts, one hoped to find more than one witness to a single historical event, but all too often that was not the case. Reflection on this problem led Bayle and Locke to insist that one person's account could not be deemed reliable merely because it had been accepted and repeated for centuries; one had no option but to examine the testimony of the single witness.[87]

Jean Mabillon's *De re diplomatica* (1681) was, and is, the classic handbook on authenticating texts. Dealing with medieval charters, he examined matters ranging from handwriting to vocabulary and customs, so that, for example, misuse of a title like "count" suggested that a particular document was forged or had suffered alteration. To know what constituted "misuse" in the period of his main concern, Mabillon had had to become familiar with virtually all surviving Merovingian manuscripts and post-Merovingian copies. Comparing a suspect document to this body of material still might yield only tentative judgments of authenticity because, as Mabillon knew, customs and vocabulary do not remain static, although a degree of stability in diction is guaranteed by the fact that charters are legal documents employing standard formulas. The closest Mabillon could come to the character of authors was by asking whether signatories to a charter had any reason to falsify its terms. Collusion and private bias usually seemed improbable because many charters were public documents, entailing such

---

[86] Much of the 1643 preface is quoted in Collis 1920–21. On the Bollandists, see Knowles 1963a, chap. 1, and the detailed analysis in Delehaye 1959.

[87] Arnauld 1964, IV, xiii–xv. Bayle in Labrousse 1963, vol. 2, chap. 1. Leibniz in Davillé 1909, 465–70. Locke, *Essay*, IV, xvi, esp. 10–11. Mabillon 1691, 232–42. In the huge literature on Muratori, which I have only sampled, I have found especially valuable the article by Righi (1933); more comprehensive is Bertelli 1960.

public consequences as the transfer of lands. If a charter had passed other tests of authenticity, questions of bias became superfluous.[88]

Mabillon knew that his verdicts on particular charters were tentative— he had to accept "moral" rather than demonstrative certainty—and he would later sum up the process of reaching conclusions in history as essentially dependent on the historian's own good judgment ("bon goust").[89] His focus on the *public* document, however, was far from unusual in this period, bringing us to the third kind of response to historical pyrrhonism: the value of so-called "monuments" as especially reliable historical evidence. The word became a generic one not only for tombs, temples, and arches, but for any evidence devoid of personal opinion or bias. Thus, monuments included calendrical systems, law codes, coinage, bridges and roads, and languages, all of which provided information about how whole populations had lived. Just as tombs told the historian about burial customs, so could linguistic filiations be used to trace the paths of human migrations. Like Mabillon's charters, a few kinds of monuments could be forged—notably coins and statues—but most could not. Above all, such material had not been intended to mislead and could not incorporate the personal views of witnesses or of admired historians like Livy and Polybius.[90]

The need for skill and discrimination in interpreting monuments was made clear in 1665 when Charles Patin distinguished between the inscriptions and the images to be found on coins and medals. As Patin noted, inscriptions might well exaggerate the virtues of a ruler, but images inadvertently conveyed information about the customs of that ruler's era. How to explain monumental facts, however, remained far from easy, as the several interpreters of Stonehenge discovered. Another danger was pointed out by a Dutch antiquarian in a letter to a fellow scholar.

> Of course the study of medals is very useful, but it will do more harm than good if one thus thinks one need no longer pay attention to the testimony of the best historians, and if one multiplies the number of emperors

[88] Leclercq 1953, vol. 1, chap. 8; see also Knowles 1963a, chap. 2, and 1963b, chap. 10. Mabillon's Italian disciples are treated by Momigliano 1966, 135–52. Mabillon's awareness of the moderate Cartesianism of Arnauld is indicated by Borghero 1983, 299–303. On the importance of the *public* document, see also Leibniz 1972, 169, and Davillé 1909, 131, 395, and some older examples in Grafton 1991, 92–93.

[89] Mabillon 1691, 236, and Leclercq 1953, 1:174.

[90] Indispensable both for monuments and evaluating witnesses is Momigliano 1950. See also Momigliano 1960, 255–71, 463–80. Studies of John Aubrey by Hunter 1975 and of John Woodward by Levine 1977 also have much to say about the value of monuments.

every time one finds variations in the depictions of their facial features or expressions.[91]

Gisbert Cuper did not invent this difficulty, since he himself had recently refuted the work of one such inventor of an ancient emperor.

The study of monuments antedated by some decades the formulation of a rationale for their value to the historian—initially, antiquarians like William Camden seem to have been bent on recording anything old, sometimes as a way of preserving the memory of objects in a state of decay, more often to celebrate a local or national past or to illustrate details in ancient texts. Nor did Edmund Gibson, editor of an expanded version of Camden's *Britannia* in 1695, seem to need any reason for reissuing a work by then classic, although Gibson acknowledged that readers would find here something quite different from the stylish but superficial narrative histories. By then, however, Charles Patin had employed a phrase that would often be repeated: monuments were "the proofs of history."[92] To Patin and others, this no longer signified merely providing concrete illustrations of what ancient authors had said, but rather corroboration of the mere opinions found in ancient texts. If no one expected that monuments would alter Livy's history of Rome, details would nonetheless be confirmed, modified, or even added. On occasion, too, as Robert Sibbald pointed out, when no written records were discoverable, history could still be reconstructed; indeed, "the only sure way to write History, is from the Proofs [which] may be collected from such Monuments" as "Triumphal Arches, Temples, Altars, Pyramids, Obelisks . . . and Medals."[93] Referring to the work of Mabillon and others, Leibniz informed one of his correspondents that the study of monuments and manuscripts had progressed toward "a completed critical method," even if, as he told another

[91] Cuper (Cuperus, Cuyper) to Huet, 1700, in Lombard 1913, 37–38. Patin 1665, chap. 2 on uses of medals. Early interpretations of Stonehenge are in Atkinson 1956, chap. 7; Hunter 1975, 178–91; and Piggott 1989, 36–38 and chap. 4.

[92] Camden 1695, editor's preface. Patin 1665, 9. Post-1665 uses of "proofs of history" are in Pomian 1976, 1688–89, and in the title of Francesco Bianchini, *La Istoria universale provata con monumenti . . .* (1697). Compare *JS* (2 February 1665), 52–54 on the importance of medals, with *JS* (3 October 1707), 631 where they have become the "principal appuy" of history. A gentlemanly appreciation of this scholarly activity is in Addison [1721] 1914, 1:281–393, and in Whitehead's discussion of Addison on the value of statuary, in Barocchi 1983, 1:287–307. Cf. Rowlands [1723] 1766, viii–ix; he was more comfortable with written records than with monuments, the latter being hard to interpret with any confidence.

[93] Sibbald, 1707, quoted in Mendyk 1989, 217. On the reliability of Swedish monuments, see the discussion of Rudbeck's *Atlantica* in *Philosophical Collections*, no. 4 (1681/82), at p. 119.

friend, historical truth still stood in need of a science of weighing evidence.[94]

The scholars who developed critical techniques and precepts did so largely in connection with ancient and medieval history. To supply a glimpse of how their methods worked in practice, the next pages will focus on selected topics in ancient history. The crucial importance of antiquity was summed up by Leibniz: "the truth of revealed Religion is based on the facts of ancient history, which cannot be better proved than by the monuments of antiquity."[95] When he wrote this, in 1697, the conventional outlines of sacred history in particular had been challenged by the biblical commentaries of Isaac de LaPeyrère and Baruch Spinoza and by the news sent to Europe by missionaries in China. For such reasons—and indeed because ancient history, both sacred and profane, was generally familiar to the educated as well as the learned—the remote past aroused far more discussion than did the Middle Ages.

Certain assumptions and limitations had long characterized the study of antiquity. Fundamental among these was the belief that Genesis was the oldest written document; one thus could not find independent corroboration of the first chapters in particular, and other means of confirmation had to be sought. For pagan antiquity, most notably for Egypt and Chaldea, the chief written sources were the Old Testament and several Greek and Roman texts, by such authors as Herodotus, Plato, Plutarch, Cicero, and Pliny. For the most remote periods of Greece and Rome, evidence came in large part from the myths in Ovid's *Metamorphoses,* Livy's history of Rome, and such authors as Plutarch.

To the modern reader, the myths recounted by Ovid can hardly be interpreted as a source of "fact" as seventeenth-century writers understood that word. Rather, Ovid recorded belief-systems or what Fontenelle described as primitive mentalities. In their overwhelming desire to find out more about the remote past, however, Fontenelle's contemporaries revived the Euhemerist tradition dating back to the Greeks themselves. Early in the century, mythographers, including Francis Bacon, typically had seen moral and natural allegories in ancient myths, but Euhemerists, in rejecting such hidden meanings, substituted another variety of hidden meaning: the historical. In this interpretation, the gods and heroes of myth had once been real mortals, eventually deified and their exploits magnified. Properly deciphered, myths could thus help to reconstruct the history of ancient rulers

---

[94] Letters to Nicaise, 30 April/10 May 1697, and Burnett, 1/11 February 1697, in Leibniz 1875, 2:567; 3:193–94. For Leibniz on monuments, see Davillé 1909, 384, 386.

[95] Letter to Nicaise, cited above, n. 94.

and other eminent persons. Furthermore, for the earliest history of mankind, myths were also interpreted as fragmentary and distorted recollections of the true facts recorded in Genesis. The perils of this method were especially obvious in Huet's *Demonstratio evangelica* (1679); if many applauded the purpose announced in Huet's title, it was still clear that he had gone too far in identifying so many ancient figures as pagan memories of Moses. As Paolo Rossi has pointed out, all such couplings of the Bible and myths could well suggest to the reader that the Bible, too, contains myths. Writers like Huet generally had no such subversive intent.[96]

From this elastic handling of myth, it was a short step to consider pagan gods and heroes not as disguised kings and patriarchs but as personifications of natural processes. Robert Hooke was one of the pioneers of this approach, as for some years he tried to persuade Fellows of the Royal Society that many myths actually referred to earthquakes. Far better known was Fontenelle's much-reprinted *Plurality of Worlds* (1686), in which there occurs one vivid example of the same kind of interpretation; not sympathetic to Euhemerism, Fontenelle here declared that he could readily believe that natural processes had opened the Straits of Gibraltar, and that ancient peoples, out of ignorance or "love of marvels," had credited Hercules with this operation.[97] As with Huet, dangers lurked here, too, for had not the dreadful Spinoza argued that Old Testament miracles had really been natural events, misunderstood by ignorant Hebrews? Despite this menace, Fontenelle's treatment of Gibraltar seems to have become a commonplace, although this variant did no more than add a detail to the prevailing form of Euhemerism.[98]

The critical canons of scholars like Mabillon can hardly be reconciled with the less than critical Euhemerism of the same and later decades. As the example of Huet suggests, however, these writers had in mind more than filling in the details of pagan antiquity: they wanted also to confirm Genesis. When Pierre Bayle remarked in his *Dictionary* that critical methods

---

[96] Rossi 1984, 152–57. Fontenelle's critique of myths was not published until 1724, but earlier hints are in his very popular *Dialogues des morts* [1683] 1971, 139–44, and the modern editor's remarks, 51–52. The dangers of Huet-like procedures were perceived by the *Trévoux* journalists, in Desautels 1956, 206–16. Especially valuable studies of the interpretation of myth in this period are Allen 1970; Manuel 1967; and parts of Walker 1972. Deshayes 1963 deals primarily with the latter half of the eighteenth century. King and Lynn 1980 is broader than the title suggests.

[97] Fontenelle 1973, 130. For Hooke, see Birkett and Oldroyd in Gaukroger 1991, 145–70, and Rappaport 1986.

[98] For example, Banier 1739, 1:30–32, where this dedicated Euhemerist allows that some fables, but not many, have their basis in natural processes. Pluche [1739] 1748, 1:vij–ix, 148, 255–58; 2:2–3, was sometimes a Euhemerist, sometimes like Huet, but also explained some myths as personifications of natural processes.

were being reserved for profane texts and not applied to the sacred, he was only half right.[99] Bayle's contemporaries had begun this task—erratically, nervously, and amid heated argument. And one of their assumptions differed markedly from that of historians dealing with profane texts: the Bible was true history. At the same time, aware of critical rules, and aware of the need to defend Christianity against a range of enemies, the same scholars who assserted that the Bible was unimpeachable also argued that the text needed confirmation.

However one interpreted the Bible on matters of doctrine (such as free will and predestination), the text was also obviously and predominantly *historical*. To confirm that history, as Don Cameron Allen showed in his classic *Legend of Noah* (1949), became a widespread preoccupation in the seventeenth century. Allen believed this trend to be peculiarly Protestant, fueled by a desire to understand the plain words of Scripture, but Catholics also engaged in the same enquiry. One whole class of problems, for example, involved the animal population of Noah's Ark. As Europeans became familiar with hitherto unknown species and genera in the Americas, Asia, and Australia, writers such as Athanasius Kircher and John Wilkins drew up lists of those animals deemed hybrids who need not have been among the Ark's passengers. How had the kangaroo migrated to Australia without leaving any progeny between Mount Ararat and the southeast Asian coast? How had peculiarly American species crossed the Atlantic? One could imagine that humans had found ways to cross the ocean, but such early migrants surely would not have taken with them noxious animals like the rattlesnake. John Ray indeed considered solutions to the problem of faunal migration so doubtful that he was prompted to ask whether the Flood had been universal. In these earnest explications of the Ark and migration, scholars had no desire to impeach Moses, but believed that to fill in details in a plausible way would enhance the credibility of Genesis.[100]

Baffling questions about the Old Testament had, of course, been debated since pre-Christian times and were still being discussed by early Fathers like Augustine. Among the more obvious difficulties, could Moses, as author of the Pentateuch, have written the account in Deuteronomy (34:5–6) of his own death and burial? How could one measure the "days" of Creation when the sun was created only on the fourth day? To these traditional questions, still alive in the seventeenth century, must be added the three provocations

---

[99] In Bayle's famous article "David," for which see Labrousse 1963, 2:340, 346 n. 1.

[100] In addition to Allen 1963, see Bligny 1973 and Emery 1948. For postdiluvial migrations, see also Gliozzi 1977; Huddleston 1967; and Northeast 1991, 123–25. Calmet 1707, 1:176–9, and 1720, 1:104–5, is a good example of a reluctant adherent to naturalistic explication.

alluded to earlier: LaPeyrère; Spinoza, and China. LaPeyrère denied that Adam and Eve were the first humans; Spinoza raised questions about the language, authorship, and interpretation of a range of biblical books; and the Chinese "annals" seemed to rival Genesis in antiquity. In reply, scholars adopted a variety of techniques to defend the veracity of the Bible.

LaPeyrère's *Prae-adamitae* (1655) may be described as a rationalist attempt to solve certain textual problems, among these the circumstances attending the exile of Cain (Genesis 4:12–17). At a time when the world's population consisted of Cain and his parents, Cain found a wife and built a city. Because Genesis contains two creation stories, LaPeyrère argued that one of the two must be a generalized reference to pre-Adamic peoples among whom Cain found a wife and so on. As a result, Genesis, far from being a history of the world, really recounted only the history of the Jews. The same conclusion he extended to Noah: the supposedly universal Flood had been a local event, of importance only to the Hebrews.[101] Not surprisingly, refutations of LaPeyrère began to appear at once. By destroying all biblical basis for calculating the age of the world, he laid himself open to charges of advocating eternalism, and some critics strove to show that the world had had a beginning in time. Sir Matthew Hale, for example, calculated how long it would take, starting from a single pair, for the world's human population to reach its then-estimated size; he concluded that the increase had been so rapid that the world could not have been in existence for more than a few thousand years.[102] According to Pierre Bayle, other critics engaged in calculations designed to show that there had been enough time for Adam and Eve to have had many children to supply the inhabitants of Cain's city.[103] (This effort, as Bayle pointed out, did nothing to dispel the impression that Cain referred to strangers, not to this hypothetical family, when he said, "whoever finds me will slay me.") Still others addressed themselves to evidence for the universality of the Flood. Among arguments used here, it was pointed out that a local flood would not have required the building of an Ark—the Hebrews could simply have moved to higher ground to escape the waters.

The replies sampled here have little to do with critical historical methods, except that Genesis could be rendered more plausible as history if one could fill in details that harmonized with and helped explain the text. More pointedly scholarly were issues raised in Spinoza's *Tractatus*

---

[101] Allen 1963, 86–91, 133–37; Pastine in Bertelli 1980, 305–18. Most valuable in Popkin 1987 are chaps. 2, 4, on LaPeyrère himself.

[102] Hale's work (1677) is an exercise in showing Moses to be in accord with reason and nature. Rossi 1969, chap. 3, is especially good on replies to the threat of eternalism.

[103] Bayle [1697] 1734, art. "Cain," rem. (A).

theologico-politicus (1670). A Hebraist himself, Spinoza insisted that the precise meaning of many biblical passages would always be in doubt, our knowledge of the ancient language being irremediably defective. Furthermore, we lack historical contexts for the several books: we cannot date them, and we either do not know who wrote them or cannot identify the persons whose names are attached to them. We can be sure, however, that the Old Testament records the history of an ancient, primitive, ignorant people, whose inability to understand nature had filled the text with references to miracles. (Alternatively, the hyperbolic "oriental" style of ancient prose meant that at least some miracles should be interpreted figuratively, not literally.) What remained, then, as the chief value of the Bible were moral precepts similar to those discoverable by unaided human reason.[104] When, a few years later, Richard Simon produced his *Histoire critique du Vieux Testament* (1678), some readers took the French Oratorian to be a Spinozist. On major points the two actually differed, since Simon did not question miracles and insisted that interpretative problems arising from the nature of the texts had their definitive solutions in the traditions and authority of the Catholic Church. At the same time, Simon, like Spinoza, denied the Mosaic authorship of the Pentateuch, arguing that the repetitions, confusions, and chronological inversions in the text pointed to compilation from older documents no longer extant.[105]

Reactions to Spinoza and Simon were mixed, the majority hostile. For Spinoza, outrage centered not on scholarship, but on his denial of miracles: his view of God as bound by His own natural laws was greeted as determinism and atheism. Not uncommon was the sentiment later expressed in the *Journal des savants,* reviewing a biography of Spinoza: "We will say nothing here [about Spinoza's writings], so as not to discuss matters which we would like to see wholly erased from the memory of men."[106] In a milder vein, some contemporaries, including Henry Oldenburg, found it baffling that a man so upright in his own life could express religious beliefs so wholly unacceptable. Again outside the realm of scholarship, Bishop Bossuet repudiated Spinozism and had the first edition of Simon's *Histoire critique* destroyed (some copies had already been

[104] *Tractatus,* esp. chaps. 6–9. in Spinoza 1951, vol. 1. A valuable analysis is in McKeon 1928, pt. 2, chap. 4, the focus not on scholarship but on the meaning of the Bible to Spinoza.

[105] See Auvray in *Religion* 1968, 201–14, for Simon's possible debt to Spinoza. The best modern study is by Steinmann 1960, esp. 100–116, but still useful is Molien's article in the *Dictionnaire de théologie catholique,* vol. 14, cols. 2094–2118. The tendency of Simon's work to remind readers of Spinoza is illustrated in Vernière 1954, 1:110, 294.

[106] *JS* (30 April 1707), quoted in Barber 1955, 30 n. 2. Lagarrigue 1990, 82, offers the example of an Amsterdam proofreader who in 1712 refused to work on a new edition of the *Tractatus* and a life of Spinoza.

distributed). Bossuet never could understand scholarly interest in what he considered to be minutiae, the message of the Bible being clear without such dangerous pedantry.[107] To at least some Protestant observers, this official hostility to Simon seemed puzzling and shortsighted, for they recognized in Simon a defender of the authority of the Roman Church and an enemy of the Protestant principle that the Bible is the sole "rule of faith."[108]

One of the earliest readers of Simon, the scholar-diplomat Ezechiel Spanheim, pointed out the net effect of the *Histoire critique:* it had undermined the authority of the Old Testament by exposing the text's human, historical, and linguistic problems and weaknesses.[109] Much the same was being said about Spinoza's *Tractatus*. Few European scholars, however, commanded the knowledge of ancient languages displayed by both writers or Simon's knowledge of surviving manuscripts; nor was there any direct way to respond to Spinoza's questions about the dating and authorship of biblical books. Indeed, even Bossuet had to admit that Old Testament books in particular had suffered alterations in ancient times. (This had long been evident because of the textual differences visible in the Hebrew, Greek, and Samaritan versions.) Typical replies to Spinoza and Simon thus tended to focus on ways to defend the truth and integrity of the Bible *despite* whatever alterations it had undergone.

Among scholars who showed awareness of and inability to deal with Spinoza and Simon, Edward Stillingfleet, future bishop of Worcester, expressed outrage at Spinoza's denial of Mosaic authorship of the Pentateuch. At a later date, preaching against Simon, Stillingfleet could only repeat arguments he had developed as early as 1662: God had protected the integrity of the biblical text, and a public document could not have suffered substantial change without provoking public outcry. More accommodating was scholar-journalist Jean LeClerc, who insisted that more systematic historical study than Simon's should be undertaken; eventually,

---

[107] For Spinoza's repute, see Vernière 1954; Colie 1963; and Friedmann 1962, 52–59. Colie points out that Spinoza and Oldenburg had been on cordial terms before the publication of the *Tractatus;* her article includes analysis of their correspondence of 1675–76, now available in Oldenburg 1965, vol. 12. For Simon, see Steinmann 1960, 124–30, and Hazard 1963, pt. 2, chaps. 3–4. The "destroyed" first edition was reprinted in Amsterdam in 1680; for rumor of a reprint early that year, see Doni Garfagnini 1981, 1:439–40.

[108] *NRL* (December 1684), in Bayle 1727, 1:190–91. For Protestant perceptions of the threat, see Bredvold 1956, 102–5; Harth 1968, chap. 6; and DuVeil 1683. The latter, addressed to Boyle, was originally published in French in London, 1678. Simon's text was placed on the Index of Prohibited Books in 1682.

[109] Spanheim 1679, 10–22. This work has variously been attributed to Ezechiel or to Friedrich Spanheim, both scholars capable of producing it.

this apparent sympathy with Simon's kind of scholarship would be subordinated to polemic.[110] Among Catholics, Huet replied to Spinoza with two axioms also pertinent to Simon: "A book is authentic when it has been believed to be so in all ages, without interruption, since its publication," and "A history is true when it records facts as they are recorded by other contemporary writers or by writers living near the times of the facts they record." Although the latter clause raises the historical question of witnesses, Huet's arguments have rightly been described as resting on a combination of faith and the concept of "universal consent."[111]

Spinoza and Simon succeeded in calling attention to the need for a scholarly approach to the Bible, even while their readers feared both the process and the results. For a long time, questions and problems of the kind they raised seemed utterly intractable where the Old Testament was concerned, and the two men had no followers of their own caliber until after 1750. As will be apparent in the next pages, types of scholarship different from theirs could be and were marshaled in these earlier decades to defend the Bible.

Another scholarly threat arose in the seventeenth century, and was debated far into the eighteenth, when missionaries reported Chinese claims to an ancient history apparently challenging calculations based on the Bible. Scholars in this highly technical field of "chronology" had long been engaged in disputing both proper methods of calculation and ways of reconciling variations presented by the different versions of the Old Testament. Familiar, too, were the classical texts claiming great antiquity for Egyptian and Chaldean civilizations. If the latter could be dismissed as products of national vanity or an "oriental" tendency to exaggerate, the Chinese annals bore a striking resemblance to Genesis: they recorded seemingly precise chronologies and genealogies, and they told of the lives and deeds of mortals, not gods or heroes. To be sure, few Chinese texts were available to Europeans before about the middle of the eighteenth century,

---

[110] Stillingfleet 1662, I, i, 16, 19 (problems of accurate transmission); II, i, 1–2 (God's care that truth be preserved); II, i, 7 (public documents). Added to the latter section in the 1675 edition was a vigorous denial of Spinoza's conclusion that the Pentateuch was not the work of Moses. Stillingfleet's remarks on Simon are in a sermon of 23 February 1682/3, printed in part by Reedy 1985, 145–55. The Carroll (1975) study of Stillingfleet has little to say about his scholarship. For LeClerc, see Voeltzel in *Religion* 1968, 33–52, and Pitassi 1987, esp. chap. 1, pt. 3.

[111] Vernière 1954, 1:126–31. Another Catholic response was an attempt to reformulate rules of historical criticism; for a devastating review of the work by Honoré de Sainte-Marie, see Chapter 1, at n. 34.

but missionaries supplied summaries of their content and commentary on their meaning.[112]

When the first chronology of ancient China was published in Europe in 1658, its author, Martino Martini, S.J., indicated that the ancient imperial annals revealed a continuous history going back to a date well before the Flood (as calculated on the basis of the Vulgate). Even if the annals could be shown not to antedate the Flood, they still began so soon after the Flood that there could scarcely have been enough time for Noah's sons to migrate, to establish colonies, and to multiply so rapidly as to produce a densely populated, highly civilized country. Nor could there have been enough time for the postdiluvial Chinese to have developed as much skill and sophistication in astronomy as the annals seemed to contain.

With minimal information, Europeans soon began to debate both chronology and such related issues as the origin of Chinese religion, science, and language. As early as 1667, Athanasius Kircher was arguing learnedly that Chinese beliefs, like those of other pagans, revealed their derivation from Adam and the Hebrew patriarchs. Shortly thereafter, John Webb, ignorant of the Chinese language, would argue that Chinese was the original language of mankind. When in 1684–85 the Academy of Sciences in Paris drew up a list of questions for the French Jesuits then due to depart for China, heading the list were questions about chronology and the text of the imperial annals.[113] Argument having already begun, it continued before any additional reports arrived. If, some scholars asked, the imperial annals were genuinely ancient, how had they survived the great book burning ordered by one emperor in the third century B.C.? Had the annals been produced more recently, the authors engaging in prideful exaggeration of the antiquity of their nation? Had Chinese astronomers really observed a solar eclipse as early as 2155 B.C., or had modern Chinese scientists, educated by missionaries, calculated when such an event must have occurred and inserted the data into the annals? Questions of such

[112] The best short treatments are by Boxer in Beasley 1961, 307–21, and Van Kley 1971. The best analysis remains that by Pinot (1932), with an important modification in Northeast 1991, 121–22. For the technical field of chronology, see Barr 1985; Grafton 1975. It would be interesting to examine the notes of editors and translators of Pliny to see how they comment on his claims for the great antiquity of Babylonian astronomical observations (*Natural History*, VII, lvi, 192–95).

[113] The Academy's questionnaire is in Pinot 1932, 2:7–9; the text, meant for Father Couplet, has no date or provenance, but Pinot (1:44–45) argues for its attribution to the Academy. In AAS, P-V, t. 11, fols, 114v–116v (16 December 1684–20 January 1685) are reports on conferences with Jesuits invited to the meetings; they focus on the natural sciences, with no allusions to Chinese history.

complexity required scholarship that only the Jesuits possessed, and the missionaries split into two groups. The Peking (Beijing) Jesuits held the annals to be genuine and reconcilable with the chronology of the Greek Septuagint. Other missionaries offended the mandarins by advocating a "figurist" (Euhemerist) interpretation of the annals, as Athanasius Kircher had already proposed. In Paris, Jesuit correspondents of the missionaries disagreed with both groups, preferred the chronology of the Vulgate, and hoped that scholars would find a way to reconcile Chinese history with traditional chronology.[114]

Lacking answers to complex questions, scholars began the process of reconciliation, generally following the route indicated by Martini himself and the Peking group: Chinese antiquity could be accommodated if one used the longer chronology of the Septuagint. In 1677, Robert Cary's massive examination of competing chronologies denied that we can be certain of the date of the Creation, but opted for the Septuagint in large part because of the Chinese dilemma. Better known than Cary's, abbé Pezron's *L'Antiquité des tems rétablie et défenduë* (1687) went through a number of editions. Announcing his aim of undermining preadamitism, libertinism, and other heterodoxies, Pezron claimed that a shortened biblical chronology, preferred by Protestants who used the Hebrew text, was quite new, and that he would "reestablish" the longer time span of the Septuagint. He went on to tackle Egypt and Chaldea as well as China, showing that all these civilizations dated from after the Flood.[115] Debate inevitably followed, but not wholly along confessional lines, as even the Jesuits were not united about the rival chronologies. Among Protestants, Leibniz found Pezron convincing, but Jean LeClerc's *Bibliothèque universelle* invited other Protestants to find ways to defend the Hebrew text. One Protestant who thought he had found such a way was William Whiston (1702).[116]

The several challenges discussed thus far—LaPeyrère, Spinoza and Simon, China—sometimes entailed a level of scholarship that few pos-

---

[114] In addition to works cited above, see Mungello 1985; Rowbotham 1956; Walker 1972, chap. 6; and Gaubil 1970. For the astronomical problem, Cassini I managed to shorten Chinese chronology by about 500 years; his early study was later published in *MARS* [*HARS*] *depuis 1666*, 8 (1730), 300–311. For Jesuit views on chronology, see also Northeast 1991, 119–23.

[115] Cary 1677, 161–62, 259–61, 270. Pezron 1687, Avertissement, pp. 10–14, and chaps. 12–14. Scholar Thomas Hearne 1698, esp. 13–14, was aware of the complexities; writing for "youth," he outlined the issues as simply as he could.

[116] Pezron was favorably reviewed by Bayle's successor, in *NRL* (June 1687), 639–62, and in Bacchini 1687, 149–53. Compare the review of Pezron's 1691 defense of his book, in *Bibliothèque universelle* (February 1693), esp. 151. Leibniz to Nicaise, 6/16 August 1699, in Leibniz 1875, 2:590, where "abbé de la Charmoye" is Pezron. Whiston 1702, esp. 60–65.

sessed, although it would be a mistake to underestimate the number of those capable of writing on a subject as technical as chronology. But the scholarship of the period was, so to speak, both broader and more fundamental than specialized studies by these not very ample ranks. The final pages of this chapter therefore return to basic questions of confirming a text, to see how scholars applied to the Bible considerations of evidence as developed by students of secular subjects.

For Mabillon and others, evidence about any text fell into two large categories: internal (such as the language of charters) and external (such as independent confirmation of the land transfers arranged by charters). Among philosophers who applied these methods to the Bible was Robert Boyle, who, in a very early essay, tackled internal evidence, describing biblical prose as clear and simple, in contrast to Spinoza's later comments about hyperbolic, "oriental" style. Boyle acknowledged that we have no equally ancient texts to aid us with Hebrew language and culture, but the Bible itself often contains the unadorned, spontaneous remarks of ordinary soldiers, shepherds, and women; the style, therefore, indicates transparent honesty, the immediacy of the reports, and the lack of subsequent tampering with the text. Because the Bible was meant for comprehension by a great variety of people in all ages, Boyle concluded not that the text was simplified for its audience but rather that it was sincere and straightforward.[117] When applied to Genesis in particular, the same kind of analysis did not always lead to conclusions like Boyle's. For Thomas Burnet, Isaac Newton, John Wilkins, and Richard Simon, the Mosaic style was indeed simple, but Burnet detected also simplification to the point of factual inaccuracy. As for Simon, he found the narrative of the Flood simple but so repetitive that he concluded: "It is very likely that if a single author had written this work, he would have explained himself in far fewer words, especially in a history."[118]

Internal evidence also included the credentials of the writer: had he witnessed the events he described, and was he an upright man? For obvious theological reasons—and perhaps, too, because much seemed to be known about the authors of the Gospels—a substantial literature focused on the witnesses to Christ's miracles; many writers on this subject were roused by Spinoza's denial of the Resurrection, as a later generation would be roused by the analysis of David Hume. This method could not be applied readily

[117] "Some considerations touching the style of the Holy Scriptures," 1661, in Boyle 1772, 2:257–62, 266, 283. Cf. Markley 1985.

[118] Wilkins [1693] 1969, 65. Simon 1685, 33. The Newton-Burnet correspondence on this matter will be discussed in Chapter 5.

to Genesis because no one had witnessed the Creation and Moses had not witnessed the Flood. In a detailed examination of these problems, Edward Stillingfleet argued that Moses learned some matters by revelation and others from oral tradition. Admitting that writing is a more accurate tool of transmission, Stillingfleet countered with a study of the character and integrity of Moses and especially with the argument that Moses reported events firmly embedded in Hebrew history and belief. Genesis, in short, was a public document, its contents undisputed by the very people who would have been able to detect inaccuracies.[119]

Other writers elaborated on these themes, arguing, for example, that so few generations separated Moses from Adam that accurate oral transmission, especially of events crucial to the history of the world, was virtually guaranteed. The problems attending oral transmission prompted some to try to determine how early writing had been invented.[120] Indeed, comparable debate was endemic during this period among those concerned with the origins of ancient Rome. Some few rejected as fabulous the story of Romulus, there being no written evidence of such ancient date; others, however, argued that indirect evidence did exist, or that tradition could be relied upon. Not until 1724 did Fontenelle's elegant attack on oral transmission appear in print, and in the same decade the subject was hotly debated in the Royal Academy of Inscriptions. Academician Lévesque de Pouilly, former editor of *Europe savante,* stirred up his colleagues when he insisted that the first four hundred years of Roman history were "uncertain," since all accounts relied merely on oral tradition. "When histories are confided only to the memory of men, they are altered in the mouths of all the successive transmitters."[121] Members of the Academy rejected this conclusion, in part because Livy's audience of educated Romans had accepted the tale of Romulus; Livy's account was thus a public document, undisputed by knowledgeable Romans. In reply, Pouilly held to his conviction that tradition is merely "popular rumor of which we do not know the source." But he also retreated somewhat, allowing that some traditions might well be accurate in their general outlines if they had relevance to large, public, and fairly simple (easily remembered) events—such, for example, would be the

---

[119] Stillingfleet, above, n. 110. For literature on the Resurrection, see Momigliano 1950; this was a crucial matter in the Spinoza-Oldenburg correspondence (above, n. 107), for Oldenburg believed the witnesses and Spinoza could not.

[120] Rossi 1984, pt. 3, passim.

[121] Pouilly [1722] 1729, 17. For the background to this debate, see Erasmus 1962. Valuable analysis is in Borghero 1983, 357–90. Historians debate the date of composition of Fontenelle's *De l'origine des fables,* Dagen (1966) arguing for the 1690s and Cantelli (1972) for the 1680s.

widespread tradition that the world was created and that all the present continents had once been covered by water.[122]

In referring to widespread tradition, Pouilly here reverted to the common method of Euhemerists: the use of pagan traditions to verify the early chapters of Genesis. On the one hand, pagan documents were "external"' evidence; on the other, they could not be considered *independent* evidence, because they were generally thought to derive from traditions dating back to Noah and even to Adam. Nonetheless, pagan sources—ranging from the Chinese annals (as interpreted by the "figurists") and the beliefs of American Indians to the much more familiar myths of the Greeks and Romans—were scrutinized for confirmation of the Creation, the Flood, the careers of the Hebrew patriarchs, and a variety of specific Christian beliefs. All this required imaginative and elastic interpretation of pagan evidence, except perhaps in the case of the Flood. Commentators on flood legends now and then paused to wonder if particular legends might refer to local events different from the biblical; on the whole, however, the worldwide occurrence of such legends could not be coincidental and must testify to the veracity of Moses.[123]

In connection with the Flood, an even more valuable kind of testimony existed: that of nature. Indeed, nature supplied the only evidence that conformed to the critical canons developed by the methodologists: nature was unfalsifiable, and it was independent of any human account of its operations. Naturalist John Woodward, who personally needed no corroboration of the truths of Genesis, understood the requirements of his contemporaries when he announced that his examination of nature would treat Moses as if he were merely a Livy or a Herodotus; in other words, the Mosaic text must be examined, evaluated, and confirmed.[124] Aspects of nature other than the geological also found a place in this confirmatory enterprise, as when Sir Matthew Hale used human reproductive rates to estimate the age of the world, and a long succession of writers enumerated the world's animal species in order to show that they could have fitted into Noah's Ark.

Despite an abundance of unresolved problems and disputes, historians had by 1700 developed critical methods that rescued the study of the past

[122] Sallier [1723] 1729, and Pouilly [1724] 1729. The Academy of Inscriptions assembled these and other writings on related subjects for simultaneous publication in a single volume of *MARI*. For Pouilly's successor, see Beaufort 1738.

[123] In addition to works on myth cited above, n. 96, the Jesuit missionary Lafitau [1724] 1974, interpreted customs and beliefs of American Indians in the same fashion as the Jesuit "figurists" did for China. The use of flood legends by naturalists will be pertinent in Chapter 5.

[124] Woodward's arguments will be examined in Chaper 5.

from the extremes of historical pyrrhonism. These methods left the most thoughtful writers aware that historical knowledge could only be probabilistic, so that Mabillon felt obliged to add to critical rules the imponderable factor of the historian's "good judgment," and Leibniz remarked that history needed a science of weighing evidence. To Nicolas Fréret, the most sophisticated member of the Academy of Inscriptions, the uncertainties of history might be lessened further by the study of "monuments," which he called "the most authentic proofs of history."[125] At the same time, critical methods enjoyed their greatest success in medieval history and in limited areas of ancient history, but were used hesitantly in connection with the Bible. That biblical scholarship required more skills than most possessed was evident in the kinds of responses elicited by Spinoza and Simon; that the results of such scholarship would be unsettling and even dangerous was also immediately apparent. It must be remembered, too, as Boyle and Spinoza both knew, that few resources existed to aid scholars in dealing with the oldest books in the world. Significantly, however, even the nervous—from the Jesuit missionaries to bishops Huet and Stillingfleet— acknowledged that the Bible, like other texts, needed confirmation.

In a larger philosophical context, Spinoza had outlined an issue crucial for students of both nature and the Bible: if we admit exceptions (miracles) to God's laws, we begin to equate God with chance and are on the road to atheism. Rephrased and reconsidered by John Locke, the argument was transmuted into a claim for the reliability of human knowledge, both sensory and rational; neither could be in conflict with true revelation. If revelation were to contradict such knowledge, we should doubt our interpretation of revelation rather than our knowledge.[126] In other words, the concept of proof was changing in the late seventeenth century, the Bible being no guide to nature, but nature being a key to proper interpretation of the Bible. At the same time, and apart from some outrageous radicals, the common expectation may be summed up as confidence that biblical accuracy would be vindicated and would not conflict with evidence supplied by nature.

[125] Fréret [1724] 1729, esp. 153–54.
[126] Locke, *Essay,* IV, xviii, 5, 8–10.

CHAPTER 3

Natural and Civil History

Philosophers who investigated the earth's history, its land-forms, or a number of related topics were aware of at least some aspects of the epistemological issues just discussed, as is evident from the appearance in the preceding chapter of such names as Steno, Hooke, and Leibniz. If few had any expert knowledge of mathematics, most could handle the language of demonstration, certainty, and probability. All confronted the difficulty of identifying the natural "fact" and finding plausible interpretations of it. And almost all, aware that their subject matter involved reconstruction of the past, found it necessary to consult the testimony of human witnesses to past events.

The present chapter is divided into two parts, the first devoted to exploring the interest of naturalists in human (or "civil") history. When the same subject enters into geological texts, it is frequently by way of allusion rather than analysis; the next pages therefore are intended to show that many naturalists had considerably more concern with and knowledge of civil history than the scientific texts alone reveal. In the second part, a few selected texts will be analyzed specifically for their attention to matters of proof and, more particularly, for the place of human testimony in the structure of geological argument.

That scientists did not regard themselves as narrow specialists is, of course, well known, and one thinks immediately of the time, intensity, and expertise Isaac Newton gave to his studies in chemistry and biblical scholarship. Less famous perhaps are the erudite activities of mathematicians such as Roberval and Viviani, who produced, respectively, editions of Aristarchus and Apollonius, or the anatomist Morgagni, who edited the works of the ancient Roman physician Celsus, or the egregious "Sir" John Hill, who edited Theophrastus. In these and other examples, one often finds the scientist combining his own specialty with a knowledge of ancient languages and some ancient history.[1]

In general, the common bond among scientists was the liberal education available mainly, but not exclusively, to men of wealth or status. Whether that education was obtained at a British university, a Reformed academy, or a Jesuit *collège*, it included familiarity with at least some of the classics of ancient Latin literature and some of the Greek as well (usually in Latin translation); in addition, Bible-reading at Protestant institutions had some parallel at Catholic ones, where students encountered commentaries by selected Church Fathers. Further exposure to various ancient texts would then be acquired by those students who trained for one of the three professions: law, medicine, and theology.[2]

The word "exposure" is used advisedly, for no curriculum dealt with history as a discipline. Rather, historical texts had other uses: they provided opportunities to learn Latin and to sharpen dialectical and rhetorical skills; they offered a gentleman acquaintance with great literature; and they contained an arsenal of authoritative information, as well as a range of philosophical speculation. In Daniel Roche's phrase, history was "un ornement de l'esprit," and studies of private libraries reveal a considerable public for those readable narrative histories that placed little strain on the intellect.[3] Quite a different matter, *erudite* history called for special skills. If schooling provided the language enabling one to read Livy, only self-education could teach the scholar subjects like epigraphy, paleography, and numismatics.

Erudite history not only flourished in the seventeenth century, but it was also the kind of learning that especially attracted naturalists during that period and for a time thereafter. To explain the general phenomenon, modern scholars have discerned a number of motives that induced so many intellectuals to devote themselves to such studies. One motive was doubt-

---

[1] *DSB*, s.v. Hill, Roberval, Viviani. Morgagni 1964, 1:62, 92, 99, and the bibliography compiled by Premuda in *Storia*, 5/II, 243. According to Fontenelle 1709, 1720, 1:230–31, abbé Gallois was a Hellenist as well as a mathematician, and he hoped to produce an edition of Pappus. This is the same Gallois who succeeded Denis de Sallo as editor of the *JS*.

[2] Among works on the history of education, see esp. Brockliss 1987 and Heyd 1982. See also Debus 1970; Dooley 1984.

[3] Roche 1988, 174–75. The various studies of libraries and of book-review journals, cited in Chapter 1, show a preponderance of subjects like history, geography, and travel.

less national or regional pride, at times coupled with the search for legal (historical) information of value to landowners, institutions, and ruling houses.[4] It is thus not surprising that Edward Lhwyd should have developed an interest in Welsh history or that Leibniz's *Protogaea* should have been intended as a prelude to a history of the House of Brunswick. Another motive may be summed up as "curiosity," or the mania for collecting either information or objects. Many collectors, to be sure, merely followed whatever fashion seemed good to them, and to amass ancient objects, whether coins or "figured stones," did not always signal any intellectual purpose. That naturalists so often assembled both human artifacts and fossils suggests that, for this group of collectors, the two kinds of specimens were mentally linked.

In Britain, the distinctive county studies, combining natural and civil history, drew much of their inspiration from Camden's *Britannia* (1586). Camden had proceeded county by county, assembling treasures of miscellany about great houses, local charters, old customs, notable features of the landscape, and much more. Pillaging from and adding to Camden, Joshua Childrey would select mainly material about nature to form his *Britannia Baconica* (1660). When Robert Plot in 1677 produced what he called a natural history of Oxfordshire, he actually once again combined the natural with the civil, now sorted topically rather than in the fashion of Camden's itinerary. Although a few of his contemporaries had already been planning to write works comparable to his, Plot's *Oxfordshire* quickly became a model in the anglophone world. Henry Oldenburg promptly saw this as a first step "to building the structure of an universal history of nature," and he urged Continental correspondents to undertake or to encourage similar enterprises.[5] But Plot's chief imitators were British writers, including such migrants to North America as John Banister and Mark Catesby. Their imitations often took the form of aping Plot's organization, and sometimes his ideas and language as well.[6]

Some readers of the *Oxfordshire* having asked if a chapter on antiquities in a work of natural history were not "altogether forraigne to the purpose,"

[4] On British scholarship put to service in a turbulent period, see Pocock 1957. See also Douglas 1939 and Levine 1987.

[5] Oldenburg to Hevelius, 22 May 1677, to Drelincourt, 24 May 1677, and to Leibniz, 12 July 1677, in Oldenburg 1965, 13:286–87, 287–89, 318–24 (quotation taken from p. 323). For John Aubrey's pre-Plot studies of Wiltshire, see Hunter 1975, esp. 192–93.

[6] Esp. the revised edition of Catesby (1771) which contains prefatory material not in the original ed. (1731). For both Banister and John Clayton, see Ewan 1970, 59–60, 80–84. Other local historians of this kind include John Beaumont, Charles Leigh, John Morton, and Robert Sibbald. I do not know how long this pattern of county histories continued, but it is still visible in Wallis 1769.

Plot's next book, a study of Staffordshire (1686), explained his principle of selection: he dealt not with persons or institutions, but mainly with things,

> such as ancient *Medalls, Ways, Lows, Pavements, Urns, Monuments* of *Stone, For-tifications, &c.* . . . Which being all made and fashioned out of *Natural* things, may as well be brought under a *Natural History* as any thing of *Art:* so that *this* seems little else but a continuation of the former *Chapter* ["Of Arts"]; the subject of *that,* being the *Novel Arts* exercised here in this present age; and of *this,* the *ancient ones.*[7]

That objects were made of natural materials seems peculiarly idiosyncratic as a rationale, but buried in Plot's prose is the idea, encountered in the preceding chapter, that objects are inherently neutral, lacking the elements of bias that might creep into a discussion of people and institutions. In short, Plot dealt with "monuments."

So, too, did other British writers, whether or not they emulated Plot, focus on objects and on the public aspects of culture. In John Ray's *Collection of English Words* (1674), the impersonal subject matter was followed by appendices giving catalogues of English birds and fish, plus an account of the refining of such ores as are found in England. Edward Lhwyd's unfinished *Archaeologia Britannia* (1707) was eventually supposed to include material on natural history as well as "the languages, histories, and customs of the original inhabitants of Great Britain."[8] Both Ray and Lhwyd also enlarged the natural history component of Camden when the *Britannia* was printed in a new edition in 1695.

In these and other instances, naturalists examined not only "monuments" but almost exclusively ancient ones. As Joseph Levine has shown in his study of John Woodward, to combine natural and civil antiquities demanded some of the same critical skills and seemed but two halves of the same problem: to reconstruct the remote past.[9] If few writers troubled to explain their aims, their activities nonetheless illustrate the persistence of this combination. Both John Woodward and Hans Sloane, for example, formed extensive private collections of natural objects and ancient artifacts, and comparable assortments could be found in the Repository of the Royal

---

[7] Plot 1686, 392. For a useful if Whiggish study of Plot, see Mendyk 1985.

[8] The quotation is from the title page of Lhwyd 1707. In addition to Gunther 1945, the memoir of Lhwyd in Owen 1777, 129–84, is still worth consulting. Modern studies by Campbell and Thomson 1963 and Emery 1971 are informative about his historical researches. Raven [1950] 1986, 167–71, gives details about Ray's *Words* and his *Collection of English Proverbs* (1670).

[9] Levine 1977, 3 and passim.

Society and at the Ashmolean Museum in Oxford, where Lhwyd succeeded Plot as Keeper of the collection. In addition, some collections featured exotica not especially ancient, such as American Indian artifacts. Their presence may be readily understood as an aspect of love of the rare and curious, but some of these objects, examined by naturalists familiar with antiquities, reminded observers of curious rocks found in Europe; Plot, Lhwyd, and others could identify the odd rocks as flint weapons used by pre-Roman Europeans, for they resembled the arrowheads fashioned by American natives.[10]

In his defense of the Royal Society, Thomas Sprat had sought to calm the critics when he announced that experimental philosophy would not undermine "the *Moral*, and *Political* Rules of ordering mens lives." History, too, "can from hence receive no dammage, or alteration . . . seeing the Subjects of *Natural*, and *Civil History* do not cross each other."[11] Consciously or not, Sprat here was mistaken, for many Fellows were already displaying a lively interest in civil history. One of the most active in this respect, Martin Lister contributed many articles to the *Philosophical Transactions* on the Roman remains at York. During his sojourn in Paris in 1698, Lister visited libraries and antiquarian collections, and he conversed familiarly with Mabillon and other historians; he also tried to arrange a scholarly correspondence between Lhwyd and another celtophile, the abbé Pezron. Among Hooke's innumerable presentations to the Society, several dealt with ancient Roman technology and one with Chinese antiquity. Halley contributed a lengthy account of ancient Palmyra and "the Inscriptions found there." And Henri Justel, long a faithful correspondent of the Society, sent to London a detailed description of a prehistoric burial site discovered in France.[12]

Not all aspects of antiquity were welcomed by the Society. When, in 1687, the *Philosophical Transactions* reported on Father Couplet's *Confucius Sinarum Philosophus* (1686), the reviewer ignored the moral and political content as "foreign to our purpose," but discussed at length the evidence for the antiquity of the Chinese Empire. Far from neutral or harmless, the subject was both delicate and controversial.

[10] For interpretations of these rocks called *cerauniae*, see F. D. Adams [1938] 1954, 118–25; Bertrand 1763, s.v. Ceraunites; Heizer 1969, 14–21. The missionary Lafitau, familiar with Indian stone axes, recalled seeing comparable European rocks in French collections. Lafitau [1724] 1974, 2:71.

[11] Sprat [1667] 1958, 325.

[12] Hooke (on Chinese writing) and Justel, in *PT* 16 (1686), 63–78, 221–26. Halley in *PT* 19 (1695), 160–75. Lister [1699] 1967, esp. 98–99 (and editor's notes) and the bibliography on pp. 298–308. Among the books Lister indicates he purchased in Paris was Pezron's defense of the chronology of the Septuagint, discussed above, Chapter 2.

'Twill be needless to advertise, that this Account places the beginning of the *Chinese* Empire long before the *Deluge,* according to the Holy Scriptures; wherefore if this be to be wholly rejected, as fabulous; or if not, how it is to be reconciled with the sacred Chronology, belongs more properly to the Disquisition of the Divines.[13]

In other words, as in the sciences, the Royal Society would take no corporate stance on the meaning or implications of such research. Decades later, the *Philosophical Transactions* would publish accounts by two divines—one English, the other a Jesuit "figurist"—who rejected all claims to the great antiquity of China.[14]

If Camden's *Britannia* appeared in Continental editions, the various county histories, written in English, seem to have been less well known. Some were indeed reviewed in the journals, but it proved understandably hard to convey to readers more than a glimpse of the odd miscellany in these volumes.[15] Antiquarian studies abounded on the Continent, but there apparently existed no concerted movement to combine these with natural history. When, for example, John Locke traveled in France in the 1670s, he acquired a copy of Jacques Deyron's little study of the antiquities of Nîmes; what Locke apparently found of interest—the claim that Nîmes had once been a seaport—was a mere detail buried in Deyron's text. Conversely, so to speak, an account of the natural history of Vesuvius did pay some attention to ancient descriptions of the volcano but not to the antiquities of the region.[16] If these impressionistic comments are accurate, there nonetheless did exist among Continental collectors the same balanced attention to natural and civil history so visible in Britain, and certain Continental naturalists also exhibit this combination of interests.[17]

---

[13] *PT* 16 (1687), 377, 378.

[14] Costard in *PT* 44 (1747), 476–93. Foucquet, S.J., in *PT* 36 (1730), 397–424. One of the two texts by Foucquet was sent from Rome by Sir Thomas Dereham, regular informant of the Royal Society. For antiquarian research by Fellows of Royal Society, more examples than I have indicated here are in Schneer 1954 and especially Hunter 1971.

[15] A French translation of Childrey was reviewed in *JS*, Amsterdam 12mo (13 June 1667), 152–57, but the editors had trouble getting a copy of Plot; see *JS*, Amsterdam 12mo (16 August 1677), 261–62, and *JS*, Amsterdam 12mo (6 September 1677), 288–89. The Roman *Giornale* tried to keep up with English books, but Plot was not reviewed; see Gardair 1984, appendices, and Rhodes 1964. For the difficulty of reviewing such works, compare the short treatment of Morton 1712 in *JS*, Amsterdam 12mo (July 1715), 62–70, with the long one in *Trévoux* (April 1714), 673–92, and (May 1714), 830–49.

[16] Deyron 1663, 61, and Lough 1953, 230. Paragallo 1705. I have not seen Paragallo's earlier treatise on earthquakes, but the review in Bacchini 1690, 169–74, refers to antiquity only for the author's discussion of ancient theories of earthquakes.

[17] My impressions are based on the fact that such studies of Continental antiquarianism as I have encountered—e.g., Stark 1880; Loewe 1924; Widera in Winter 1962, 162–68—make no mention of natural history. This may well stem from the modern notion of

Most familiar in this connection is the career of Leibniz, who was already both naturalist and historian when Oldenburg informed him of Plot's *Oxfordshire*. Now and then, to be sure, Leibniz told his correspondents that he preferred natural to civil history, God's laws to man's, certainties to probabilities. Nonetheless, in addition to his skill as a medievalist, Leibniz maintained the liveliest interest in ancient China; on the one hand, he made a long tour of Italian archives, and, on the other, he corresponded with at least one of the Jesuit missionaries. In ancient Chinese culture, Leibniz perceived a number of issues of worldwide significance, including the need to accommodate Chinese history to biblical chronology. He also thought the ancient Chinese possessed a "natural theology" second in antiquity only to the Bible and presumably derived at least in part from the Hebrew patriarchs. Furthermore, Leibniz believed that he detected in the ancient writing (trigrams) supposedly invented by Emperor Fu Hsi a system of rational communication equivalent to the binary arithmetic that Leibniz himself had "rediscovered."[18]

Leibniz saw his own studies of natural history as a necessary prelude to the history of ancient peoples. The surviving evidence about pre-literate European cultures consisted mainly of those physical (natural) conditions with which they grappled, and such information as one could detect in ancient languages. Impatient with those who sought to identify the very first language, the "Adamic" language which he believed no longer in existence anywhere, Leibniz thought that the filiations of languages would be a trustworthy guide to the migrations of ancient and early medieval peoples. Natural and linguistic evidence lacked both the bias of early writers such as Tacitus and the vagueness of local traditions transmitted orally.[19]

Among Leibniz's correspondents, Swiss naturalist Louis Bourguet shared the older man's interests, and Bourguet found time to investigate the evidence for Chinese antiquity and to attempt an analysis of Etruscan inscriptions. In Bourguet's own circle was J. J. Scheuchzer, the Zurich physician and naturalist who translated into Latin a French treatise on numismatics and who also wrote but never published a "Chronological

---

specialization that led Schnapper 1988, an art historian, to insist that early collectors did not separate paintings from other items in their collections. For information on Continental collections, see works cited above, Chapter 2, n. 42.

[18] For the early dates of Leibniz's interest in both history and natural history, see Roger and Scheel in *Leibniz* 1968. His preferred kinds of study are in Leibniz 1875, 3:61, 182. Details of his year in Italy are in Robinet 1988. For his interest in China: Lach 1945; Mungello 1977, chaps. 1–3; and Leibniz 1977, introducton and 59, 107, 115, 157–58.

[19] Heinekamp in Parret 1976, 518–70. Aarsleff 1982, 42–83. See also Davillé 1909, 222–23, 385–86, 407–8; Leibniz 1903, 224–29.

History of Switzerland" apparently as massive as his folios on natural history.[20]

In Padua, Antonio Vallisneri, a friend of both Bourguet and Scheuchzer, greatly admired Scheuchzer's expertise as both naturalist and antiquarian, as the three friends shared the latest news about recent books and a variety of discoveries. For Vallisneri, however, the great models of scholarly mentality and activity were medievalist L. A. Muratori and Muratori's own mentor, Benedetto Bacchini. If the Vallisneri-Muratori letters allude to but rarely discuss erudite matters, they do contain instances of Vallisneri's aid in gaining access to manuscripts needed by Muratori.[21] In his own early explorations, an account of which he sent to the Royal Society, Vallisneri did not limit himself to observations of nature, but also paid attention to local customs and to those ancient texts recording place-names and descriptions of landscape. As a collector, Vallisneri emphasized natural objects, using his museum as a teaching instrument; he also owned human artifacts in abundance, having acquired the collections of two older contemporaries. That his knowledge of antiquarian subjects was not merely casual and fashionable is further implied in a letter from a friend who, having read Vallisneri's geological treatise of 1721, asked his views on the competing claims to antiquity of Egypt, Chaldea, and China.[22]

Italian intellectuals ever since the Renaissance had almost inevitably developed at least some interest in the antiquities surrounding them. Most versatile of all, perhaps, was the Jesuit polymath Athanasius Kircher, whose many folios dealt with everything from the "subterranean world" to the antiquities of Rome, Egypt, and China. The ultimate curioso with a philosophy all his own, Kircher assembled in Rome a museum that was a remarkable hodgepodge of natural and artificial objects, ranging from the remains of a "siren" to laboratory equipment to demonstrate optical

---

[20] The Leibniz-Bourguet letters are in Leibniz 1875, 3:544–96; see also Sticker 1971. A detailed bibliography of Bourguet's publications is in *DSB*, 15:56–58, with indication also of some of the riches of extant manuscripts. For Scheuchzer, see Hans Fischer 1973, 23, and Jahn in Schneer 1969, 200.

[21] Muratori 1978, 119–20, 168. Most of the letters in this collection are from Vallisneri to Muratori, and the former's admiration is obvious throughout. The early letters of Vallisneri to Scheuchzer (before 1710) are now available in Vallisneri 1991; this edition does not include the few extant replies located among the Scheuchzer manuscripts at the Zentralbibliothek in Zurich. I am indebted to the Zentralbibliothek for microfilming these texts, and I regret that I could decipher only isolated phrases.

[22] The manuscript sent to the Royal Society seems no longer to be extant, but a version was later published in *GLI*, suppl. 2 (1722), 270–310, and 3 (1726), 376–428. For Vallisneri's museum, see Pomian 1987, 123–25; Caylus 1914, 56–57; Montesquieu 1949, 584, 586–87. Letter from Porcia to Vallisneri, 24 July 1721, in Brunelli 1938, 230.

illusions.[23] More restrained than he, his younger contemporaries illustrate clearly the perceived compatibility of studying natural and civil history. Filippo Buonanni, for example, one of Kircher's successors as curator of the museum, published monographs on numismatics and biological generation—the latter text incorporating a discussion of fossils. Best known as a painter, Agostino Scilla published a volume on fossils and seems to have contemplated writing a history of Sicily based on its coins and medals. In the north, Scipione Maffei became an expert on the antiquities of Verona, and he also assembled a volume of "letters" on natural subjects, including an informed discussion of the earth's history. And one of Vallisneri's disciples, Giovanni Bianchi of Rimini, had in his home a noted collection of natural objects and of Greek, Roman, and Etruscan antiquities.[24]

One more Italian example, the Marsili brothers, merits attention for its importance in Bolognese cultural life. The older brother, Anton Felice, in the 1680s established in his home two short-lived academies, one for history (especially ecclesiastical) and the other for experimental science. Although Luigi Ferdinando Marsili shared such modernizing views, he was a soldier and an engineer, not a cleric, and he evidently had little interest in a scholarly approach to Church history. His own interest, as shown in his great folios (1726) describing the basin of the Danube, lay in natural and civil antiquities. Here he displayed remarkable expertise, examining ancient texts, medals, and inscriptions; using his military experience to explore terrains where now-vanished ancient monuments might have been built; and studying the rocks, soils, fossils, and watercourses of the area. An assiduous collector, Marsili donated his holdings to the new Institute he had long worked to create at the University of Bologna. This endowment included laboratory equipment, a museum of plant and animal specimens, "fossils" of various kinds, a library of the best editions, some Greek and Arabic manuscripts, some ancient statuary, and models of fortifications.[25]

Relatively less versatility can be found in the writings of Paris academicians, although a succession of astronomers—Cassini I, Louville, Delisle—gave some attention to the reports from China in which it was alleged that

[23] Rivosecchi 1982 is especially valuable for both analysis and illustrations. Also, Findlen 1994; Godwin 1979.

[24] *DBI*, s.v. Bianchi, Buonanni. For Scilla, see the sketch with bibliography in Morello 1979b, 148–51; his natural history collection was later purchased by John Woodward, in Purcell and Gould 1992, chap. 6. Maffei 1747; his letter on geology in this volume will be discussed below, Chapter 7.

[25] The range of his collections is indicated in Marsili 1728, xxi–xxii. The various skills employed in his research may be seen, for example, in Marsili 1744, 2:25–34, where he discusses Trajan's bridge.

Chinese astronomical skills dated back to a period close to the Flood. To Delisle, in fact, the Academy in 1750 assigned the task of corresponding with one of the most scholarly of the missionaries, Antoine Gaubil, S. J. Fontenelle, however, considered the combination of history with mathematical sciences to be theoretically incompatible. As a member (inactive) of the Academy of Inscriptions, and as eulogist of Leibniz and Francesco Bianchini, Fontenelle found occasion to express admiration of the skills needed by the erudite, but he also announced his doubts in the *éloge* of Bianchini, who had been mathematician, astronomer, and historian. As Fontenelle put it, mathematical truth and historical scholarship were "opposed" to each other, were indeed mutually exclusive, and certainly required different kinds of talent. He was nonetheless aware that mathematicians did do these curious things, as when he reported that Cassini I and Viviani, employed as hydraulic engineers by the Grand Duke of Tuscany, had combined these activities with the examination of fossil shells and of "sepulchral urns and Etruscan inscriptions."[26]

In contrast to Fontenelle, one of the first generation of academicians, Claude Perrault, began his *Mémoires pour servir à l'histoire naturelle des animaux* (1671) with a discussion of history. Making no distinction between the natural and the civil, he declared that there were two kinds of history, distinguishable by method and trustworthiness. One kind entailed compiling information from various sources, so that the compiler, however committed to truth, could not vouch for the accuracy of his material. Preferable was "the Narrative of some particular Acts, of which the Writer has a certain knowledge." Accounts of the latter sort will inevitably be incomplete, but they have "this Advantage; that Certainty and Truth, which are the most recomendable Qualities of History, cannot be wanting . . . provided the Writer be exact and sincere."[27] Although he then moved on to a discussion of the trustworthiness of various accounts of animals, Perrault captured here what his contemporaries considered to be the most valuable traits of all writing, ancient and modern, natural and civil. In the post-1700 generation of Parisian naturalists, however, one finds little of Perrault's interest in evaluating texts. At least, this is true of the Academy's leading naturalists, Réaumur and Antoine de Jussieu, as well as Fontenelle. That Réaumur had a nodding acquaintance with such activities can hardly be

---

[26] Eulogy of Bianchini, 1730, in Fontenelle 1825, 2:232; of Viviani, 1704, in Fontenelle 1709, 1720, 1:73. The valuable Gaubil correspondence (1970) unfortunately does not include any extant replies to his letters. For a degree of competition between the scientists and members of the Academy of Inscriptions in tackling ancient sciences and technology, see Maury 1864, 59–65.

[27] Claude Perrault [1671] 1688, preface.

doubted, but his apparent separation of the natural from the civil in his publications aroused some discomfort among contemporary readers of academic memoirs.[28]

More conventional than the Parisian naturalists, provincial French academicians now and then undertook projects resembling the work of Robert Plot and the British county historians. The chief modern historian of such groups, Daniel Roche, has shown that these academies typically concerned themselves with local history—usually ancient and medieval, sometimes erudite—and with such observational, relatively nontechnical sciences as physiology, anatomy, and natural history.[29] In Bordeaux and Montpellier, academicians planned to assemble natural and civil histories of their regions. The Bordeaux project of 1719, announced in the *Journal des savants,* called on volunteers to send information to Montesquieu, then an active member of the Academy. Twenty years later, however, one member was still urging his colleagues to take up the work seriously, pointing out that regional studies were precisely within the realm of expertise of a local academy.[30]

The Montpellier project had no better success. Two members later claimed that the Society had intended, ever since its founding in 1706, to produce a natural history of the province, but twenty years later had not yet managed to get funds for such a collaborative enterprise.[31] In the event, something akin to the Montpellier plan was carried out by one man, physician Jean Astruc. A scholar who would later produce a masterly work of biblical erudition, Astruc had in 1707 presented to the Montpellier society a memoir on fossils that then circulated in the detailed accounts published in the *Mémoires de Trévoux* and the *Journal des savants.* The journalists appreciated Astruc's combination of talents, as he not only examined terrains but also combed ancient writings for descriptions of the Mediterranean littoral. When Astruc's *Histoire naturelle de la province de Languedoc* appeared in 1737, however, the "natural" content was remarkably slender, even omitting most

---

[28] Torlais 1936 has no great interest in subjects not scientific, but see his chapter on "Réaumur et sa société." Réaumur also corresponded with Louis Bourguet, and some of these letters will be cited in Chapter 5. For his correspondence with Crousaz, see LaHarpe 1955. The "separation" of natural from civil is treated in detail in Chapters 7 and 8.

[29] Roche 1988, chaps. 6, 7.

[30] Barrière 1951, 92–96, 208. According to Masson, in Montesquieu 1950, 3:87, the Bordeaux project was published in both the *JS* and the *Mercure de France.* Varloot in *La Régence* 1970, 138, reports that it was also carried in *Europe savante.*

[31] Plantade in Montpellier, *Mémoires,* vol. 1 (1766), 267; the text published here dates from 1726. Also, letters from Gauteron, secretary of the Montpellier society, to abbé Bignon in Paris, 1726–27, cited above, Chapter 1, n. 69. Gauteron complains that the Estates of Languedoc had expressed liking for the project but had voted no funds.

of the details of Astruc's own earlier memoir.[32] For whatever reason, Astruc had shifted the balance toward the antiquarian, including much information of the neutral and monumental kind—such as place-names, local dialects, and topography, as these were recorded in ancient texts.

These Continental examples, like the British, do not exhaust the field—one might add, for example, Linnaeus's concern to copy Runic inscriptions and to note the location of barrows during his travels in 1741—but they are sufficient to suggest that naturalists often had a lively interest in and knowledge of ancient texts and artifacts.[33] Few attained a level of scholarship comparable to that of Leibniz or Lhwyd or Astruc, but even those who, like Vallisneri, collected but rarely wrote about antiquities gave evidence of their belief that natural and civil history were compatible and indeed complementary areas of research. It is not surprising, therefore, that when naturalists discovered fossil bones apparently belonging to elephants, they turned first to ancient literature for traces of these animals accompanying the Roman legions in their march through Europe.

Thomas Sprat's assertion to the contrary notwithstanding, natural and civil history did "cross each other" in more than one way. Methodologically, practitioners of both sought to eschew personal bias by a focus on facts. Just as historians valued "monuments," geologists regarded fossils and rocks as unbiased evidence about the past. Indeed, the historian's vocabulary permeated geological texts that refer to fossils as the earth's monuments.[34] At the same time, neither historians nor scientists imagined that ancient texts could be ignored, for these contained much concrete information. If historians wondered about personal bias in such texts, scientists had less cause to worry because it seemed unlikely that ancient authors would have had any reason to lie about or to distort accounts of floods, earthquakes, or changing shorelines. In short, geological evidence was part of what constituted, in Robert Plot's apt phrase, "the Records of time," and ancient texts were part of the same records.[35]

Because naturalists did refer to ancient writings, it is obvious that they used the same timescale for both geology and human history. Questions of time will be treated in detail in a later chapter; here it is more pertinent

---

[32] *Trévoux* (March 1708), 512–25, and *JS* (suppl. for 1708), 119–24. The two texts are identical in all but spelling; *Trévoux* includes, but *JS* omits, footnotes to ancient authors. It seems likely that *JS* cribbed its account from the Jesuits. Astruc's full text was published in Montpellier, *Mémoires*, vol. 1 (1766), 48–74. See also *HWL* 10 (March 1708), 163–64.

[33] Linnaeus 1973, 20.

[34] Rappaport 1982, 27–31. Gohau 1987, 9, 61, 72–74; 1990, 115.

[35] Plot 1677, 113.

to ask why they sought human witnesses to natural events. After all, in the common rhetoric of the decades around 1700, nature was more reliable than human beings. Nature obeyed laws; men were self-interested and prone to bias. In Fontenelle's pithy phrase, geologists had the inestimable advantage of studying "histories written by the hand of nature itself."[36]

Many of Fontenelle's contemporaries were less optimistic than he. They certainly knew that a chief way of minimizing reliance on human testimony was by replicating natural processes in the laboratory. But when Fontenelle himself advocated this method, he inadvertently implied one of its limits: in the laboratory, one manipulates *known causes* in the design of experiments. In other words, experiments did not disclose new causes but allowed further study of those already known. Even so, as Vallisneri argued, we have no guarantee that we have selected those causes actually used by nature. Furthermore, as Leibniz remarked, igneous activity in the early history of the earth far exceeded "in intensity and duration the heat of our furnaces," so that the limits of technology made replication impossible.[37]

As another and critical reservation about the geological past, several writers pertinently wondered about the extent to which one could project present processes *uniformly* into the past. Leibniz expressed this worry in more than one text, remarking that those who judge the past by the present would never conclude that the sea had once covered virtually all land masses—and yet the ubiquity of marine fossils showed that this had clearly happened. In a different context, Réaumur wondered whether changing shorelines implied a uniform past rate of marine movement, measurable by the observable (modern) rate; so, too, did Celsius wonder whether the supposed diminution of the Baltic Sea had occurred at the same rate in the past as in the present. Such writers did not question the uniformity of nature's laws, but they were acutely aware, as was any experimenter, that constancy of effects can be expected only when conditions are also constant.[38]

For these several reasons, investigation of the earth's past was deemed an uncertain enterprise. Collecting facts was not at issue, but how could one be sure that one had correctly described the physical conditions obtaining in the past? Could one properly infer past conditions and causal explanations merely from an examination of objects? How might inferences be

---

[36] *HARS* [1722] 1724, 4.

[37] *HARS* [1700] 1703, 51–52. For Vallisneri, see M. Baldini 1975, 60. Leibniz [1749] 1993, chap. 3, p. 17.

[38] Leibniz [1749] 1993, chap. 26, p. 91, and Leibniz to Jakob Bernoulli, 1703, in Daston 1988, 237. For Réaumur and Celsius, see below, Chapter 7. Rossi 1984, 43–44, remarks on the view that, after the Creation, "laws of nature . . . have no history."

transformed from the tentative to the more reliable? In the texts to be examined below, words such as "conjecture" and "hypothesis" deserve our attention, for neither was considered desirable as more than a temporary answer to questions about nature. If, as Robert Hooke phrased it, we have no helpful human records of geological events, the results "at best will only afford us occasions of Conjecture." According to Emanuel Swedenborg, "it is only by conjectures and experiments that we can obtain any insight into [the] causes" of events dating from remote antiquity. When Leibniz dealt with natural conditions antedating all "histories," his results were described as "conjectures based on natural monuments." As one writer would eventually remark, "an historical truth is . . . preferable to any hypothesis whatsoever."[39] What we are seeing here is apprehension that study of the past will be merely conjectural and hypothetical. To avoid such a result as far as possible, scientists hoped to discover human evidence to supplement and confirm that of nature. Far better to find human testimony than to rely solely on what Hooke, referring to fossils, characterized as "dumb Witnesses."[40]

As a fine example of the procedure just outlined, Thomas Molyneux revealed his methods to the Royal Society in his "Discourse concerning the *Large Horns* frequently found under Ground in *Ireland*." These antlers— belonging to what we now recognize as the extinct Irish elk—were, as his title indicates, numerous, and no such large animal still lived in Ireland. Molyneux tells us that he consulted an antiquarian to find out if human beings had recorded seeing the animal; furthermore, since he identified this animal with the American moose, he wanted to know when the Irish survivors had migrated westward. When the antiquarian reported a lack of written references to the elk, Molyneux concluded: "seeing it is so many Ages past, that we have no manner of Account left to help us in our Enquiry, the most we can do in this Matter is to make some probable Conjectures about it."[41]

In comparable fashion, antiquarian William Somner produced a "Brief Relation of some Strange Bones" in which he argued that England and France had formerly been connected by an isthmus. Admitting that neither history nor tradition mentioned such a land-bridge, Somner proposed a

---

[39] Hooke 1705, 334, text probably dating from 1686–87. Swedenborg [1722] 1847, 16; also, p. 28. Eulogy of Leibniz, in Fontenelle 1709, 1720, 2:286; similar remarks are in *Europe savante* (November 1718), 139, based on Fontenelle's text but written by editors skilled in matters of historical scholarship. The final quotation is from the *Gentleman's Magazine* 27 (July 1757), 299.

[40] Hooke 1689 in Gunther 1930, 7:712.

[41] *PT* 19 (1697), 489–512, at 490, 499.

line of argument for which he suggested a fragment of proof. The discovery of marine fossils near Canterbury could be explained by the one-time presence of the sea in that area. When the sea retreated from Kent, it destroyed the hypothesized isthmus, leaving behind not only fossils but also a bit of linguistic evidence: the valley of the "Sture" (modern Stour) could be construed as "Esture," a corruption of "estuary." All this seemed to Somner so persuasive that "were I of the Jury, I should more than incline to concur with them, who find for the [existence of the] *Isthmus.*"[42]

John Wallis immediately produced his own addition to Somner's evidence, namely,

> the *Unity of Language* between the Ancient *Gauls* and *Britains;* and . . . the great Intercourse between those in *Gaul,* and the *Druides* in [Britain]; (of which Ancient Writers take notice:) Which is not likely to have been, if there had not been an easie Communication between the one and the other. Which, though it be not a *Physical* Argument (as are those of Mr. *Camden,*) is a good *Moral* Inducement, in Confirmation of them.[43]

Even this evidence was not wholy satisfying, albeit "morally" so, because no historian mentioned an isthmus or its destruction, "which being a thing remarkable, might have been thought worthy to be reported." Wallis therefore suggested further that the Atlantis myth might refer not to an island, but to an isthmus.[44] In this fashion, Wallis tried to add to the admittedly sparse human testimony.

The case of Robert Hooke is more complex, Hooke himself having been willing to do without human witnesses to the many earthquakes he thought had shaped the earth's history. As he argued, many historical events, including such striking ones as the building of Stonehenge, had not been set down in writing; one could thus legitimately assume that modern earthquakes testified to the way nature had also worked in the past. In general, the argument was reasonable, appealing to the common conviction that

[42] *PT* 22 (1701), 882–93, at 888, 891.

[43] *PT* 22(1701), 967–79, at 968.

[44] Ibid., 973–75. The temptation to reduce a large myth to an event or person of local significance was not peculiar to Wallis. According to Davillé 1909, 527–28, Leibniz succumbed to Teutonic temptation in identifying Ulysses with Odin. Better known at the time, Olaus Rudbeck's *Atlantica* identified the cultured society of Plato's Atlantis with ancient Sweden, omitting Plato's disappearance of the Atlantean continent! Rudbeck was reviewed in the *Philosophical Collections,* no. 4 (1681/2), 118–21. There were also many attempts to show that one or another of Noah's sons had founded particular nations, and to show that languages ranging from Breton to Hungarian had been the language of Adam and Eve. In the latter connection, John Webb was unusual in arguing for Chinese, not English, as the Adamic language.

nature worked in lawful ways. Still, as John Wallis and others noted, even the dullest chroniclers could hardly have failed to observe and record at least some of the earthquakes posited by Hooke, and Fellows of the Royal Society, as well as members of the Oxford Philosophical Society, demanded that Hooke confirm his theory with the use of human testimony. This Hooke tried to do, with results as sparse as the fragments later employed by Somner and Wallis himself.[45]

Among Continental naturalists, in addition to Leibniz and Swedenborg, a noteworthy French example comes from the travel account of botanist Joseph Pitton de Tournefort. During his voyage to the Levant in 1700–1702, Tournefort wondered whether the Black Sea had once been wholly landlocked. If so, then rivers emptying into it would have caused the sea to overflow, flooding the countryside and ultimately carving an outlet into the Sea of Marmora and hence the Mediterranean. When Tournefort discovered that an ancient text did report a tradition of flooding in the very region in question, he expressed his relief in the following terms: "what we have just suggested as a scientific conjecture becomes an historical truth."[46]

With these examples in mind, one may better understand why Jean Astruc, examining the coast near Marseilles, could not rest his case for changing shorelines wholly on physical evidence, however persuasive. Seeking support for his "simple conjectures," he wished the ancient natives had left written descriptions of the area; lacking such, there was a limit to how far back in time he could go, as he had to be content with the accounts by Greek and Roman immigrants.[47] Similarly, when Louis-Bertrand Castel, writing in the *Mémoires de Trévoux,* criticized the theory of marine transgressions proposed by Antoine de Jussieu, he complained that such large events "should have been . . . noted in all the Histories." Commenting on Jussieu and Castel, Antonio Conti in Paris wrote to Vallisneri in Padua:

> Our most authentic histories do not go back further than Cyrus, i.e., about five centuries before the birth of Christ. That is when the centuries of myths end. But in order for the movement of the seas to be detectable (*sensible*),

[45] Rappaport 1986, esp. 136. For a more sophisticated treatment of the silences of texts, see Bayle in Labrousse 1963, 2:20; also Lange 1966.

[46] Tournefort 1717, 2:403–9, quotation from 406. See also J. G. Gleditsch, cited in Chapter 4, at n. 22.

[47] References given above, n. 32. The version published in 1766 (p. 57) is clearer than the journalistic accounts about "simple conjectures" and the regrettable need to use accounts by immigrants.

could it not have a period longer than that of thirty centuries, for example?[48]

Conti's query does not explicitly refer to a pre-human past, but only to a time, "the centuries of myths," that cannot be trusted for "authentic histories."

As a final example of these conventional modes of thought and argument, the writings of Nicolaus Steno merit analysis for the author's awareness of his own methods. Like his contemporaries, Steno knew that the "geometric mode" produced certainty, that "matters of fact" should be accompanied by causal explanations, and that such explanations could at best be probabilistic. To lend greater weight to probabilities, he, too, sought confirmation in ancient writings.

Two years before the famous *Prodromus* (1669), Steno's treatise on muscles showed his attachment to the geometric mode as an effective way of conveying certainty. Appended to the treatise, his "digression" on *glossopetrae,* or fossilized sharks's teeth, made no use of such precision; instead, Steno began with a series of descriptions—the section called "historia"—of strata and of the physical condition of the fossils found in them. There followed a long list of "conjectures" to explain the causes of these phenomena. If Steno had no doubts about the origin of *glossopetrae,* he was notably cautious in claiming more than probability for the organic origin of all marine fossils.

> The same phenomenon may be interpreted in various ways; indeed, Nature in her processes gains the same end by various means. Hence it would be unwise to regard just one method of them all as true, and condemn all others as false.[49]

In short, some fossils might well be the products of unknown natural powers.

In the *Prodromus,* some former "conjectures" became "propositions"—for example, the key statement that bodies enclosed in the strata must have been buried there before the strata became consolidated.[50] Indeed, the

---

[48] Conti to Vallisneri, n.d. (but shortly after the account by Castel), in Conti 1972, 390. Castel in *Trévoux* (June 1722), 1093.

[49] Steno [1667] 1958, 43.

[50] Ibid., 15; compare Steno [1669] 1916, 218. The next several citations from the *Prodromus* will all be from the 1916 translation, the page numbers inserted in the text.

whole structure of the book entails ranging Steno's observations under the rubric of the proposition. The very title of the treatise offers a taxonomic statement in highly abstract terms: how to analyze "a solid body enclosed by process of nature within a solid." And the chief aim is expressed in traditional mathematical form: "given a substance possessed of a certain figure, and produced according to the laws of nature, to find . . ." (209). Historians have customarily and rightly agreed with Steno's own contemporaries that the *Prodromus* has a solid empirical foundation; at the same time, some modern readers have noticed that the work is so abstract that it is virtually impossible to identify what terrains in Tuscany Steno actually examined.[51]

To look at the structure of the *Prodromus* is to see the priorities and epistemological pronouncements of the late seventeenth century. First comes the indubitable theoretical analysis: those propositions which conform to Cartesian clear and distinct ideas because their opposites are obviously impossible. Each proposition is illustrated by a range of supporting examples. There follows a brief application of the several principles to the actual structure of a part of Tuscany, and then the tidying up of the proofs by asking whether human witnesses confirm the whole analysis.

That this apparently neat structure is artificial becomes evident if one examines the odd array of material assembled under the propositions. Steno's analysis of the behavior of fluids, for example, is essential to his discussion of the deposition of horizontal and parallel sediments (227–31); but this analysis is wholly irrelevant to the very next pages, where he attributes subsequent changes in the position of strata to either "the violent thrusting up" caused by a number of possible forces or "the spontaneous slipping or downfall of the upper strata . . . in consequence of the withdrawal of the underlying substance, or foundation" (231–34). Similarly, propositions concerning the behavior of particles and fluids work well when applied to the patterns of accretion of various types of crystals (237–49); but the same propositions are strained to the utmost in the very next pages when he discusses the vital processes involved in the growth of the shells of mollusks (249–56). And these propositions are irrelevant when he moves from shells to other fossils: wood, bones, and teeth. Here he simply compares living forms with fossil specimens (257–62).

Artificial, too, is the search for confirmation in ancient texts, since the whole structure of the book involves the application of propositions deemed certain to the phenomena they can explain. On the face of it, further confirmation would seem to be unnecessary. If one looks carefully

---

[51] E.g., Azzaroli in *Niccolò Stenone* 1988, 78, and Gohau 1987, 69.

at Steno's final pages, however, it becomes clear that he perceived some gaps in his arguments. Below are excerpts from his recapitulation of the six stages of Tuscan history, the key phrases italicized (263–65).

> In regard to the first aspect of the earth Scripture and Nature agree in this, that all things were covered with water; *how and when this aspect began, and how long it lasted,* Nature says not, Scripture relates . . .
>
> Concerning the *time and manner* of the second aspect of the earth, which was a plane and dry, Nature is likewise silent, Scripture speaks . . .
>
> *When the third aspect* of the earth, which is determined to have been rough, *began,* neither Scripture nor Nature makes plain. Nature proves that the unevenness was great, while Scripture makes mention of mountains at the time of the flood. But *when those mountains . . . were formed, whether they were identical with* mountains of the present day . . . , neither Scripture nor Nature declares.

Much more detail is offered in these pages, but the emphasis on "how and when," "time and manner," reveals Steno's sense of the limited information that he had been able to extract from nature itself.

The only text sufficiently reliable and ancient to supply at least some of the missing details was Genesis. Although Steno also took care to reassure readers that his science did not contradict the Bible, one should note that Genesis was not his only historical text. He was cautious about interpreting ancient myths, arguing that "the history of nations regarding the first ages after the deluge is doubtful in the view of the nations themselves, and thought to be full of myths" (267). Nonetheless, the "authority of history" (259) could not be limited to the Bible, and Steno thought that some ancient myths and traditions incorporated a few credible elements. Among these were the myth of Atlantis, geographical details dating from the journeys of Bacchus, Ulysses, and others, and the Greek tradition that early men descended from the mountains when the plains became dry and habitable (268–69). If these tales were somewhat less credible than later texts—such as Roman accounts of the expedition of Hannibal, or Tacitus on earthquakes—Steno still thought them potentially useful in reconstructing aspects of the geological past.[52]

The curious structure of the *Prodromus* did not disguise valuable features of the book, and readers such as Paolo Boccone, Francesco Nazari, and

---

[52] Hannibal is in Steno [1669] 1916, 259. That Steno in 1667 considered Tacitus sufficient even without natural evidence is indicated by Noe-Nygaard in Poulsen and Snorrason 1986, 175. Unusual among writers on Steno, Greene 1961, 58–59, notices that he and Hooke sought confirmation from ancient witnesses.

Henry Oldenburg ignored the organization, commenting instead on specific matters of interest to the "curious," such as the discussions of fossils and crystals. Certainly, the arrangement of propositions, illustrative phenomena, and confirmation would have been familiar to such readers.[53] If we compare for a moment Rohault's popular *Traité de physique,* published two years later, major sections of the text begin with the equivalent of propositions, followed by illustrative experiments. When he dealt with the earth, Rohault, unlike Steno, singled out "timeless" phenomena (like the nature of matter) and thus had no need of historical confirmation; for him, the repetition of experiments sufficed. Indeed, later in the decade, Mariotte's *Logique* (1678) coupled a defense of empiricism with the acknowledgment that a quasi-deductive form of presentation is more persuasive.[54]

The methods employed by Steno and others may be fruitfully compared with those developed by the erudite historians analyzed in Chapter 2. To numismatists such as Charles Patin and Ezechiel Spanheim, monuments were "the proofs of history" but could not substitute for the text of Livy's history of Rome; rather, coins could flesh out details, add some information, and correct textual errrors or confirm textual accuracy. When relevant texts were lacking, confronting a monument like Stonehenge left investigators struggling to find a suitable, provable historical context. For naturalists, fossils and strata were monuments and proofs of history, but texts, too, added to the stock of facts. Having no interest in syntheses of human history, naturalists nonetheless resembled historians in seeking bias-free statements in ancient texts. As we have seen, such texts were scrutinized for concrete information—on earthquakes, floods, land bridges, elks—about which the witnesses had no reason to lie.

Like historians, naturalists found texts to be in short supply for the most remote past, so that myths and legends became as important to natural as to civil history. (The opening chapters of Genesis were not regarded as myth or legend, except by some outrageous freethinkers.) How to explicate myths remained problematical, as historians and naturalists struggled to separate tales intended to be fiction from those thought to convey useful information, however disguised. As Robert Hooke explained it, he believed the myths recounted by Ovid to be constructed "partly from the Theory of

<hr/>

[53] Boccone 1671, 54–66. Nazari 1669, 111–14. *PT* 6 (1671), 2186–90. Oldenburg's introduction to his translation of Steno 1671. Silence on the structure of the *Prodromus* is reminiscent of the silence about Jüngken's later "demonstrative" chemistry text (above, Chapter 2, n. 14).
[54] Rohault [1671] 1728, pt. 3, chap. 1, for his nonhistorical approach to the earth. Coumet in *Mariotte* 1986, 302–4; cf. Dear 1990, esp. 665–69.

the best Philosophy; partly from Tradition, whether Oral or Written, and partly from undoubted History." In a moment of caution, Hooke added that diversity of interpretation prevailed because "In these Matters Geometrical Cogency has not yet been applied."[55] In fact, when asking himself whether myths are reliable sources of data, Tournefort, like Steno, had to admit that most were not. But he did not hesitate to accept Plato's tale of Atlantis.[56] Unlike historians, however, naturalists generally did not engage in Euhemerist explications of myth, for their aim was not to identify the real persons behind ancient figures of gods and heroes; understandably, they instead combed through myths for naturalistic elements.

How to synthesize their materials into a coherent narrative proved more difficult for naturalists than for historians. In one of his perceptive texts, Fontenelle reflected on the reconstruction of ancient (human) history, but his remarks apply to geology as well.

> If one found the wreckage of a great palace, the pieces scattered over a large terrain, and one were sure that no pieces were missing, it would be an enormous labor to reassemble the whole—or at least, without [physically] doing so, by studying the pieces, to arrive at a correct idea of the whole structure of the palace. But if pieces were missing, the work of mental reconstruction would be more difficult, and the more pieces missing, the more difficult the task; and it would be quite possible to draw up various plans of the structure with little in common among themselves. This is the condition in which we find the history of the most ancient times. Innumerable texts have vanished; those remaining are rarely complete; numerous little fragments which could be helpful are scattered here and there. . . . But worse yet—and not true of material objects—the texts of ancient history often contradict one another; and one must either find a way to harmonize them, or settle for making a choice that one always suspects is a little arbitrary.

And, Fontenelle concluded, historians merely do what they can to combine the available materials into a synthesis that will inevitably be challenged by someone who claims to have done a better job.[57]

Although Fontenelle thought geology to be in a less parlous state than civil history—nature and "objects" in general being more trustworthy than human documents—one could scarcely hope to find a better statement of

---

[55] Hooke 1705, 384, 391, texts dated 1688 and 1693, respectively.

[56] Tournefort 1717, 2:407–9.

[57] Eulogy of Bianchini, 1730, in Fontenelle 1825, 2:230. This passage would later be cited by the French translator of Warburton 1744, 1:xij–xiv, and by d'Alembert, "Chronologie," in Diderot, *Encyclopédie*, vol. 3 (1753), 390b. See also Lombard 1913, 395, 398.

the problems confronting both disciplines. To historians living at that time, however, syntheses probably seemed quite feasible. Some, like Livy's history of Rome, already existed, and others could be achieved by taking an important theme, such as the history of the French monarchy, to provide the essential continuity of a national history. Naturalists had no such easy options. They knew the earth had a history, in which the biblical Flood was one episode. But to understand even trustworthy objects proved more difficult for naturalists than did ancient coins for historians. Every attempt at synthesis, every "theory of the earth," led to what Fontenelle summed up as a challenge from someone who claimed to have done a better job.

If syntheses came and went, a general framework, shared by historians and geologists, remained more stable, and it was biblical. Much has been written about the fear of heterodoxies, but to such fear one must add the dread of losing another kind of certainty: the known outlines of world history. To imagine, as LaPeyrère did in 1655, a pre-Adamic world of men was not liberating, but plunged one into a history without dates, without human records, and without landmarks. The "authority of history," in Steno's phrase, provided a bulwark against unverifiable conjecture.

CHAPTER 4

Fossil Questions

So much has been written about "the fossil question" of the later seventeenth century that another examination of the issues may seem supererogatory. Historians have tended, however, to concentrate on the arguments of some notable naturalists—chiefly Nicolaus Steno and certain of his British contemporaries—and lesser figures, both British and Continental, have received only sporadic attention. To be sure, debate in Britain was lively, but Continental naturalists discussed the same arguments, sometimes with, sometimes without knowledge of British publications. Understandably, historians have also focused on the most common and problematical fossil forms, namely, marine shells. Here I argue that due attention to fossilized wood, bones, and teeth suggests an illuminating shift of emphasis from identification of specimens to issues of transport and deposition. The different classes of fossils posed somewhat different problems of transport, and only in the case of marine shells did some naturalists refuse to consider them as organic remains unless their transport could also be explained.

To summarize the fossil debate in its most familiar form, one must begin with the shells that so many naturalists were discovering buried in peculiar locations: at high altitudes, deep in the earth's crust, and far from modern seas. Naturalists could identify such finds as organic if they could satisfy themselves on three counts: form, matter, and position.[1] Did a fossil resemble in form a known living organism? Did it show traces of organic material, or had it been completely lithified or mineralized? How did it come to occupy its particular position in the strata? No consensus could be reached quickly on these subjects. If, for example, the fossil shells examined in Italy and southern France could readily be identified with living genera and even species, the giant ammonites in English strata could not. If marine deposits near modern coasts and at low altitudes could be explained as the result of occasional flooding or episodic variations in sea level, comparable deposits in other locations could not as readily be accounted for.

The debaters, then, divided into three groups. One, which included Steno, Hooke, and Leibniz, unequivocally identified fossil shells as the remains of organisms; the same men, however, differed among themselves on how fossil-bearing strata had been deposited. A second group, led by Robert Plot and Martin Lister in Britain and by other men on the Continent, denied the organic origin of most fossil shells, arguing instead that nature, in some mysterious way and for some inscrutable purpose, had produced rocks bearing an uncanny resemblance to the shells of living organisms. Three phrases regularly occur in the writings of Plot, his school, and his critics: the specimens themselves are *lusus naturae* (sports or tricks of nature) or *lapides sui generis* (rocks of a unique kind), and the cause that produced them is a *vis plastica* (plastic force or virtue). Competing with these two views was the theory commonly associated with the name of Edward Lhwyd, but also suggested independently by Tournefort in Paris, that "seeds" lodged in soil or rocks had there grown into what Lhwyd called "mock shells." For the Plot and Lhwyd groups therefore, no problem of transport and depositon arose, since seemingly marine forms had actually originated in the very places where they are now found.

The various concepts employed by these writers not only seemed to resolve particular difficulties, but also had a respectable intellectual ancestry in ancient philosophies and in Renaissance and post-Renaissance revivals. Essentially Aristotelian was the idea that some forms of life arise by spontaneous generation; seriously questioned by Francesco Redi in the

---

[1] The best discussion is by Rudwick 1972, chaps. 1–2, whose book has rendered earlier studies of the fossil question obsolete. For form and matter, see also the discussion of Fabio Colonna by Morello 1979b, 1981.

1660s, the idea survived because even his admirers recognized that a larger range of experiments would have to be undertaken to supplement Redi's.[2] First cousin to spontaneous generation, so to speak, the Neoplatonic notion that nature in general and the earth in particular possessed "generative" powers had never entirely vanished from philosophy. In the Neoplatonic revival then associated with Henry More and Cambridge, such "plastic" powers became a sort of intermediary between God as Lawgiver and the mechanistic conception of the universe advocated by Descartes and others.[3]

As is evident from this brief survey, ancient survivals and revivals could hardly be confined to Britain, and Continental naturalists shared the same philosophical heritage as their British colleagues. Less evident from this summary, writers often declared that their answers to the fossil question depended not so much on form and matter as on finding a satisfactory explanation of transport and deposition. This stark choice can be found, for example, in a letter from one of Henry Oldenburg's correspondents, mathematician R. F. de Sluse (Slusius): he could readily believe that fossils were "little creatures . . . hardened by a petrifying juice," but their location so far from the sea was baffling, "unless indeed they are the results of nature's playfulness."[4] In the same vein, Roman journalist Francesco Nazari in 1676 began a review of two new publications by declaring that the authors were debating the key issue of whether fossil shells had been generated in situ or had been transported by the sea. One of Nazari's authors, Giovanni Quirini, unable to solve the transport problem, had opted for generation within the rocks; the other, Jacopo Grandi, gave the Flood a greater role than did the man he admired, Nicolaus Steno.[5] Indeed, the question of transport instantly and inevitably leaped to mind when one encountered objects far from their natural environment. Thus, when physician and botanist Sir Andrew Balfour asked a friend to collect some fossil shells for him, he indicated the location of a good deposit on a hill near Rome where there were

---

[2] Compare the Whiggish view that Redi "refuted" spontaneous generation (*DSB*, s.v. Redi) with Vallisneri's early publication intended to add to Redi's experiments, in Porcia 1986, 52–53, 56–57. For analysis of the long persistence of spontaneous generation, see Mendelsohn 1971.

[3] The best studies of Henry More et al. are now A. R. Hall 1990 and Hutton 1990. Other useful discussions include W. Hunter 1950; Hutchison 1982; Henry 1986.

[4] Slusius to Oldenburg, 24 October 1668, in Oldenburg 1965, 5:90.

[5] Nazari 1676, 8–12. Quirini is sometimes "Querini." The Quirini and Grandi tracts were published simultaneously in Venice.

> Shells of all sorts *Petrified,* or if ye please to call them Stones, resembling all the Species of Shells. By what means they came there, I leave you to find out.[6]

Equally succinct was the Huguenot traveller F. M. Misson who compared shells found at high altitudes with kidney stones: both resemble "monsters" because their locations are so unnatural.[7] When astronomer G. F. Maraldi reported in 1703 on assemblages of marine fossils he had seen in Italy, he concluded by announcing to the Paris Academy that he had limited himself to facts, without seeking to explain "how these Fish and Shells find themselves on Mountains, apparently contravening the order of Nature."[8]

Such doubts and choices proved to be unnecessary for those who examined fossilized wood, bones, and teeth. Although individual specimens might be hard to identify precisely, rare was the naturalist who thought them *lusus naturae,* in large part because the transport question could be so easily answered for these objects.

Petrified Wood

Probably the most exhaustive analysis of petrified wood occurs in Robert Hooke's *Micrographia* (1665), where Hooke begins his examination by comparing his specimen with "rotten Wood" from "a huge great *Oak.*"

> I found, that the grain, colour, and shape of the Wood, was exactly like this *petrify'd* substance; and with a *Microscope,* I found, that all those *Microscopical* pores, which in sappy or firm and sound Wood are fill'd with the natural or innate juices of those Vegetables, in this they were all empty.[9]

If other writers did not imitate Hooke's use of the microscope—and few Continentals could read his English—comparison with living specimens proved persuasive. As Steno put it, a petrified sample could be shown to be real wood "by the knots of its branches and by its bark, whose fissures

---

[6] Balfour 1700, 272. This is a posthumous publication by Balfour's son, the text dating from perhaps 1668 when the recipient of these letters was about to embark on his grand tour (ibid., iv–vi).

[7] Misson [1688] 1717, 2:315–16.

[8] AAS, P-V, t. 22, fol. 366r (12 December 1703); Fontenelle expanded on this statement, as will be evident below.

[9] Hooke [1665] 1961, 107. Petrified wood, both process and transport, were being discussed in 1663 at the Royal Society: Birch 1756, 1:247–48.

had been filled with mineral matter."[10] In Paris, astronomer Philippe de La Hire informed the Academy of Sciences that although criteria such as weight and hardness showed that samples of a petrified palm trunk were genuine rock, other criteria, including fibrous structure, proved that this rock had once been living wood.[11]

Hooke, La Hire, and other naturalists also thought it significant that one could almost observe directly how wood became petrified, namely, by the agency of "petrifying waters" or a "petrifying spring." This phenomenon had so long a history, dating back to Pliny, and was apparently so much a commonplace, that Robert Boyle once reported with surprise that there existed "a place in England, where, without petrifying Water, Wood is turned into Stone."[12] Like Steno, Hooke never attributed petrification to a single causal agency, but he attempted some experimental replication in the case of wood. Although, as he admitted, he never succeeded in petrifying a wooden stick "throughout," he seems to have been satisfied with the results. For La Hire, the reported effects of a petrifying spring proved even more decisive than the comparison of fibrous structures; observation of the spring, he declared, "decides the question and leaves no doubt at all" that petrified objects are not *lusus naturae*.[13]

"Petrifying waters" turn up regularly in the decades around 1700, often, but not exclusively, in association with petrified wood. On one occasion, for example, La Hire showed the Academy a petrified shark's tooth, and the minutes of the meeting simply contain the remark: "the waters of that place [where the tooth was found] are petrifying."[14] Some writers tried to explain the causal process further, usually suggesting that particles in the water had entered the pores of the organic material. Others would eventually begin to doubt that such encrustations were the first stage in producing petrification "throughout."[15]

The combination of comparative structures and an effective causal agency convinced Robert Plot that petrified wood was not among the *lusus naturae* that he otherwise defended with such vigor. Specimens he encountered retained "some course [*sic*] representation of the very lineaments of the wood it self," and he offered some analysis of how salts or a "saline prin-

---

[10] Steno [1669] 1916, 261.

[11] *MARS* 1692, 122–23.

[12] *PT* 1 (1665), 101–2. Pliny was still being cited on this subject by Feuillée 1714, 1:329–30. See Pliny, *Natural History*, XXXI, xx.

[13] Hooke 1705, 294, text dated ca. 1668. *MARS* 1692, 125, later cited approvingly by Astruc 1766, 70–71.

[14] AAS, P-V, t. 12, fol. 107r (11 December 1688); *HARS depuis 1666*, 2:43. See also Clayton 1693 in Berkeley 1965, 56, on the petrifying waters of Virginia.

[15] Doubters included Dezallier 1742, 84–85.

ciple" found in petrifying waters might enter the pores of organic bodies. John Beaumont, a disciple of Martin Lister, offered no explanation of his own reasoning, but when, in 1685, he presented to the learned societies of both London and Oxford a plan for a natural history of Somersetshire, he divided what are now called fossils into two categories: on the one hand, "*Stones* curiously wrought by Nature" and resembling plants and animals, and, on the other, "*pieces of wood, fossile teeth,* or other *bones* and the like."[16] In the years before he developed his theory of "seeds," Edward Lhwyd debated with himself and his friends about the origin of fossils, petrified wood being one of the cases he examined. Was this real wood or merely stone with a woody texture? More cautious than most, Lhwyd declared: "For my part I incline at present, very much to ye former, but will not yet conclude but that there may be both in rerum naturâ."[17]

Lhwyd was under the impression that naturalists disagreed on this subject, but it seems that remarkably few believed petrified wood to be a mere rock resembling wood. One of these few, Nehemiah Grew, may have attempted to compare the petrified specimens in the Repository of the Royal Society with living wood, but he records only what he considered to be a decisive chemical test: the fossil specimens would not react with any acid, as wood ought to do.[18] Lost irretrievably, it seems, are the names of participants and details of the debates in Bourdelot's "academy" in Paris, but one day a rock "entirely similar to a piece of oak" revived discussion of "whether bones, wood, or other substances can petrify." Some members supported the negative, but the editor of an account of these "conversations" chose not to report their views. Instead, he emphasized the more persuasive opposition: petrified wood resembles real wood in structure, and the behavior of a petrifying spring is cited.[19]

If most naturalists agreed about the nature of petrified wood, to the standard arguments based on structure and the agency of petrifying springs must be addded a third consideration: no unusual mechanism of transport was needed to explain how specimens came to be buried where they were found. As a case in point, when Childrey's *Britannia Baconica* (1660) reported on the several instances of "buried trees" discovered in Cornwall and elsewhere, the author offered two possible explanations: they had been toppled by coastal storms (when found in coastal areas), or they had been

---

[16] Plot 1677, 32–34. Beaumont in Birch 1756, 4:379–80, and in Gunther 1939, 12:276.

[17] Lhwyd in Gunther 1945, 14:161.

[18] Grew 1681, 270, also 253–54 on the nature of fossils, and 257, 262, 265 for the examination of particular specimens.

[19] Le Gallois 1674, 1:123–24. The text also alludes to stalactites as showing that water itself can turn to rock. On this point, see Plot 1677, 33.

chopped down and left to decay or petrify by either Romans or ancient
Britons. Cornish miners, he added, believed that they had lain underground since the era of the Flood. A few years later, reviewing a French
translation of Childrey, the *Journal des savants* mentioned the Flood and the
additional possibility that such trees had actually grown underground, but
the reviewer declared that we need not have "recourse to causes so distant
and so extraordinary." It was simple and plausible to say that the sea or
rivers had flooded the areas in question, and also that some trees had been
felled by ancient inhabitants.[20]

From the early years of the Royal Society various Fellows had now and
then discussed how trees could be buried and the wood turned to rock. By
the time William Derham tackled this subject in 1712, interpretation had
become easy. The deposit examined by Derham had been exposed in the
aftermath of a local flood, and Derham could readily imagine that the trees
had originally been uprooted either by storms or by earlier floods. Indeed,
Derham remarked, the soil covering these buried trees closely resembled
sediments still being deposited by the Thames.[21] Some thirty years later, in
1745, Jean-Etienne Guettard examined a comparable deposit and concluded as readily as Derham that the local river had been responsible for
burial. In the Berlin Academy, J. G. Gleditsch would introduce his own
analysis with the remark: "If we can credit ancient monuments, this whole
area [the Marches of Brandenburg] was, in the days of our ancestors,
covered with huge forests," so that it was hardly surprising to find deposits
of petrified wood.[22]

These and other writers sometimes explicitly acknowledged that the key
problem of transport was more readily resolved for petrified wood than for
other fossils. Derham and Antoni van Leeuwenhoek, for example, attributed marine shells to the Flood, but had other explanations for petrified
wood. As one anonymous commentator noted, one did not need the Flood
in such cases.[23] John Ray did wonder if the petrified wood found in English
fens could be identified with fir trees not native to England, and John
Morton would later argue that such examples required the agency of the
Flood. To Morton's claim a reviewer for the *Bibliothèque raisonnée* would
respond that one need only consult travel accounts for persuasive evidence

[20] Childrey 1660, 6, 142–43, 150. *JS*, Amsterdam 12mo (13 June 1667), 153–54. Discussion also at the Royal Society in 1663, in Birch 1756, 1:247, 248, and in Ray 1673, 7.
[21] *PT* 27 (1712), 478–84.
[22] Guettard travel journal, MHN, MS 2187, fol. 258r–v; the handwriting is too poor for
deciphering the precise location, but Guettard also noted that these sediments contained
what appeared to be freshwater fossils. Gleditsch 1748, 33–34.
[23] Leeuwenhoek 1800, 1:145–53, text originally published in Dutch in 1702. *Universal
History* (1740), 1:97.

that rivers and ocean currents could transport logs for long distances, without need of the Flood.[24]

Buried Bones and Teeth

Like petrified wood, fossil bones and teeth were readily identified as organic remains, although not without occasional dispute. Exactly what animals these fossils belonged to, a problem distinct from acknowledgment of their organic origins, aroused in naturalists an awareness that they needed to know more about comparative anatomy. Conscious of their limited knowledge, naturalists nonetheless knew that some of their fossil specimens did not correspond to any living animals native to Europe. The transport problem, seemingly acute, was actually easy to solve by resorting to ancient historical texts.

To modern historians, the most famous early remarks on fossil "elephant" bones are Steno's, but in the late seventeenth century the stirring text was a modest brochure by W. E. Tentzel, an *Epistola* first published in 1696. This letter was addressed to that prince of communicators, Antonio Magliabechi, in part because Tentzel hoped for information about comparable specimens in the collection of the Grand Duke of Tuscany. Magliabechi dispatched copies of the tract to at least one Neapolitan correspondent and probably to Modena as well, where Bacchini's *Giornale* gave the pamphlet a detailed review. In Paris, too, the *Journal des savants* considered Tentzel's work worthy of serious attention. Tentzel himself sent both the epistle and some bones and grinders to the Royal Society, wanting the Society's judgment. The *Philosophical Transactions* reprinted the whole text, with the comment—inadequate from Tentzel's point of view—that the specimens were "found agreable to his Description."[25] In the decades that followed, Tentzel's little text would be cited repeatedly and certain of its arguments considered authoritative.

Tentzel rejected the idea that his specimens could be sports of nature, in part on the grounds that no mere accidental concretion would reproduce in detail even the hollows formerly containing bone marrow. Using published reports of the anatomy of elephants, he also argued against the

---

[24] Ray 1673, 9. Morton 1712, 258–61, and *BR* 18 (1737), 82–89. In the same periodical, see also the review of Bourguet, 30 (1743), 162–67.

[25] Tentzel [1697?]. Quondam and Rak 1978, 1:461–62. Bacchini 1696, 251–53. *JS* (20 August 1696), 393–95. *PT* 19 (1697), 757–76, and note appended on p. 776. For Tentzel's efforts to get the Society's judgment, see Ray 1848, 322–23, 366, 472. A bibliography of works on elephant fossils is appended to Guettard 1768, 1:29–79.

popular belief that such bones belonged to giant human beings. Once he could say with assurance that the bones, tusks, and grinders were those of an elephant, he tackled the more difficult matter of how the animal had been buried under more than twenty feet of soil and rock. Since the superincumbent strata appeared to be undisturbed, the bones had not been interred at some time later than the formation of the strata themselves. Nor had the animal accompanied the Roman legions into Thuringia, because, Tentzel argued, the Romans would surely not have buried the valuable tusks. (Contemporaries found this point especially striking.) The obvious conclusion could only be that the Flood had transported the animal as well as the sediments in which it was buried. This "conjecture" Tentzel offered the learned world, asking for reactions and commentary.

Tentzel posed clearly the two questions adumbrated by earlier writers: how to identify specimens, and how to explain transport. Although many thought the two questions had to be answered simultaneously, they will here be treated separately for the sake of clarity in analysis. As many revealed in their writings, naturalists knew that these questions required different approaches—knowledge of comparative anatomy on the one hand, knowledge of history on the other.

In pre-Tentzel decades, one finds a good many references to fossil bones and teeth, chiefly to show that these are the remains of organisms, rather than to identify the animals. Steno certainly devoted little attention to such remains, noting that they might be the bones of human giants and offering no explanation of why he thought some specimens could belong to elephants brought to ancient Rome by the army of Hannibal. In the same year as Steno's *Prodromus,* the Roman journalist Nazari would discuss, in connection with some recent finds, whether one should interpret Genesis literally on the existence of giants.[26] Some discussion also took place in Paris at Bourdelot's meetings, where various options were canvassed: *lusus,* elephants, giants—or even the possibility that ordinary bones might grow larger when buried in the earth.[27] Rare in these decades was the fullness of detail and analysis offered in Plot's *Oxfordshire* (1677). After examining what seemed to be part of a femur as well as other bones and teeth, Plot noted that these were all too large to belong to an ox or a horse. Were they the remains of an elephant? Plot answered in the negative, pointing out that there was no evidence that the Romans had ever brought elephants to

[26] Steno [1669] 1916, 258, 259. Nazari 1669, 23–24.
[27] Schnapper 1986b, esp. 182–90. The growth of buried bones would later be defended by a Toulouse anatomist, J.-J. Courtial; reviews in *JS* (5 January 1705), 3–14; *HWL* (January 1705), 12–23; *Trévoux* (April 1705), 611–26.

England. As the clinching argument, the fossil teeth differed significantly in size and shape from those of a living elephant. (Plot reports that he had seen an elephant on display in Oxford in 1676, and he took the opportunity to examine the teeth of this "very young and not half grown" animal.) He then suggested that the fossils might be human, providing the reader with an enchanting list of giants, from Goliath to the sons of the Titans to some more recent cases.[28]

The puzzling issue of human giants lingered for many decades, in part because some large bones had in early times been entombed in churches— Plot mentions this—as relics of antediluvian men of large stature. Would anyone have buried *elephant* bones in churches? Clearly not.[29] By 1696, Tentzel could struggle through such entrapments by consulting reports on the dissection of elephants. But Tentzel's analysis did not satisfy Leibniz, who wrote at length to Tentzel about his own conjecture, formulated earlier in connection with other fossil finds, that an elephant-like creature, perhaps marine, had once inhabited Thuringia. To give his views more currency than a private letter, Leibniz also informed the Royal Society of his thinking.[30] In that milieu, Hans Sloane eventually considered the further study of elephant anatomy so important for the identification of fossils that he encouraged surgeon Patrick Blair to prepare a long report that filled two numbers of the *Philosophical Transactions* for 1710. Sloane himself would ultimately produce two long articles, analyzing various finds, including some from Russia and Siberia, as well as written accounts ancient and modern. In Sloane's articles, elephants won an easy victory over giants, although Sloane also conscientiously reported a Siberian superstition that such bones belonged to a still-living animal, the "mammuth" (said to be "very like the Elephant"), that roamed around in subterranean caverns.[31]

Sloane's analysis, published almost simultaneously in London (1728) and Paris (1729), did not end the discussion but may well have succeeded in establishing as "fact" that some great fossil bones belonged to elephants.

---

[28] Plot 1677, 131–39. Cf. Delair and Sarjeant 1975, 6–7. Martin Rudwick has suggested to me that Plot's "very young" animal was an Indian elephant, its teeth differing significantly from those of both African elephants and mammoths.

[29] C. Cohen 1994, chap. 2, and Schnapper 1986b.

[30] Aiton 1985, 208–9, and letters in Leibniz 1875, 3:184; 1926, ser. 1, 13:204–5, 293–94, 345–47. Royal Society, JB, under date 17 November 1697. Leibniz later included this proposal in a memoir sent to Paris (AAS, P-V, t. 25, fol. 336v, under date 4 September 1706), but the passage was omitted from the version in *HARS* [1706] 1707, 9–11.

[31] *PT* 35 (1728), 463–64. For earlier use of "mammoth," see C. Cohen 1994, 88. Sloane's two articles were translated and published as a single memoir in *MARS* [1727] 1729, 305–34.

Commenting on Sloane's papers, Fontenelle pointed out, however, that a good many fossil bones and teeth could not be identified with elephants, and naturalists needed more knowledge of comparative anatomy than they yet possessed. On another occasion, in connection with yet another mysterious fossil bone, Fontenelle remarked: "It is thought that this bone could be the lower end of the humerus of some large animal different from the elephant. But what animal? This fact would apparently have important implications if one could investigate it thoroughly."[32]

Bones and teeth clearly not elephantine left the door open to a curious parade of large animals—domestic (the horse, the ox), exotic (the hippopotamus), and human (giants). One journalist, writing in 1705, went so far as to claim that such remains were commonly being attributed to giants.[33] That this was not the case may be suggested by the contrast between Robert Wodrow and Cotton Mather. Wodrow, after examining a large tooth that he thought "of the marine kind," commented: "That ther are and wer some persons of larger stature then others is past doubt, but such a proportion of body as this tooth would require can scarce be digested with me." In Puritan Boston, Mather had no such doubts, as he assembled evidence for his "Biblia Americana." When he sent to the Royal Society a report on specimens discovered in the Hudson Valley, interpreting these as giants, the editor of the *Philosophical Transactions* complained that he had neither sent drawings nor explained his reasons for considering these bones and grinders to be human.[34]

To complicate matters further, the problem of giants tended to merge with debates about human "monstrosities." Participants ranged from Thomas Molyneux in Dublin to Claude-Nicolas LeCat in Rouen to members of the Paris Academy. If some focused on anatomy and biological generation, the larger issue, in Molyneux's words, was to determine "how far the power of Nature may reach . . . beyond her usual bounds." This consideration prompted some writers to reflect on God's freedom and man's ignorance of nature.[35] Metaphysical discussion being distasteful to Fontenelle, he remarked in his elegantly acerbic way: "it is more reasonable to attribute great bones to great animals that one is familiar with than to huge men of whose existence one is uncertain." One of his successors

---

[32] *HARS* [1727] 1729, 3; [1738] 1740, 36. For other hard to identify teeth and bones, see *MARS* [1715] 1717, 182–83, and *HARS* [1719] 1721, 23–25.

[33] *HWL* (January 1705), 19, probably drawn from *JS* (5 January 1705), 12.

[34] Letter of 1702, in Wodrow 1937, 234–35. *PT* 29 (1714), 62, 63.

[35] The quotation is from Molyneux in *PT* 22 (1700), 488; see also *Trévoux* (September–October 1701), 289–302. LeCat, 1747, in Gosseaume 1814, 1:73–79. See also Ehrard 1963, 1:213–15; E. Martin 1880, 125–26; and Roger 1993, 404–18, 555.

as secretary of the Academy dismissed the existence of giants even more quickly, as one of those things people used to believe.[36] But the dismissal seems to have been a bit premature. In any event, fossil bones did only a little to clarify and much to confuse the issue of giants.

To virtually all writers who tackled fossil bones and teeth, the explanation of transport could not be ignored but was less problematical than identifying specimens. One looked, in the first instance, for historical witnesses to the presence of the various animals in Europe, their possible migration, and their death and burial. For exotic animals native to warm climates, one searched in the same records for evidence of European climatic change. Diligent investigation had remarkably poor results, except for some ancient Roman texts.

In his analysis of elephant fossils, Sloane offered a review of the possible agents of transport that may be taken as typical of this period. For some southern Russian specimens, had they accompanied the army of Alexander the Great? Had others been transported by the Romans? (Echoing Tentzel, he commented here that the Romans would have saved the ivory when the animals died.) Had some perhaps lived in Siberia in antediluvian times, when the earth's climate had been temperate? Without developing the latter question, Sloane argued that his fossils antedated ancient Greece and Rome and thus were probably transported to their present sites "by the Force of the Waters of an universal Deluge."[37] Predictably, Fontenelle was less than enthusiastic about this explanation, which struck his contemporaries as very persuasive. British writers argued in terms like Sloane's about elephant bones found in Britain: if the Romans had not brought them— and this point was debated—then the Flood was the obvious alternative. For L. F. Marsili, discoveries in the Danube basin prompted a Roman interpretation, the soldiers having thrown the dead animals into Hungarian lakes. When another Bolognese, Giuseppe Monti, discovered a jawbone he thought that of a walrus, he could not attribute the transport of so exotic a creature to any human agency, and he made this mandible the centerpiece of his treatise on the Flood.[38]

To some writers, the transport of exotic animals seemed stronger evidence in defense of the biblical Flood than then-controversial interpreta-

---

[36] *HARS* [1727] 1729, 3; [1743] 1746, 50. Giants nonetheless "persisted," as Schnapper 1986b shows; see, for example, Maillet 1750, 53.

[37] *PT* 35 (1728), 510 for the quotation. Fontenelle's summary, in *HARS* [1727] 1729, 4, omits the Flood and refers to "grands bouleversements" and "grandes inondations" in the earth's history.

[38] For British writers, see Stokes 1969, 29–34. Marsili's interpretation is in Sloane, *PT* 35 (1728), 509–10. Monti, 1719, in Sarti 1988, 56 and fig. 41.

tions of the transport of marine shells. This certainly was the message conveyed by J. P. Breyne who wrote from Danzig to the Royal Society about Siberian fossils, as it was also for Louis Bourguet who explained his reasoning in detail in his last publication.

> One cannot reasonably attribute the burial of so many bones in so many ways and in so many places to local floods, to upheavals and earthquakes, to the assembling and enclosure of bones in materials brought to the site long after [the death of the animals] and by some particular event, [or] to the rituals of superstitious men. . . . I prefer, at the risk of seeming ignorant to some moderns, to hold to a general flood, which I hope one day to prove, God willing. Besides, such a proof seems to me so natural, if one does it properly, that I am constantly amazed that several great men in our century prefer having recourse to any other cause.[39]

More succinctly than Bourguet, naturalist J. F. Esper would later explain that transport must have been accomplished by a universal flood—if not the one mentioned in the Bible, then one would have to assume another such event, "without being able to base it upon History."[40]
As Bourguet's allusion to "great men" indicates, there did exist nondiluvial explanations for the transport and burial of great animals. Leibniz had provided a speculative alternative by suggesting, in reply to Tentzel, that elephant fossils had not been transported but represented earlier forms native to Germany. More than a decade later, Leibniz would repeat this proposal, acknowledging the paucity of evidence about animals and events so distant in time.[41] Doubtless unfamiliar to Bourguet was the way that Robert Hooke tried to summarize the nondiluvial options he detected in Leibniz and in Molyneux's analysis of the Irish elk. Prompted by Tentzel, Leibniz, and Molyneux, and concerned to defend his own theory, Hooke canvassed possibilities that he admitted were conjectural: the migration of great animals (the elk), fossils as ancestors of living species (Leibniz), and his own preferred explanation of burial, namely, during earthquakes.[42] If Bourguet did not read Hooke, he certainly did examine Antoine de Jussieu's nondiluvial explanation of exotic fossils in French strata, Jussieu arguing for marine currents that had transported plants and animals from the

---

[39] Bourguet 1742, 1:150–51. Cf. Breyne in *PT* 40 (1737), esp. 129.
[40] Esper 1774, 81.
[41] Leibniz 1710, 118–20.
[42] Hooke 1705, 436–39, texts dating from 1700 and 1697, with explicit reference to Tentzel. For Molyneux and Leibniz as noticed by Hooke, see Royal Society, JB, under dates 2 June and 17 November 1697.

Indies to Europe. Like Hooke and Leibniz, Jussieu knew his views to be conjectural, for he lacked historical evidence that such events had occurred.[43]

That these proposals could not be confirmed by the "authority of history" was plain, too, in the reflections by Henry Baker, F.R.S., published in 1745 in the *Philosophical Transactions*. Prompted by the discovery of "an extraordinary large fossil Tooth of an Elephant," Baker began in the usual way, with the Romans, only to deny their responsibility. The tooth had been found in a cliff constantly being undermined by tides, and no Roman would have chosen such a burial site. Ignoring the Flood, Baker announced that "this Discovery seems a convincing Demonstration, that the Earth has undergone some very extraordinary Alterations." Indeed, Baker assumed, for no apparent reason, that elephants had once been native to England, and that that could only have been true if the English climate had in the past resembled that of "very hot Countries." Having begun with a single tooth, Baker rapidly reached his chief supposition: if we imagine even a small shift of the earth's axis, a remarkable chain of consequences would follow, namely, a change in the earth's center of gravity, climatic alterations, changes in the positions of seas and land masses, the breakup of continents, and so on. Not only would the "antient Bed of the Sea . . . be changed into dry Land," thus explaining how marine shells are now found buried far inland, but the elephant and other tropical animals could once have inhabited the very regions where their fossil remains are discovered.[44]

By the time Baker produced these reflections, not a single element in his tissue of suppositions was new. Among British writers alone, Hooke and Edmond Halley, Thomas Burnet and Abraham de la Pryme had discussed all these conjectures. What does distinguish Baker—along with Hooke, Leibniz, Molyneux, and Antoine de Jussieu—is the admission that ancient texts did not contain the requisite information, and that physical conjecture was a legitimate substitute. Such a procedure made contemporaries uneasy, and they continued to discuss the two options outlined by Tentzel: the Flood and the Romans. Indeed, diluvial transport of large animals long remained the most persuasive explanation, even among naturalists otherwise averse to diluvial geology, and even when, in the years around 1800, the Flood itself had been transformed into a recent event in the lengthened history of the earth.[45]

---

[43] The point was made clear by Fontenelle's summary of one of Jussieu's memoirs, in *HARS* [1721] 1723, 1–4, where he remarks on "floods unknown in Histories," and researches enabling scientists to unravel (his word is *deviner*) the history of ancient changes. Jussieu's first article on this subject is in *MARS* [1718] 1719, 287–97.

[44] *PT* 43 (1745), 331–35.

[45] See, for example, the analysis of William Buckland by Gould 1985, chap. 7.

From the foregoing discussions of wood, bones, and teeth, one may extract a number of suggestive conclusions. Most obviously, these were relatively "easy" fossils, and rarely did anyone argue that they might be *lusus naturae*. Precise identification of specimens did provoke debate, but even those writers, such as Plot and Lhwyd, who doubted the organic origins of most fossil shells, could not deny that petrified wood, bones, and teeth had once been organic material. A few writers even proposed that the same arguments for organic origins be transferred from these easy forms to the more problematical marine shells.[46] As for transport, fossil wood usually required little or none, but elephants required a great deal. Still, historical solutions could be found for the latter, unless one were willing to enter the realm of physical conjecture.

FOSSIL SHELLS

Unlike wood, bones, and teeth, fossil shells posed additional problems stemming in part from their ages and modes of preservation. Some, as Martin Rudwick has pointed out, did occupy the "easier" end of the spectrum: they were sufficiently recent to resemble living forms, some retained traces of organic material, and many could be found in coastal or other low-lying deposits where it was possible to imagine the former presence of the sea. Other shells, however, had undergone complete transformations of substance or had vanished entirely (leaving casts or moulds), many had no identifiable living analogues, and deposits were located far inland, both at high elevations and buried deep within the earth's crust. Further complexities arose when some fossils were found to resemble forms still living in non-European seas; such exotica posed in acute form the problem of transport.[47]

Without minimizing such difficulties as the identification of specimens and the worrisome matter of extinction, it can be argued that transport was as much a key issue for fossil shells as for Tentzel's elephants—and an issue much less readily resolved for shells than for elephants. For Hooke, Steno, and their contemporaries, an organic interpretation of fossil shells generally *had to be accompanied by* an explanation of their burial on land. In addition to the examples of Slusius, Quirini, and Grandi, mentioned earlier, there is the more familiar case of Robert Hooke, who, as early as 1667–68, was already emphasizing transport: unless this puzzle could be satisfacto-

[46] Spener, 1710, as reported in *JS* (15 December 1710), 659–60. See also the comment in *Trévoux* (April 1714), 687. Behrens [1703] 1730, 23–24, reversed the argument: since fossil shells are not organic, nor are fossil bones.

[47] Rudwick 1972, chaps. 1–2.

rily resolved, naturalists would resort to interpreting fossil shells as *lusus naturae*. For many years, Hooke lectured at Gresham College on these twin issues. At times, despairing that his theory of earthquakes would be accepted, he would focus on the fossils themselves, arguing that it did not matter if one could not explain their transport. If, he declared, one could not identify the builders of Stonehenge, one still could not permissibly conclude that these megaliths were *lusus naturae* or the work of Merlin. Toward the end of his life, Hooke unhappily admitted that his views on fossils had gained adherents for the (to him) deplorable reason that their transport and deposition were being attributed to the Flood.[48]

The chief arguments on behalf of an organic interpretation of fossil shells may be readily summarized on the basis of texts by Hooke and Steno. On the whole, the essential point differed not at all from what both men said about wood and bones: structural comparisons between living and fossil specimens provided the conclusive evidence. Both distinguished natural crystals from what we now call fossils—at the time, all such bodies were "fossil objects," or objects dug up—crystals being, in Hooke's words, elemental forms "very easily explicable Mechanically."[49] Unlike crystals, the chambered shells of certain mollusks showed a different growth pattern resulting from the lifelong activity of the animal itself; the same pattern, Hooke argued, could be found in the fossil ammonites he broke open and sometimes put under the microscope. For Hooke and Steno, crystals and fossil shells both had natural, albeit very different, explanations, and neither writer had any patience with the essentially mysterious concept of *lusus naturae*.[50]

In addition to these main lines of argument, both men had to confront problems peculiar to the fossils they examined. For Steno, the huge numbers of *glossopetrae* could be readily explained: every shark has more than 600 teeth. But he was conspicuously silent about why the fossilized teeth were not accompanied by any bony remains of the animals. Since ammonites had no known living counterparts, Hooke argued for their similarity to the pearly nautilus living in distant seas. At various times, he applied to fossil shells the same analysis he used for elephant bones: local faunas might have been wholly destroyed by earthquakes, or particular species might have undergone some variation—he here used the analogy of breeds of dogs and sheep—in the course of time. For example, the giant

---

[48] Details in Rappaport 1986, and further discussion below, in Chapter 7.

[49] Hooke 1705, 281, text dating from 1667.

[50] Hooke's impatience was already visible in 1665 (1961, 111–12). Steno [1669] 1916, 211, 258. Both thought that this concept implied a purposelessness in nature; as will be evident below, this kind of criticism later became common.

ammonite, on the assumption that it had formerly inhabited tropical seas, could have become smaller as (or if) climates had changed, and Hooke invited his audience to seek in historical records evidence that British climate had once been tropical. Finally, Hooke and Steno both tackled problems of petrification and transport. On the former subject, Steno proposed a mechanical replacement of organic material by inorganic particles; Hooke said much the same, but added other possible agencies of petrification, such as compression under conditions of great heat or great cold. If Steno analyzed at length the process of sequential sedimentation, he became hesitant and tentative in trying to explain how marine deposits could have been elevated above sea level. One possibility, left undeveloped, involved "a sudden burning of subterranean gases, or . . . the violent explosion of air" released by neighboring crustal collapses. Hooke at first (in 1665) paid little attention to such matters of dynamics, but he soon after developed his theory of earthquakes which was designed to show how land masses had at various times either emerged from the sea or sunk beneath its waters.[51]

Among the chief British opponents of Steno and Hooke, Lister, Plot, and Lhwyd all allowed that certain fossils were organic in origin. In this category they placed fossilized wood and bones, as well as some incompletely petrified shells that showed traces of organic material. All three men acknowledged, too, that shells found in low-lying coastal deposits presented little if any problem, and Lister in particular did not challenge Steno's interpretation of "easy" Italian fossils. Now and then, such admitted organisms raised unexpected questions, as when Robert Plot and John Ray marveled at the sheer numbers of shells concentrated in certain localities. Both men tentatively suggested an interpretation that would later, in Voltaire's hands, achieve notoriety: such sites might have been centers of human habitation in ancient times, the shells being the debris of meals eaten by the local residents.[52]

Lister, Plot, and Lhwyd took into account the morphological resemblances between fossils and living forms, but argued that mere resemblance was not enough. For Lister, using such examples as the giant ammonite, unless one could discover the same living *species* (generic or family likeness

---

[51] The quotation is from Steno [1669] 1916, 231. On the number of teeth per shark, ibid., 257. For the variability of species and possible climatic changes, Hooke 1705, 327–28, 343, texts dating respectively from 1668 and 1686–87.

[52] As far as I know, the earliest writer to make the latter proposal was Bernard Palissy [1563] 1844, 37. See Ray 1673, 294, and Plot 1677, 118–19. See also Réaumur in *MARS* [1720] 1722, 401, and Dezallier 1742, 156–57. For Lister and Steno, see Rudwick 1972, 61–63.

was insufficient), it would be pointless to ask how fossils had been "transported" from the sea to inland quarries. No transport had taken place.[53] Plot adopted much the same attitude, with an additional argument of importance. Like other Fellows of the Royal Society who had heard Hooke's lectures, Plot insisted that notable geological events—floods, landslides, earthquakes—should have left some trace in written history. As John Wallis would later say, in connection with the disappearance of that hypothesized isthmus linking England with France, no ancient witness could have failed to record events so striking. Wallis scrambled to find fragmentary evidence for a theory he thought plausible, but Plot was more militant: if ancient records were silent about large events, then the events themselves had not taken place.[54] Without a method of transport confirmed by historical records, fossil shells had not been transported, could not be called organic remains, and were produced in situ by a plastic virtue.

For both Lister and Plot, the conclusion that many fossil shells had never been organisms was further supported by their observations that such remains occurred in curious groups in the strata. To Lister, the fact that "Quarries of different stone yeild us quite different sorts or species of shells" meant that the different rocks had *produced* the specimens found enclosed in them. Plot arrived at the same conclusion by another route. Noting that deposits of cockles resembled "breeding places, where they [the animals] had aboad for some considerable time," he could imagine no way that geological changes could produce tranquil conditions allowing organisms "to get together and sequester themselves from all other company, and set them down, *each sort,* in a convenient station." Population clusters and tranquil surroundings led Plot to conclude that most fossil shells had been produced in situ by a plastic virtue, the resemblance to living organisms being fortuitous and inexplicable from a human point of view, but doubtless rational and purposive to God.[55]

Familiar with such arguments, John Ray found them untenable. Or, more precisely, Ray vacillated on the subject of fossil shells and their transport for much of his adult life. In his last years, he would prepare a visual argument in response to Lister's views on ammonites in particular (see Plate I); by showing an ammonite found in association with a variety of shells, Ray informed readers that all the depicted specimens must have the same

[53] Lister's views, published in 1671, would be briefly repeated more than twenty years later in Lister [1699] 1967, 230–31.

[54] Plot 1677, esp. 113 for "the Records of time."

[55] Lister in *PT* 6 (1671), 2283, and Plot 1677, 113, 120–21. Plot explicitly rejects Hooke's complaint that human reason can detect no purpose in the operation of plastic powers.

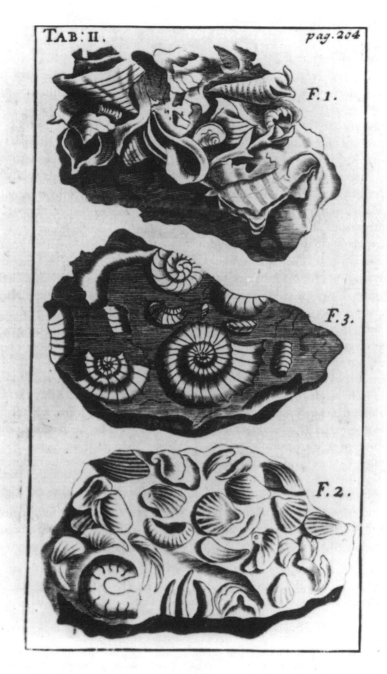

PLATE I. John Ray, *Three Physico-Theological Discourses,* 3d ed. (London, 1713), Tab. II. Ray intended fig. 3 in particular to show that ammonites, found in association with marine shells, must also be considered organic in origin. (Courtesy of the History of Science Collections, University of Oklahoma Libraries)

organic origins.[56] As for the transport of shells and their discovery at high elevations, Ray on one occasion threw up his hands: "For when as we see the thing done, it is vain to dispute against it from the Unlikelihood of the doing it." Implicit in this passage is Ray's long struggle with himself and his correspondents to see whether the Flood might provide the solution to transport, or the means "of the doing it."[57] One of Ray's most intimate correspondents, Edward Lhwyd, also gnawed at these problems, his commitments changing more than once in a period of about a dozen years. In an early letter (1686), Lhwyd adopted Plot's views and phrases, suggesting that fossil shells are "lapides sui generis yt owe their forms to certain salts whose property 'tis to shoot into such figures as these shell-stones represent." A few years later, in 1692, Lhwyd had become increasingly convinced that at least certain fossil forms resembled true organisms, but he now worried that to identify these fossils as organic would imply their extinction. For a short time, Lhwyd flirted with diluvial explanations of transport and deposition, but he soon, thanks in part to Ray's arguments, found this mechanism subject to serious objection. Eventually, in 1698, Lhwyd would devise his theory of "seeds" as a sort of compromise. Some fossils did have organic origins, but the adult forms of most marine fossils had not actually lived in the sea; instead, their waterborne or airborne "spawn" had penetrated into soil and rocks and had bred there into the adult fossils. Lhwyd did not much like his own theory, admitting that it was bizarre to think of organisms as growing within rocks, but in no other way had he been able to solve the problem of transport.[58]

If Lhwyd's theory of seeds had a respectable philosophical ancestry, the British anti-Hooke (anti-Steno) school derived much inspiration not from Lhwyd but from Lister and Plot. In the pages of the *Philosophical Transactions* and in separate publications, such writers as John Beaumont, John Clayton, Nehemiah Grew, G. Hatley, "Charles King" (Henry Rowlands), Charles Leigh, Robert Sibbald, and the anonymous "L.P." debated the

---

[56] Ray 1713, plate II and explication on 205. In the preface "To the Reader" (xvii–xix), William Derham explains that Ray had long since sent these materials to the printer, but various delays had retarded publication for several years.

[57] Ray 1718, 166; see also Ray 1848, 151–56. His views on the Flood will be treated in Chapter 5.

[58] The several texts by Lhwyd are in Gunther 1945, 14:79–80 (1686), 167–68 (1692); *PT* 17 (1693), 753–54; Gunther 1945, 14:381–96 (1698, esp. p. 393 for the admission that having bodies grow within rocks is bizarre). Lhwyd described himself (ibid., 396) as a one-time diluvialist, but his text of 1693 actually shows only that he hoped John Woodward would clarify this issue; as will appear in Chapter 5, his objections to diluvial geology crystallized when Woodward published his theory in 1695. The oddity of Lhwyd's 1698 theory was duly noted in *Acta* (July 1699), 335.

issues and declared their allegiances. Their arguments run the gamut from peculiarities of preservation (broken specimens interpreted as forms in the process of growth within the rocks) to the problematical idea of extinction. Although the occasional writer did explicitly conclude in favor of Lister or Plot rather than endorse the possibility of extinction, it was also realistic in that period to suggest that apparently extinct forms might still be found alive in unexplored oceans. No such ready solution could be found for the difficulties of transport, as followers of Lister and Plot pointed out

Good examples of Plot's language and ideas are provided by the writings of John Clayton and John Banister, both of whom explored the natural history of the North American coast. Upon his return to Britain, Clayton reported to the Royal Society in a series of articles his own ambivalence about fossil shells: those found in coastal deposits were the remains of organisms, but deposits farther inland should probably be explained in Plot's language of the "shooting" of salts into figures resembling shells.[59] Precisely the same alternatives can be found in the manuscripts Banister never had an opportunity to publish—he died in a hunting accident. In the same vein, the Huguenot traveller F. M. Misson, in his much-reprinted and translated *Nouveau voyage d'Italie* (1688), stated what he thought was obvious: without a plausible means of transport, fossil shells found far from the sea could only be interpreted as *lusus naturae*. In a tract of 1705, Charles King would announce that the identification of fossil shells as the remains of organisms "appears very plain and rational," until one begins to consider transport; with obvious reluctance, he admitted that the lack of an acceptable method of transport would oblige naturalists to remain attached to theories that he found inevitable but disturbing: either the concept of *lusus* or Lhwyd's "seeds."[60]

Comparable debates took place during the same decades in Italy, where one can detect clusters of writers whose works were sometimes known north of the Alps. One such cluster consisted of Maltese physician G. F. Buonamico, Sicilian botanist Paolo Boccone, and Sicilian painter Agostino Scilla. All were familiar with elements of Aristotelian and Neoplatonic philosophy incorporated into the writings of Athanasius Kircher, whom Buonamico also knew personally. In the writings of these three men, one sees the same remnants of ancient philosophies, the same concern with modern obser-

[59] Clayton 1693, in Berkeley 1965, 58.
[60] Banister in Ewan 1970, 323, 332–33. Misson [1688] 1717, 2:312–16. King 1705, 4–16. I shall continue to refer to King by his pseudonym because his treatise can be found under that name in standard library catalogues.

vations, and the same attention to historical verification that characterized British writings of this period.[61]

Buonamico almost inadvertently launched the public debate that Scilla and Boccone were already conducting in private correspondence. Asked by Boccone to express his views, Buonamico in 1668 produced an array of arguments, doubts, and proposals strikingly similar to those of British writers. Admitting that he had formerly believed marine fossils to be the remains of organisms deposited by changes in the locations of the sea, Buonamico proceeded to formulate the same questions asked by John Ray: Where was the great shark population needed to supply so many fossilized teeth? Why did the strata contain no shark skeletons? (Adding a question ignored by Ray, he also wondered why the teeth of the living animals did not possess the medicinal virtues attributed to the fossils.) Further anticipating Ray, Plot, and Lister, Buonamico noted that fossils were not assembled pell-mell, as if in the aftermath of a flood, but instead occupied strata in an orderly fashion—here a deposit of *glossopetrae,* there a bed of sea urchins. Nor did Buonamico think fossil shells, ammonites being among his examples, sufficiently like living forms to conclude for the organic origins of such objects. Wishing to combine "modern" philosophy with acceptable elements of Aristotelianism, Buonamico concluded that God was the first cause of fossils that had been generated within the rocks; regularities of shapes and structures testified to nature's working in uniform ways.

From this mixture of modern with traditional, Boccone and Scilla salvaged the modern and some of the acute observations and problems pointed out by Buonamico. In Scilla's *La vana speculazione* (1670), the concept of *lusus naturae* was rejected vigorously. Fossils were a "trick of time, not of Nature."[62] As Scilla observed in connection with *glossopetrae,* these fossils displayed such a variety of orientations in the strata—"one is planted with roots upwards, one sideways, one in a straight position, an infinite number broken"—that they must have been deposited by ocean waves.[63] Without examining transport in detail, Scilla did think a variety of means available, ranging from the Flood to assorted fresh- and salt-water floods and other "accidents."[64] A year later, in 1671, Boccone published compa-

---

[61] For Boccone, see *DBI* and Accordi 1975. For Buonamico, *DBI* and Morello in Maffioli and Palm 1989, 131–45. For Scilla, Accordi 1978 and esp. Rudwick 1972, 56–59. Accordi needlessly insists on the "modern" elements in his authors, but he also provides useful quotations from their writings. The next paragraph is based on the Morello article just cited. Kircher's views on fossils, published in 1665, are analyzed by Ellenberger 1988, 2:74–76.

[62] In Morello 1979a, 65.

[63] In Accordi 1978, 137.

[64] Scilla 1670, 17; Morello 1979a, 66; 1979b, 56–57.

rable arguments, rejecting *lusus* mainly on the grounds that only the substance of fossils, not their form, had undergone change. Considering, too, reports that the sea in some coastal areas had built up shorelines, he saw no difficulty in concluding that fossil shells, however far from modern seas, had been deposited "by floods and by storms which occur at sea."[65]

News of Scilla's tract eventually arrived in London, but the book itself proved hard to obtain and was not reviewed in the *Philosophical Transactions* until more than two decades after its publication.[66] Boccone, however, traveled to both Paris and London, delivering lectures, displaying fossil specimens he carried with him, and having some of his writings published in French. In Italy, further debate would be provoked by a more formidable writer than Buonamico: Filippo Buonanni, S.J., who in 1698 would become curator of Kircher's museum in Rome. Buonanni's first substantial publication (1681) would be cited for decades because of its copious illustrations of shells, and his later catalogue of the Kircher museum (1709) doubtless appealed because the museum itself had long been famous.[67] In his *Ricreatione dell'occhio e della mente* (1681), Buonanni devoted a chapter to the fossil question, rehearsing some of the arguments of Steno and others. Despite the fact that Aristotle and other ancient authors had described fossil shells as marine organisms and had commented on changing shorelines, Buonanni found the mechanics of all this to be obscure and dubious. The Jesuit pondered, as had Quirini and Buonamico, whether water could transport heavy shells, and he considered this so unlikely that he preferred to think that fossils had been produced in situ. As an Aristotelian, he preferred spontaneous generation to Neoplatonic plastic powers, but he also could not rule out the possibility that floods had deposited on land the seeds capable of sprouting into fossil forms.[68] As familiar as Hooke and Plot with the Aristotelian maxim, "nature does nothing in vain," Buonanni inadvertently echoed Plot's reply to Hooke: flowering plants do not always bear fruit, not all fruits come to ripeness,

---

[65] Boccone 1671, 18–20, 44–45. Guettard 1768, 6:146–81 considered the writings of Steno, Scilla, and Boccone to be so persuasive about *glossopetrae* that he was surprised by persisting dissent; he classified interpreters as *dentaires, conciliateurs* (only some specimens are sharks' teeth), and *plastriens*. As examples of non-*dentaires*, see Reiske 1687, 64, and Leigh 1700, fig. II and explication (pages following bk. 1).

[66] Pulleyn to the Royal Society, in Birch 1756, 4:42, under date 10 June 1680. Boyle, 1686, in Adelmann 1966, 1:497. William Wotton in *PT* 19 (1695/6), 181–201. That the latter account was intended as an attack on John Woodward as possible plagiarist became clear a year later, in Arbuthnot 1697, 65.

[67] For the usefulness and the criticisms of Buonanni's illustrations, see Chapter 2, at n. 49. The fame of the museum is discussed by Findlen 1994.

[68] Buonanni 1681, 80–81.

nature can thus produce (fossil) teeth without jaws and fossil shells that had never housed living animals.[69]

Some twenty years later, Buonanni had become more sophisticated and discriminating, as he divided fossils into two groups: the remains of organisms (fossil teeth included here) and the products of natural powers. But his earlier book had established his reputation as a non-modern who adhered to what the *Philosophical Transactions* scornfully called the "old and antiquated" doctrine of spontaneous generation.[70] In a period when Francesco Redi's experiments had aroused international interest, similar sentiments were expressed even more forcefully by Marcello Malpighi who declared that Buonanni's work "corrupts the true method of philosophizing a posteriori and renders everything uncertain and every possible extravagance credible."[71] Significantly, these critics focused on spontaneous generation, not on fossils or on plastic powers. Aristotelianism was being condemned, but plastic powers remained respectable for two or three decades after 1681.

Such debates as occurred in France have left only slight traces for the last decades of the seventeenth century, and one is tempted to conclude that defective records, rather than lack of interest among savants, explains the dearth of materials. Certainly Paolo Boccone's visit to Paris did provoke or add to discussion when he appeared at the "conferences" of J.-B. Denis and at Bourdelot's "academy." Although he did not speak French, Boccone is said to have conveyed his ideas effectively because he had brought with him "all the things which are the subject of his observations."[72] Boccone's message can be detected in the content of his books, published in French in 1671 and 1674, and from the specimens—including Maltese *glossopetrae*—that he carried from Paris to London and then donated to the Repository of the Royal Society. He apparently argued forcefully in Stenonian terms against the concept of *lusus naturae,* made ample use of Steno's analysis of *glossopetrae*, and declared that "floods and storms at sea" were quite sufficient to explain how marine shells had been transported to burial places on land.[73] That Boccone himself was regarded as having an exper-

---

[69]   Ibid., 82–83.

[70]   *PT* 14 (1684), 507. For Buonanni's views in 1709, I am relying on the review in *GLI* 7 (1711), 258–60.

[71]   Adelmann 1966, 1:636. By 1697, Vallisneri and his friends thought the debate on spontaneous generation resolved sufficiently to poke fun at the older view; see Cuaz in *Storia,* 5/I, pp. 115–16.

[72]   Denis 1672, 4–5, and Le Gallois 1674, 1:124–26.

[73]   Boccone 1671, 18, 44–45, 65–66; in general, pp. 38–66 for Steno. He also (51–52) summed up the role of the Flood as an agent of transport, as earlier advocated by Fabio Colonna, but did not himself adopt this explanation. The specimens he donated to the Royal Society are listed in Birch 1756, 3:116–18.

tise not (yet) available in Paris is implied in the surviving account of the Bourdelot meetings, where, on one occasion, discussion of *glossopetrae* was briefly undertaken and then postponed until a date when Boccone might attend the meeting.[74]

That some discussion also took place in the Academy of Sciences is evident in the minutes of meetings, as members now and then brought fossils to show to their colleagues. What was said on these occasions is only rarely recorded, but signs of debate and some indication of issues become clearer for the Academy after 1700 and for French provincial academies at a slightly later date. Among the early traces of this developing interest in Paris is a memoir presented to the Academy in 1702 by Tournefort, in which the author digressed from his main topic by proposing a theory similar to, but independent of, that of Edward Lhwyd. Almost casually in these pages, Tournefort suggests that the growth of plants (he was a botanist) might serve as a model for the "growth" of hard bodies like rocks and fossil shells. All could stem from "seeds."[75] This text may have attracted the attention of astronomer G. F. Maraldi who, the very next year, brought from Italy some rocks containing "dried fish" and other fossils; he had also observed specimens of this kind at high elevations far from the modern sea. But it was Fontenelle, not Maraldi, who decided to broach the crucial problem of transport.

> What can have carried these fish and shells into the earth and up to high altitudes in mountains? It is likely that there are subterranean fish, as there are subterranean waters, and these waters, which . . . rise as vapors, perhaps carry with them the very light eggs and seeds [of organisms]; when the vapors condense and become water again, these eggs can begin to hatch there [within the earth] and become fish and mollusks.

Eventually, according to Fontenelle, soils containing these organisms may have dried out, for one reason or another, with the result that the organisms and the soil itself hardened into rock. And he cautiously concluded: "If all rocks were formerly fluid, as capable scientists believe, then this kind of theory becomes more acceptable."[76] In his capacity as interpreter, Fontenelle here tried to use Tournefort's theory to explain Maraldi's observations.

The Tournefort theory did not last long in the Academy, in part because

---

[74] Le Gallois 1674, 1:124–26.

[75] Tournefort in *MARS* [1702] 1704, 222–23.

[76] *HARS* [1703] 1705, 22–24. The text by Maraldi was transcribed into AAS, P-V, t. 22, fols. 365r–366r (12 December 1703). Fontenelle's reflections were chiefly confined to the final paragraph in the published version.

Tournefort himself did not pursue the matter, although a few years after his memoir of 1702 it was rumored that he intended to tackle the fossil question. His sudden death aborted whatever projects he may have had in mind and have left only the merest trace, buried in the minutes of the Academy's meetings: here it is reported that in 1708 Tournefort once more suggested that marine organisms grow to adulthood after being embedded in rock.[77] If we cannot know what Tournefort intended, Fontenelle's messages in the annual *Histoire* are clear. Initially impressed by Tournefort's suggestions—"growth" or "vegetation" might be a powerful principle in all areas of nature—Fontenelle rapidly abandoned such ideas when he began to receive communications from the Scheuchzer brothers in Zurich and from Leibniz. From about 1706, "seeds" and "eggs" vanish from the publications of the Academy, where they had had only a short lifetime.[78]

If the "seed" theory enjoyed only brief popularity in the Academy, the concept of *lusus* seems never to have gained a foothold either in the Paris institution or in the academies of Bordeaux and Montpellier. The Neoplatonic revival in England, which attracted Plot, Ray, and Newton, aroused no enthusiasm among French savants and was subject to heavy criticism in the writings of Nicolas Malebranche. Whether or not naturalists read Malebranche—and he was considered more readable then than he is now—they assuredly read Fontenelle's annual *Histoire,* where "mysteries" were treated with impatience. In 1703, Fontenelle was already rejecting *lusus* as a way of explaining the birth of "monsters." In his usual epigrammatic style, he insisted that nature does not play games and that any such philosophy merely cloaks our ignorance of natural processes.[79] By the time that Leibniz sent an account of some geological observations to the Academy in 1706, Fontenelle was probably quite comfortable in rejecting *lusus* as "a poetic idea" unsuitable to philosophy. If nature played games, she would surely play freely and not confine herself to reproducing in fossils the sizes and even the smallest traits of living organisms. The *Journal des savants* would also dismiss as poetry the "seed" theory as it appeared in the work of one of Lhwyd's Swiss disciples, K. N. Lang. In a more critical tone than usual, the *Journal* summed up Scheuchzer's attack on theorists who had "recourse

[77] AAS, P-V, t. 27, fol. 33v (4 February 1708). The rumor about Tournefort's intentions reached Astruc in Montpellier, as reported in his letter to Tournefort, 9 July 1708, MHN, MS 1391. Tournefort died on 28 December 1708.

[78] The seed theory was still visible in *HARS* [1705] 1706, 35. The next year, in a summary of Leibniz's views, both seeds and plastic powers would be dismissed with minimal discussion, *HARS* [1706] 1707, 11.

[79] *HARS* [1703] 1705, 28. Fontenelle's locutions would later be repeated by abbé Bellet of the Bordeaux academy; see below, n. 82. In general, see Ehrard 1963, vol. 1, pt. 1, chap. 2, and Tocanne 1978, pt. 1, esp. chap. 6 on Malebranche.

to images or phantoms resembling those described by the poet Lucretius. [They] claim that these phantoms fly through the air and then, falling into the bowels of the earth, produce the [fossil] shells and fish one encounters there."[80] Clearly conveyed here was that such philosophy was silly; also conveyed was the danger signaled by reference to Lucretius, the most notorious of ancient atheists. How the pious intentions of naturalists became transmuted into a threat of atheism will be explored below.

In Montpellier, two of the earliest discussions of fossils, based as they were on "easy" (recent) forms and on "easy" (Mediterranean littoral) locations, rejected *lusus* without any elaborate argument. One of the two naturalists, Jean Astruc, received detailed attention in two of the most prestigious book review journals, but the reasoning behind his formal presentation appears most clearly in a letter he wrote to Tournefort.

> The question of the origin of fossils entails many large problems that I could not deal with in the lecture I delivered, but that I would like to discuss here. One cannot say that these forms, so often found, are sports of nature, the similarities they usually have to living shells being too notable and too regular. . . . The problem is to figure out what could have scattered them in so many different places and so far from the sea.[81]

In this very pointed text, Astruc found the fossils themselves easily explained—no mystery, no *lusus* here—and transport the chief problem. The Bordeaux academicians seem to have had more difficulties than the Montpelliérains, for the Bordelais were still offering the fossil question as a subject of prize competitions as late as the 1740s. Early in the century, however, one of the most active members of the Bordeaux academy, the abbé Bellet, insisted that Fontenelle was right: nature was lawful, not mysterious or erratic. If Bellet did not address himself explicitly to the origin of fossils, this academician clearly had no patience with mystery or mystification.[82]

Those who attacked *lusus* and "seeds" considered their targets to be illegitimate forms of philosophizing. Needless to say, neither Lhwyd nor Tournefort thought it illegitimate to use the analogy of plants in order to say that some fossils might grow from seeds. Nor did Robert Plot think it

---

[80] *HARS* [1706] 1707, 11. *JS*, Amsterdam 12mo (suppl. to September 1708), 585, and (7 January 1709), 12. *Trévoux* (April 1709), 729. For Scheuchzer, below, at n. 85.

[81] Letter cited above, n. 77. Also, Rivière [1708] 1766.

[82] Barrière 1951, 167, 209–10. One of the contestants in the 1740s relied heavily on Fontenelle and the Paris academicians for arguments and prose; see Barrère 1746, esp. 38–41, 50–51.

mystery-mongering to assert that some fossil shells seemed analogous to crystals and concretions; if chemical processes in nature could produce the latter, why not also the former? But advocates of *lusus* in particular had long insisted that nature really does produce some uniquely mysterious forms, ranging from "landskips" depicted on rocks to pious images, crosses, and even the Crucifixion.[83] If critics found it easy to dismiss such imaginative interpretations of specimens, they took more seriously the supposed "powers" that had produced them.

Hooke and Steno, as well as Bayle and the La Hires, had long seen plastic forces as the equivalent of unpredictable purposelessness in nature. John Ray, too, uneasy about such philosophizing, eventually rejected it in strong terms:

> Now that Nature should form real shells, without any design of covering an Animal . . . gives great countenance to the Atheists Assertion, That things were made or did exist by chance, without counsel or direction to any end.[84]

In 1708, J. J. Scheuchzer made plain his view that those who used *lusus naturae,* as he himself had done a few years before, were "soldiers of Epicurus," seeing "haphazard" processes, "random jumbling," and chance combinations of atoms, instead of the Divine order in nature.[85] Such attacks continued in the next decades, sometimes with less natural theology and teleology, and more emphasis on naturalism as opposed to mystery. If Vallisneri would eventually dismiss with scorn the use of "ancient occult faculties of generation," he and his friends actually took the need to reply to Neoplatonists more seriously than such scorn implies. Apparently at Vallisneri's urging, Antonio Conti in 1712 used the pages of a learned journal to rebut at length the theories of an Italian disciple of Henry More and Ralph Cudworth. Although Malebranche thought Conti had wasted his time on so bad a book, others considered the conflict not yet over.[86] As late

---

[83] For example, Kircher in Ellenberger 1988, 2:74–76; Beaumont in *PT* 11 (1676), 739; Feuillée 1714, 2:531; and the image of the Crucifixion sent by Lang to the Royal Society in 1736, above, Chapter 1, n. 45.

[84] Ray 1693, 133–34; compare the more tentative statements in Ray 1673, 127–31, 294. In addition to works by Steno and Hooke, cited above, see Hooke in Hunter and Schaffer 1989, 16–17. La Hire in *MARS* 1692, 122, and La Hire *fils* in AAS, P-V, t. 17, fol. 131r (19 March 1698): both texts reject "chance" (*hazard*). Also, Bayle in Labrousse 1963, 2:215 n. 114.

[85] These passages by Scheuchzer are quoted in the editor's notes to Beringer 1963, 175. *Trévoux* found these arguments worth reporting in January 1713, pp. 66–68.

[86] Vallisneri 1728, 16, for the scorn. Conti's critique of Nigrisoli is in *GLI* 12 (1712), 239–330. Vallisneri was one of the founders and editors of this publication. See also Badaloni 1968, 38, and Garin's introductory remarks in Fardella 1986, 10–12.

as 1725, one of Louis Bourguet's correspondents was still urging Bourguet to abandon incomprehensible metaphysical ideas—in this case, both plastic powers and Leibnizian monads.[87]

For fossils in particular, J. J. Baier's treatise of 1708 indicates in some detail what was happening to *lusus*. He rejected vigorously any notion that,

> after the manner of superstitious vulgarity, I am conjuring up some agent distinct from God, which . . . directs and modifies corporeal creatures, sometimes toying idly with them, often fashioning absurdities and monstrosities.

Clearly no advocate of plastic powers, Baier nonetheless retained *lusus* as a category into which he could relegate

> those stones which either display a unique figure or, if they imitate the forms of other bodies, exhibit them somewhat less than perfectly, and rather fall short in bulk or extension or some other characteristics, so that they can in no way be shown to have received their origin from the things to which they are somehow likened.[88]

Removing from this category fossils he thought clearly organic in origin, Baier used *lusus* for a curious assortment of objects he could not explain—such as belemnites, dendrites, and stalactites. Apart from failure to imitate organisms "perfectly," what these and other objects also had in common was a symmetry, a geometrical regularity (round, conical, star-shaped) that Baier thought atypical of true organisms (see Plate II). Ironically, when John Morton in 1712 discussed some of the same specimens, he argued that only organisms, not minerals or rocks, could produce such characteristics as symmetry.[89]

Journalists in France and England found Baier sufficiently interesting to review, and his little treatise is certainly instructive and suggestive. In his effort to be a modern, he assigned no mysterious powers to nature, merely placing in the category of *lusus* some inexplicable forms. In making his selection, he seems to have been willing to contract the size of that cate-

---

[87] In Crucitti Ullrich 1974, 87. See also the argument of Mallatt 1982 that Beringer's curious (factitious) fossils could only have been explained by a plastic force; when Beringer in 1726 rejected this concept, he had no explanation at all.

[88] Translations by the editor in Beringer 1963, 181–82. Baier was reviewed in *JS*, Amsterdam 12mo (4 March 1709), 378–87, and *HWL* (February 1709), 70–74.

[89] Baier 1708, esp. 30. Morton 1712, 236–37 and plate 10, esp. figs. 14–20. Conrad Gesner's earlier illustrations of some of the same forms that puzzled Baier are reproduced by Rudwick 1972, 29, 32.

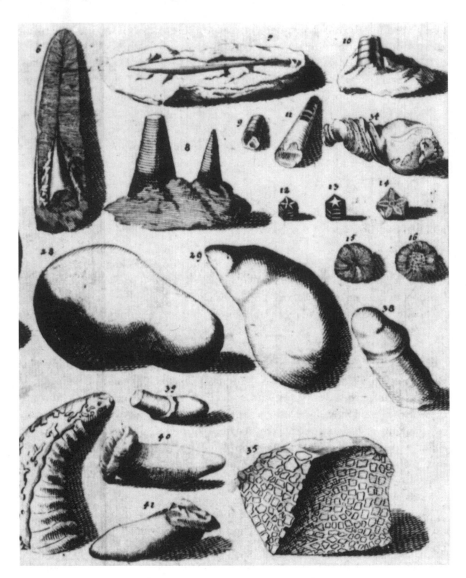

PLATE II. Detail from Johann Jacob Baier, *Oryktographia norica* (Nuremberg, 1708), Tab. I, showing some of the "geometric" shapes that he classed among *lusus naturae*. His interpretation of "star-stones" (stalked crinoids, figs. 12–14) was disputed by John Morton (1712). (Courtesy of the History of Science Collections, University of Oklahoma Libraries)

gory as various objects became more understandable in naturalistic terms. The same attitude can still be detected in the 1740s, when naturalists such as Dezallier and Da Costa, baffled by belemnites in particular and unable to accept existing interpretations of these fossils, retained the category of *lusus* as a convenient hold-all for the inexplicable.[90]

Several of Baier's contemporaries tried to estimate when the *lusus* controversy had come to an end. One journalist implied a turning point as early as about 1705, another suggesting 1710; Réaumur in 1715 thought most naturalists favored the organic interpretation of fossils, while in 1719 Giuseppe Monti thought it no longer worthwhile even to discuss such antiquated issues. What such writers did not realize is that the concept was dying, even while the word and the category survived. If no one could "refute" the concept, it could be—and was—gradually replaced by more acceptable forms of explanation, as Fontenelle noted when he announced that the abandonment of mystery signaled the emergence of science from its "infancy."[91] Accompanying this change, however, was a development Fontenelle himself did not like: the decline of *lusus,* the increasing acceptance of the organic interpretation of fossil shells, and the triumph of naturalism were aided and indeed promoted by naturalists who used Noah's Flood to explain the transport of marine shells. It would seem, then, that the philosophical "modernization" detected by Fontenelle was only part of the story, the Deluge being the other part. Like Tentzel's elephants, fossil shells required a mode of transport. If the diluvial explanation proved durable for large animals, it had a more controversial career in connection with shells.

[90] Dezallier 1742, 64–65; he was familiar with Baier, cited on page 29. DaCosta in *PT* 44 (1747), 397–407, esp. at p. 404. Guettard 1768, 6:215–96 offers a review of thirteen different interpretations of belemnites.

[91] *HARS* [1722] 1724, 2. The several estimates of when the controversy ended are in *HWL* (April 1705), 249–51; *Bibliothèque italique* 1 (1728), 143; Réaumur in *MARS* [1715] 1717, 181; Monti, 1719, in Sarti 1988, 53. Also, *Dictionnaire de Trévoux* (1721), s.v. fossile. I here agree with Porter 1977, 52–53, that the decline of *lusus* was accompanied by increasing acceptance of both the mechanical philosophy and diluvial geology, although the latter also occasioned much criticism.

CHAPTER 5

Diluvialism, For and Against

In the preceding chapter, I mentioned diluvialism as a solution to the pressing problem of transport. More precisely, as some historians have noted, John Woodward's *Essay* of 1695 persuaded many readers both that fossil shells were organic in origin and that the Flood could account for their transport and deposition. The history of diluvial geology has not yet been written, historians, with few exceptions, having as little patience with such "fantasies" as did Archibald Geikie. Indeed, in traditional analyses of pre-Huttonian geology, it has been customary to label any naturalist who refers to the Flood as a diluvialist and even to suggest that alternative theories, such as Abraham Gottlob Werner's universal ocean, were but disguised, euphemistic forms of diluvialism. In the interests of better historical understanding, it is imperative to distinguish between diluvial geology and the belief that Noah's Flood took place.

Naturalists writing in the period 1665–1750, and later, generally did accept the Flood as a fact of history.[1] The latter phrase requires emphasis, for it signifies that Moses, the presumed author of Genesis, had written the early history of the world, and that the Mosaic text, like other historical works, could be confirmed using a variety of evidence. As internal evidence,

---

[1] In an otherwise valuable article, Dean 1985 regards the Flood as speculation eventually defeated by increasing observation and fieldwork.

the text of Genesis itself showed that the ancient Hebrews had kept careful records, as illustrated by the detailed genealogies which could be contrasted with the mythological fancies of other ancient peoples. As not quite external evidence, at least some of the many flood stories in different parts of the world were generally held to be distorted folk memories of the original event itself; in the common locution of the time, the Bible and pagan traditions "agreed," and the pagans thus "confirmed" Genesis. Genuinely independent support could be found in the fossil evidence, nature testifying to the great upheaval by the universality of marine sediments and their organic contents.

To every sentence in the preceding paragraph there existed significant contradictions in the decades around 1700. A few bold spirits doubted the authorship of Moses, relegated Old Testament miracles to the realm of myth, and took seriously the possibility that other peoples might be older than the Hebrews. Such extremism had only limited appeal, and philosophers were more commonly divided about the Flood itself. Was it a miracle or a natural event? If miraculous in its cause, did it have natural effects? Some queried the Flood's universality, unable to account for the amount of water necessary and for its subsequent diminution, or unable to explain satisfactorily the postdiluvial migrations of animals to distant continents.

A preliminary word about miracles seems suitable here. It is usually said that naturalistic, rational explications of the Bible were a peculiarly Protestant concern, whereas Catholics more readily resorted to miracle and allegory. In a large sense, this is true enough, as the Protestant program involved a return to "the plain words of Scripture," rather than to allegedly fanciful Catholic interpretations. But although Protestant polemicists had long charged Catholics with needlessly multiplying miracles, such charges applied mainly to post-apostolic events or to the centuries since St. Augustine. In other words, confessional divisions did not apply to the Flood, all parties agreeing that miracles had occurred in biblical times. The one position perhaps peculiar to Catholic naturalists was the refusal of some few to consider the Flood as anything but a miracle, both in its causes and its effects. More often, however, Catholics and Protestants agreed that the Flood as a fact of history had had natural effects; its causes, whether natural or miraculous, aroused much debate, but not along confessional lines.

To be sure that any single event actually was miraculous posed insuperable problems of interpreting both nature and Scripture. As a conventional blending of things natural and divine, a learned French monk had no difficulty in declaring: "all these arguments which establish the natural causes for such effects do not destroy the truth of this proposition, that God's justice is their first cause."[2] This reminder was not intended to discourage the study of nature, and the *Mémoires de Trévoux* would later castigate the abbé Pluche, that popular expositor of natural history, for his anti-intellectualism in counseling his readers not to be too curious about the secrets of the natural world. The potential dangers of such curiosity had long been apparent, but the advice of St. Augustine was echoed by Protestants and Catholics alike: the natural world is open to human study, and natural truths cannot contradict the divine. When and if seeming contradictions arise, the fault probably lies in an erroneous interpretation of the Bible, which, after all, was not intended as a textbook of physics.[3]

Augustinian wisdom proved difficult to apply in an age of enthusiasm for progress in natural knowledge. According to Pierre Bayle, "we live in a period of philosophy which explains everything by natural causes, as much as possible, and . . . I rather like this method."[4] Bayle's "as much as possible" was precisely at issue. Naturalistic explanation had been extended very far indeed by Descartes in his hypothetical explanation of how the universe had evolved, but one did not have to be a Cartesian to ask what natural causes God might have used either during the six days of Creation or at the time of the Flood. (Even Isaac Newton, vigorously opposed to a naturalistic cosmic evolution, suggested chemical processes that might aid in explaining the creation of the earth's crust during the six days.) As Jacques Roger once said in describing the technique of naturalists, they tried "to make the Bible say what it does not say but does not explicitly contradict." They also tried to explain naturally what the Bible says God did directly, and miraculous elements consequently diminished almost to vanishing point.[5] This process did not go unobserved by naturalists themselves.

---

[2] Placet 1668, 73. This is a simpler way of saying what Bishop Bossuet and others worked into their view of world history.

[3] *Trévoux* (March 1733), 416. The Augustinian position was, of course, expressed by Galileo; for echoes of it, examples include Bishop Wilkins, 1675, in Shapiro 1969, 237–39; P. Perrault [1674] 1721, 2:758; and Vallisneri, 1721, in Rappaport 1991b, 84.

[4] *NRL* 1685, quoted in Labrousse 1963, 2:11.

[5] Roger in *Leibniz* 1968, 141. The Newton text alluded to here is a letter to Burnet, discussed below, at n. 18. The classic argument, pitting physical law against the Newtonian version of Providence, occurs in the Leibniz-Clarke correspondence of 1715–16. Cotton Mather tried to find a middle course, combining miracle with God's use of natural causes; as shown by Middlekauff 1971, 288–90, the result was difficulty in deciding among competing versions of the causes of the Flood.

The Burnet Controversy

If "diluvialism" is used in the restricted sense of attributing to the Flood *a major role* in producing the earth's landforms, then this school of thought should probably be dated not from Burnet but from one of his critics, John Woodward. Although Woodward promised repeatedly to produce a longer treatise to answer questions posed by disciples and critics, his *Essay* of 1695 took into account a greater array of scientific evidence than had any of his predecessors. Furthermore, Woodward claimed to be arguing in the "modern" mode: he would rely on natural facts, and he would treat Moses as a historian whose text required confirmation.

What Woodward produced had been seen by others as a major desideratum for three or more decades. Serious treatises had already been tackling the question of whether the Flood had been universal. Most inflammatory among these was Isaac de LaPeyrère's *Prae-Adamitae* (1655), in which the limited extent of the Flood was a relatively minor issue compared with the author's thesis that there had existed men before Adam. Relatively less controversial were scholarly works by Isaac Vossius (1659), Edward Stillingfleet (1662), and Abraham van der Myle (1667). These and other writers defended the veracity of Moses, even while they found naturalistic objections to the universality of the Flood. When two such works were reviewed in 1668 in the *Journal des savants* and Nazari's *Giornale,* the reviewers detected an obvious flaw in physics: How could water cover mountains in biblical lands and yet not spread to the rest of the earth? As a later journalist pertinently remarked, this would be a greater miracle than a universal flood.[6] Nevertheless, John Ray and Robert Plot were among the few naturalists to conclude that a local flood obviated a number of baffling problems. As long as all humans, except Noah's family, perished, a chief message of Genesis would be preserved.[7] In a different milieu, the Roman Congregation of the Index would later, in 1686, consult the great Mabillon about one of the works of Vossius. The judicious scholar allowed that to defend a local flood was as permissible as a nonliteral reading of the six "days" of Creation; such views, albeit irrelevant to faith and morals, might

[6] *JS,* Amsterdam 12mo (12 November 1668), 144–47, and (19 November 1668), 169–71; translated with modifications in Nazari (28 December 1668), 174–76, and (29 January 1669), 15–16. One of the works reviewed is by Myle (or Milius), the other an anonymous tract sometimes attributed to the same author. Cf. *Trévoux* (May–June 1701), 17. For one of the disputes stimulated by Vossius, see Rossi 1984, 145–52.

[7] Ray 1693, 122. Plot 1677, 112. Ray was quite hesitant on this subject, seeming to lean ultimately towards universality. The only effort to study this question of non-universality seems to be the article by Mangenot 1890.

still merit "reproof" because they conflicted with Genesis and the consensus of the Fathers.[8]

For naturalists interested in marine fossils, the universality of the Flood offered an instant and obvious explanation of the ubiquity of such forms. This, at least, was the proposal of Sir Matthew Hale, writing in the 1670s, in the aftermath of some of the treatises mentioned. Others did not see the matter in quite this way. Nicolaus Steno and Agostino Scilla, for example, allowed that the Flood could be responsible for some fossiliferous deposits but not for the whole succession of sedimentary strata. Various writers doubted that Flood waters could have lifted heavy shelis to any altitude, whereas others argued that the Flood had consisted chiefly of fresh water and thus could not have been responsible for sediments marine in origin.[9] For some authors, as indicated in the preceding chapter, these insoluble problems led them to interpret fossil shells as *lusus naturae.*

These selected examples do much to explain why John Woodward believed he had the solution to a whole range of controversial issues. To Woodward, however, the most provocative among earlier books and tracts was Thomas Burnet's *Sacred Theory of the Earth,* the crucial part of which—containing his methodological principles and his diluvial theory—appeared in Latin in 1681 and in a modified English version, produced by Burnet himself, three years later. Burnet nowhere mentioned fossils, but he addressed himself in detail to the earth's history and to biblical exegesis. A brilliantly imaginative and engaging work, and a masterpiece of prose, the *Sacred Theory* has too often been described as mere fiction, uncontrolled speculation, and biblical literalism. In the late seventeenth century, it was immediately recognized as modern and daring, in large part because Burnet, far from being a literalist, had ranged very freely indeed in his interpretation of Genesis.[10]

Burnet shared the then common view that Genesis contained the only reliable ancient history of the earth, but he was equally aware of traditional and contemporary latitude in biblical exegesis and of long disagreement on specific matters of importance to him, such as the origin of mountains. By no means denying miracles, and indeed insisting that nature does not act without Providence, he also declared that to resort to miracle was a con-

---

[8] Leclercq 1953, 1:408–10. Several works by Vossius were placed on the Index in 1686.

[9] Allen 1963, 93–94; Levine 1977, 63–65. Misson [1688] 1717, 2:312–13; Buonanni 1681, 73–74; Accordi 1978, 137–38. Steno's use of the Flood will be discussed below, Chapter 7.

[10] Older interpretations of Burnet and his era have now been superseded by such works as Allen 1963, chap. 5; Nicolson 1963; Pasini 1981; Porter 1977, chap. 3; Roger 1973. But Ewan 1970, 53 could still call this Cartesian rationalist "medieval."

fession of ignorance of nature.[11] Modeling his theory in part on Descartes, Burnet also rejected an essential feature of Cartesian cosmogony: if Descartes was hypothetical and "counterfactual," the real facts being in Genesis, Burnet aimed to produce a genuine history, documented by both nature and Scripture.[12] At the same time, because the opening chapters of Genesis contain two accounts of Creation, Burnet argued that one of the two must be an allegory, a moral tale, or a simplification meant for comprehension by ordinary people. This exegetical method, applied wherever Burnet thought suitable, allowed him much flexibility in deciding which biblical statements should be considered "facts."

As Burnet pointed out, the single most difficult hurdle in the way of accepting the Mosaic account of a universal flood was the sufficiency of water—and some way for the water to diminish afterward. To defend the Flood's universality, Burnet argued that, since there could not have been enough water to cover mountains (and he refused to resort to the miraculous creation of additional water), then mountains had not yet existed. The original formation of a smooth and featureless antediluvian earth Burnet described in terms owing much to Descartes: the heaviest matter constituted the earth's core, which was enclosed in concentric layers of water, oil, and residual dust. The dust settled out of the atmosphere and consolidated sufficiently to form a fertile crust and a proper habitation for men and other creatures. Eventually, however, the soil dried and cracked, releasing the subsurface water; this natural disaster, the Flood, shattered the hitherto smooth crust, producing mountains, valleys, and sea basins as they exist today.

If the outline of this theory has become quite familiar, several points that startled Burnet's readers must be made explicit. First, the Flood here has become so much a matter of physics that virtually lost in this analysis is God's decision to punish sin. Admitted but buried in Burnet's scheme, God's foreknowledge accounted for the coincidence in time of the Flood and the prevalence of sin. Second, Genesis has been flatly contradicted on a number of points, most obviously on the existence of mountains before the Flood. Burnet knew that his nonliteral reading had a long tradition behind it, as commentators had for centuries struggled with such problems as how "days" had been counted before the creation of the sun on the fourth day. But to ponder the literal and symbolic meanings of "light" seemed a pious exercise compared with Burnet's naturalism. Finally,

[11] Burnet 1684, 303, 314–15. For exegetical traditions known to Burnet, see Nicolson 1963, chap. 2 and pp. 175–78.
[12] Roger 1973, and Roger in Lennon 1982, 95–112. See also Clarke 1989, 189.

Burnet skillfully deployed the several strategies that would make his theory almost demonstratively certain. He began with a series of deductions, moving on to inductions that confirmed his earlier chain of reasoning; biblical and other ancient texts then provided witnesses and traditions in harmony with his analysis of nature. More than once he could thus declare that the outlines of his theory, if not all the details, should be granted "more than a moral certitude."[13]

No reader of Burnet could be neutral, and the first reactions were of pleasure and joy. Even the many critics, including those shocked and terrified, had to admit that the prose was magnificent and the theory ingenious. As John Evelyn informed diarist Samuel Pepys, he had read the original Latin edition "with greate delight" and found the English version "still new, still surprizing, and the whole hypothesis so ingenious and so rational, that I both admire and believe it at once."[14] As a sober journalist, Pierre Bayle judged Burnet's book to contain "very new and very profound ideas" about the Flood, but to a friend he would later drop restraint: Burnet's *Theory*

> is one of recent books I have most enjoyed. His is a profound mind which supports his ideas cleverly and learnedly. I do not say that he always succeeds in making his hypotheses agree with the words of Moses.[15]

Lorenzo Bellini, the Florentine disciple of Malpighi, confessed to his mentor that he had "never read a *fantasia* more noble than [Burnet's], nor one more learnedly and nobly worked out." And Malpighi replied with comments on Burnet's undeniable genius.[16] As Marjorie Nicolson showed in her classic study, whether one agreed with Burnet was often irrelevant, his prose and his imagery being irresistible to travelers and poets who contemplated the Alps as the "wreckage" of a former world.[17]

Burnet himself may not have foreseen delight of this kind—even if he clearly took pleasure in his own theory—but he did know that he could expect opposition to his biblical exegesis, as he had already tested the climate by explaining his views to Isaac Newton. To Newton, the Mosaic

---

[13] Burnet 1684, 149–50; also, 65, 78–79, 113.

[14] In Pepys 1926, 1:23. Partly quoted in Nicolson 1963, 189; also, Davies 1969, 72, where the statement is attributed to Pepys.

[15] Bayle did not actually write a review of Burnet. The first statement quoted is from *NRL* (July 1685), and the second from a letter to Lenfant, 28 March 1693; both are in Bayle 1727, 1:328, and 4:685.

[16] Malpighi 1975, 3:1219–22, 1224.

[17] But Nicolson 1963, 246 errs in calling Woodward's *Essay* "in part a sequel" to Burnet, and "in part a reply." It was meant to be a refutation and a replacement.

account, however lacking in naturalistic detail, should be regarded as "philosophical" rather than "feigned." More bluntly than Newton, other critics accused Burnet of turning Moses into a mere poet. One Scottish writer even seized on the charge made by Modenese physician Bernardino Ramazzini that Burnet had departed so far from the Bible as to revive an "ancient Abyssinian" creation myth. The Scot, Dr. Robert St. Clair, apart from this curious addition to anti-Burnet literature in English, joined other critics in trying to show that Burnet was a bad exegete. Some opponents also engaged in efforts to impeach Burnet's physics. If John Keill's was the most capable work of the latter sort, Keill was equally disturbed by Burnet's exegetical liberties, and he analyzed at length the permissible limits of allegorical interpretation of the Bible.[18]

To many British critics, the existence of antediluvian mountains was a trifle when compared with more profound Christian issues challenged by Burnet. One such writer, Archibald Lovell, brother of the London Charterhouse where Burnet himself was Master, noted that by making the whole earth a paradise that lasted until the Flood, Burnet had ignored the entire scheme of sin and redemption that had begun with Adam and Eve. To Erasmus Warren, Burnet's naturalism implied that the Flood would have occurred even if man had not sinned, and Warren vigorously rejected this assault on God's justice and providence. Herbert Croft, Bishop of Hereford, discussed at length the various biblical passages Burnet had misinterpreted. Acknowledging the Augustinian wisdom in not having frequent recourse to miracle, Croft argued that Burnet seemed to exclude the very possibility of divine intervention. In fact, the octogenarian bishop did not know quite what to make of Burnet. Was he an atheist? Was his flexible approach to the Bible akin to the method of papists? Was he sincere but misguided, or did he harbor some "evil design" against Christianity? Whatever the answers, Croft judged Burnet to be arrogant not only in his attachment to his own theory—most critics commented on this—but also in his allegiance to naturalistic explanations even when such a method conflicted with what was "plainly declared" in Scripture.[19]

Scientists also had more to say about Burnet's rationalism than about such topical matters as his failure to discuss fossils. The cautious Edward Lhwyd informed John Ray: "I cannot (as yet) reconcile his [Burnet's] opin-

---

[18] Newton to Burnet, n.d. (but January 1681), in Newton 1959, 2:329–34. Keill 1699, esp. 148–53; see also Allen 1963, 110–12, and Rossi 1984, 71–73. Ramazzini (and St. Clair) will be treated below.

[19] Lovell 1696, 23–25. Warren 1690, chap. 6. Croft 1685, preface and 65–67, 73, 98. Goldgar 1995, esp. 158, points out that individual arrogance was frowned upon in the Republic of Letters in which the Republic itself was the judge of scholarly excellence.

ions either to Scripture or reason," despite his "learning and ingenuity." As for Ray, opposed to theories that strayed too far from the Bible, Burnet's was "no more or better then a meer chimaera or Romance."[20] One Fellow of the Royal Society, William Petty, had years earlier informed a correspondent that he disliked naturalistic interpretations of the Flood, "a Scripture Mistery, which to explain is to destroy." Two other Fellows, John Beaumont and John Keill, would unwittingly echo this sentiment in response to Burnet. As Beaumont put it, the Flood was "a particular Judgment, than which nothing in the Scriptures seems to me to carry more the face of a Miracle." Burnet "destroys the Miracle by lessening it, and makes it cease to be a Wonder, while he strives to make it fit to be believ'd."[21] John Locke, too, found Burnet's theory fraught with difficulties, and he would eventually couple the Flood with Newtonian gravitation as among those phenomena "which I think impossible to be explained by any natural Operation of Matter or any other Law of Motion, but the positive Will of a Superiour Being, so ordering it."[22]

If individual Fellows could pass judgment, the Royal Society could not. When classicist Thomas Gale read an account of Burnet's theory to the Society, therefore, it was "discoursed of and well approved of as to some particulars of the theory, though the proof and management thereof could not be judged of without a perusal of the discourse itself." No judgment is detectable in Gale's own account, published in the *Philosophical Collections*, although the editor of these volumes, Robert Hooke, did not himself look favorably on Burnet's theory.[23] What "particulars" the Fellows "well approved of" thus cannot be identified, but one may speculate that Burnet's originality and ingenuity startled and captivated Gale's listeners.

On the Continent, Burnet's theory evidently aroused less excitement than in Britain—that is to say, comment and criticism can be found, but in less abundance. Erratic diffusion of the original Latin edition doubtless helps to explain why reactions appeared at different times in different places. In Leipzig, the alert *Acta eruditorum* called attention to Burnet's book in 1682, and young Christian Wagner, a Leipzig pastor and occasional reviewer for the *Acta*, would the next year produce a critique of Burnet's

---

[20] Lhwyd in Gunther 1945, 159. Ray 1928, 237.

[21] Petty 1928, 21. Beaumont 1693b, 185. For Keill, see below, at n. 39. In 1740 Hans Sloane sent to abbé Bignon a sketch of Beaumont's life (BN, MS fr 22,229, fols. 258r–265v), the first pages published by Jacquot 1953, 94–96. Omitted by Jacquot, the longer part of the text comments on Beaumont's interest in fairies and spiritual phenomena and becomes a discourse by Sloane on mental illness.

[22] Locke 1693, sec. 180. Also, Locke 1976, 3:138–39.

[23] Birch 1756, 4:83 and 69. *Philosophical Collections*, no. 3 (10 December 1681), 75–76.

theology.[24] In Paris, Nicolas Malebranche received a copy of the book, probably sent by Burnet, quite early in 1681, but the *Journal des savants* did not catch up with the publication until more than two years later. In Florence, Lorenzo Magalotti, former secretary of the Cimento, was one of the first recipients, but Malpighi in Bologna managed to get a copy only in 1686, thanks to the good offices of the Italian chaplain to James II's queen. And in Modena, Bernardino Ramazzini somehow became acquainted with the book shortly before or during 1691.

Among Continental writers concerned with biblical exegesis, Magalotti expressed views similar to those we have encountered in Britain, but without the cries of alarm so visible among British critics. According to Magalotti, one could admire Burnet's ingenuity and method of argument, but not his claims to "more than a moral certitude." If a modified theory of the earth could one day be devised, it would presumably handle Moses with more discretion, for, Magalotti added, Genesis is hardly an ordinary book, and one should not complain that its content does not always conform to the dictates of human reason.[25] More sympathetic than Magalotti was Nicolas Malebranche, whom Burnet had met in Paris and to whom he wrote explaining his exegetical method. The Bible, as Burnet indicated, does transmit to us "general conclusions" about natural events, but it leaves us to seek out their causes; this is more worthy than any effort to distort the text in order to have "the Holy Spirit pronounce on the details of science." This point of view suited Malebranche very well indeed, as he recognized in Burnet a fellow Cartesian. In his *Méditations chrétiennes* (1683), the Oratorian would incorporate into his philosophy, without naming Burnet, an important part of Burnet's theory. Arguing that God's power and wisdom are more evident in the establishment of laws than in miraculous interventions, Malebranche used as a prime example those physical laws that (coinciding in time with the peak of general sinfulness) caused the earth's crust to collapse and release the subsurface waters of the Flood. He could not embrace Burnet's denial of the existence of antediluvian mountains, but argued instead that crustal collapses had pushed water up to the level of the highest peaks. What followed the disaster, however, was fully in accord with Burnet: the earth's center of gravity changed, its axis became inclined, and the era of perpetual spring ended.[26]

[24] *Acta* (March 1682), 70–76. I have not seen Wagner's treatise; he is identified in an appendix to Laeven 1986, 341–54, as having later reviewed for the *Acta* Erasmus Warren's critique of Burnet.

[25] Magalotti 1741, 92–98. Magalotti's letter, dated 1681, was first published in 1719.

[26] Burnet's letter, 24 May 1681, is in Malebranche 1958, 18:196–98; the Malebranche side of the correspondence may not be extant. For the text of the *Méditations,* ibid.,

In less rarefied realms of philosophy, brief comments in the correspondence of Malpighi with Bellini reveal an initial enthusiasm they both shared. At some stage, however, Malpighi thought it would be desirable to reply to Burnet's apriorism, and he urged Bernardino Ramazzini to do so. Ramazzini's *De fontium mutinensium* (1691) obligingly took up the issue in one chapter, in which Burnet was said to have revived an "ancient Abyssinian" myth. Nonetheless, Ramazzini had remarkably little to say about Burnet's ideas, his own approach to the earth's history being wholly different from the Englishman's. His chief concern was to study Modenese terrain and the remarkable springs of the book's title. Having examined in detail the several strata exposed during the digging of wells, Ramazzini suggested that the whole area, "in the first beginning of the World," had been a great plain submerged under the Adriatic Sea which had gradually retreated and was still doing so. Meanwhile, continual erosion had brought sediments from the Apennines and the Alps, so that

> this growing up of the Ground . . . was but slowly made, and by Slices, as it were, through length of time, as the several Lays [*sic*] of Earth do witness, which are observed in all Wells constantly in an equal Order and Distances when they are digged; so that this growing up of the Ground . . . ought to be thought rather the Product of so many Ages, than the tumultuary and confus'd Work of the common Deluge.[27]

The bulk of Ramazzini's chapter, in fact, attacked Burnet indirectly by showing the Flood to be geologically irrelevant.

Although the *Journal des savants* had initially expressed no criticism of Burnet's *Sacred Theory*, it did single out Ramazzini's chapter as worthy of special attention in its own right. The more balanced review by Benedetto Bacchini did not focus on one chapter, but Bacchini, unlike his confrère in Paris, explicitly pointed out that Ramazzini had given persuasive evidence against the diluvialism of both Burnet and an earlier Italian author, Jacopo Grandi.[28] The reference to Grandi serves as a useful reminder that

---

10:79–80. He and Burnet disagreed on the physical mechanisms which would bring about the end of the world, discussed in *Méditations*, 80–81; the letter cited above indicates that the two men had exchanged views on this subject.

[27] I quote St. Clair's translation 1697, 104–5, 107–8, 116. It seems not to have occurred to St. Clair that Ramazzini's views were not especially biblical. Malpighi's request to Ramazzini is mentioned by U. Baldini in Rossetti 1988, 223, n. 85.

[28] *JS*, Amsterdam 12mo (19 July 1683), 227–35 for Burnet; Paris quarto (29 March 1694), 150–54 for Ramazzini. Bacchini 1692, 31–42. Ramazzini's letters 1964, 98–101, 113–15, show his interest in Grandi who was not named in the publication itself; the acute Bacchini nonetheless recognized the allusions.

Continental journalists were familiar with earlier naturalistic explications of the Flood, including efforts to argue against its universality; in this perspective, Burnet's treatise was simply one more book that might arouse comment but not alarm.

On the Continent, as in Britain, the initially mixed reception of Burnet veered to the more hostile after the publication of his *Archaeologiae philosophicae* (1692). In this scholarly work, Burnet offered essentially a study of ancient concepts of the origin and earliest history of the earth. In the *Sacred Theory*, he had made relatively little use of ancient pagan writings, arguing that their ill-developed cosmogonies showed that the authors retained only a confused tradition of the true history accounted by Moses. In 1692, however, Burnet seemed intent on reducing all ancient writings, including Genesis, to the same level of authority. Burnet's naturalism in 1681 had induced deist Charles Blount to couple him approvingly with Hobbes and Spinoza, but it was still possible in the 1680s to defend Burnet, as did Malebranche in France and Thomas Browne in England. In 1692, Blount once more perceived Burnet as his ally, but his services were not necessary for readers to take alarm.[29]

Burnet's new book aroused debate in the Royal Society, and, in fact, some of the British critics discussed earlier actually were responding to both of Burnet's works. Continental theologians—such as J. Graverol, a French émigré in London, and Caspar Bussing in Hamburg—now took alarm, Graverol characterizing Burnet as both Spinozist and Socinian.[30] If the *Journal des savants* could be impartial about the *Sacred Theory*, it now cautioned its readers that Burnet was expressing "new ideas, which are hardly finding acceptance even among those of his Communion." Nonetheless, the reviewer provided an intelligent and uncritical summary even of Burnet's exegetical method.[31] Such detachment also characterized journalist Jean LeClerc in Amsterdam, but his fellow émigré Henri Basnage de Beauval found it harder to maintain his sangfroid. Basnage had not been unduly disturbed by the *Sacred Theory*, merely noting that Burnet "pushes very far the examination of these unique questions, without requiring any

[29] Blount assembled some extracts from Burnet, Hobbes, and Spinoza as *Miracles No Violations of the Laws of Nature* (1683), which provoked Browne's *Miracles Work's* [*sic*] *Above and Contrary to Nature* (1683). Blount then translated two chapters from Burnet's *Archaeologiae philosophicae* for publication in his *Oracles of Reason* (1693). See Nicolson 1963, 238–39, and Tuveson 1950, 59–62.

[30] Rossi 1984, 71, and Pasini 1981, 106–7. Debate is not apparent in the review in *PT* 17 (1693), 796–812, but the manuscript JB, 18 January 1693, indicates that Fellows engaged in "much Discourse" on a topic crucial to ancient history, i.e., the antiquity of writing (as distinct from oral transmission).

[31] *JS*, Amsterdam 12mo (18 July and 25 July 1695), 504–18, quotation on 504.

agreement on the part of his readers."[32] A year later, in 1690, Basnage encountered Pierre-Daniel Huet's *Quaestiones alnetanae,* in which the author expounded at length the common traits of sacred and profane theologies. With pious intentions, Huet wished to show that Christian truth had a foundation in "universal consent," but the message conveyed to Basnage and others seemed a dangerous equalizing of all theologies and the possibility that all religions derived from a "natural revelation."[33] By the time Basnage read Burnet's comparable arguments, he could easily recognize the dangers and describe the book to Leibniz as an attack on Moses: "No one has heretofore done this with so much freedom." In reviewing Burnet's new book, Basnage now warned readers that "sinister" interpretations—such as the possibility that the world is eternal—might be extracted from the work and that Burnet had failed to meet the case by insisting that he was merely offering conjectures.[34]

Compared with other critics, Basnage was mild and judicious. For decades, Burnet would become a symbol of incipient atheism, as when Richard Bentley, in his Boyle lectures of 1692, included Burnet among the advocates of "blinder mechanism and blinder chance." In Naples, Domenico d'Aulisio coupled Burnet with Spinoza, both having derived their "blasphemies" from Descartes; such philosophers believed that the world had been formed by mere mechanical causes and thus was "without architect."[35] To William Nicholls, as to Keill and others, Burnet displayed the "presumptuous pride" of Descartes and other "world-makers" who thought they could fathom the intimate workings of nature. Unlike Bishop Croft, who paused to wonder if Burnet might think himself a Christian, others focused not on motives but on results. As the *Trévoux* journalists later declared, Burnet had "scandalized" all Christians by "his way of explaining or rather of abolishing" biblical mysteries.[36]

The Burnet controversy continued at least into the early decades of the eighteenth century, the methodological issues, not the geological theory, being fundamental. In effect, Burnet had shown that naturalistic explication of biblical texts could have unforeseen and unpalatable consequences—consequences not really visible during earlier decades of

---

[32] For both LeClerc and Basnage, see Reesink 1931, 211, 212, 304–5. Also, Basnage, June 1689, conclusion on 335.

[33] Dupront 1930, 183–272, citing Basnage at 249.

[34] Basnage, March 1693; Reesink 1931, 304–5. Letters in Leibniz 1875, 3:98 and 92–93.

[35] Bentley and Aulisio are quoted in Rossi 1984, 73–74.

[36] Nicholls 1696, esp. 119. For Keill et al., see Tuveson 1950, 63–65. *Trévoux* (July 1721), 1222; the journalists were here reviewing a posthumous work by Oratorian Bernard Lamy who admired Burnet's naturalistic explanation of the Flood. Astonishingly, *HWL* (July 1699) 429–31, treated the Burnet affair as one of those "*Paper-Combats . . .* about trifles."

designing Arks and fitting the animals into their stalls. To critics who objected that mountains were not the "ruins" of a former world, the real issue was God's wisdom and beneficence. For those who objected to mechanical laws explaining both Creation and Flood, the real issue was conflict between determinism and God's freedom.[37] Burnet himself did reply to some critics, reiterating his belief that the Bible should be interpreted literally only on matters for which we have no natural evidence; but when such evidence does exist, the sacred text must be re-interpreted, or we encourage atheism by showing nature and Scripture to be in conflict.[38] Here Burnet echoes St. Augustine, but it is clear elsewhere that he did not fully understand the debates he had provoked, as shown in his exchange with John Keill who preferred that the Flood be considered a miracle. When Burnet asked "wherein this miracle consisted," Keill replied: "I never thought it my business to explain miracles."[39] Keill and others made a distinction that would be hard to sustain in the next decades: they wanted the *cause* of the Flood to be miraculous, but the *effects* detectable in nature. This program would be adopted by John Woodward, whose disciples and critics at times echoed Burnet, for they demanded that Woodward explain "wherein this miracle consisted."

WOODWARD AND HIS DISCIPLES

To naturalists like Ray, Burnet's most obvious omission from the *Sacred Theory* was all reference to fossils. Ray himself, as he told Edward Lhwyd, had been urged by friends to remedy this defect by writing a treatise to show that fossils had been deposited by the Flood. The result, Ray's *Miscellaneous Discourses* (1692), hardly met the case, for Ray could not even be sure that the Flood had been universal. Nor could he attribute fossils to the Flood, having long since wondered how any flood could affect the earth's crust to any depth; instead, accumulations of fossiliferous sediments seemed to him to signal "a work of time." His most substantial contribution to the Burnet controversy therefore turned out to be his effort to enlarge the role of Providence. Arguing that mountains were not "ruins" of a former world but beautiful and useful to man, Ray could deal with

---

[37] Nicolson 1963, chap. 6, puts some emphasis on the question of wisdom and hence purposefulness in nature. The issue of determinism had become associated with Spinoza, and Burnet seemed to follow in this dangerous tradition; the more general source of the danger had long been recognized as the mechanical philosophy. See Osler 1994.

[38] Reply to Erasmus Warren, in Burnet 1726, vol. 2, esp. 418–20, 478–80.

[39] Keill 1699, 4.

Design but not with miracles; he could only shift the balance somewhat by pointing to God's role in setting natural causes in motion.[40] By 1694 Ray had to confess to Lhwyd that he saw "little likelyhood of demonstrating how the Universal Deluge could lodge [fossils] so deep in ye bowels of ye mountains & rocks; yet it would be a great satisfaction to me to see it well made out; or any good attempt toward it."[41]

Ray and Lhwyd both knew that John Woodward proposed to do what Ray thought unlikely but desirable. In the event, Woodward's *Essay* of 1695 did not initially present itself as a defense of the Bible; instead the preface announced that Moses would be treated no differently from any historian. Reinforcing this program elsewhere in his book, he would declare, for example, that moderns should use ancient texts—pagan texts in particular—only for factual descriptions rather than for arguable theories or viewpoints. He himself would concentrate on natural facts, limiting his treatise to those inductive conclusions "whereof we have a plain and undeniable Certainty: those which flow directly and immediately from the Observations." The chief conclusions of this kind were that rocks must once have been dissolved in order for organisms to lodge in them, and that this dissolution occurred during the Flood.[42]

Already the owner of a considerable natural history collection, Woodward relied heavily on the fact that most fossils are marine forms; since these occurred universally, the obvious explanation of the phenomenon was a universal event, the Flood. That such an event had taken place could readily find historical confirmation in Genesis. By contrast, other alternatives—such as traditions of local floods, marine transgressions, and changes in the location of sea basins—he rejected as having no solid historical foundations. Some traditions might well be confused memories of the biblical event, but pagan texts could generally be dismissed as unreliable allusions to postdiluvial changes that he judged to be of no geological significance.[43]

Woodward's leap to the accuracy of Moses and the inaccuracy of pagans indicates a less than impartial approach to ancient writings, especially when compared with the efforts of his contemporaries to find nuggets of information even in the myths recounted by Ovid. Having settled the matter to his satisfaction, however, he proceeded to adopt the few details Moses had provided: not only had mountains existed before the Flood, but there had

---

[40] In Ray 1693, the most relevant passages on pp. 35–39, 70–73, 121–24, 144.
[41] Ray to Lhwyd, 7 September 1694, in Ray 1928, 254.
[42] Woodward 1695, preface and 62–63, 73–74.
[43] Ibid., 46, 62–63.

also been enough water to cover them, the water coming from rain and from the opening of "the abyss." Since Moses had not described the effects of the Flood, except to imply that the earth's topography had not much changed, Woodward here was on his own. He thus explained how the Flood must have dissolved the earth's crust, after which sediments settled in order of their specific gravities; fossils themselves—the organisms had not been dissolved—followed a similar pattern of deposition,

> those which are heaviest lying deepest in the Earth, and the lighter sorts (when there are any) such in the same place) shallower or nearer to the Surface, . . . the heavier Shells in Stone, the lighter in Chalk, and so of the rest.[44]

For presumably only a short time, the earth's surface was "eaven and spherical," its strata "parallel, . . . plain, eaven, and regular; . . . continuous, and not interrupted, or broken." Expansive forces emanating from "the abyss" altered this arrangement, producing earthquakes and volcanic eruptions, and allowing excess water to drain from the earth's surface. The most solid parts of the earth remained unaffected, but other areas collapsed, and the net result was the crust as we now see it, with its mountains, fissured rocks, sea basins, and so on.[45]

This summary hardly does justice to Woodward's *Essay*, parts of the volume being solidly packed with information and displaying a degree of analytical skill. Probably most impressive at the time was his discussion of marine fossils, detailing their resemblances to living organisms and responding to problems that had led other interpreters astray. Indeed, his subsequent publications confirmed his stature as an expert fossilist, even among readers who could not accept his diluvial theory.[46] In 1695, however, the key issues remained the origin of fossils and an explanation of their transport and deposition. To some contemporaries, Woodward had adequately solved both these problems. This was certainly the view of Woodward's most vigorous Continental disciple, J. J. Scheuchzer, formerly an advocate of the concept of *lusus naturae*. Other disciples do not fit comfortably into a single category, as some were persuaded by Woodward to abandon *lusus* or Lhwydian "seeds," some adopted Woodwardian diluvial-

---

[44] Ibid., preface; also, 75–78.

[45] Ibid., 80–81, 133–42, 164.

[46] Ibid., 15–33. For his collection, repute as a collector, range of correspondence, and post-1695 publications, Price 1989 is convenient, but Eyles 1971 and Levine 1977, chap. 6, are preferable.

ism, and some thought Woodward's theory required modifications both large and small.[47]

Less immediately obvious to some contemporaries, and to some modern historians, was Woodward's reliance on miracle, although he had firmly declared: "a Deluge neither could then, nor can now, happen naturally."[48] If the reviewer for the *Acta eruditorum* noticed that the Flood was not natural, he also asked the naturalistic question: why did not organisms also dissolve? More ambiguously, the *Philosophical Transactions* merely carried the comment that geological changes had been accomplished by "the Hand of Almighty God."[49] Both disciples and critics actually had so much difficulty distinguishing nature from miracle in Woodward's theory that they urged him to produce the promised larger treatise in which various questions would be answered. This Woodward tried to do, in 1714, and additional explanatory texts would be assembled and published in 1726 by disciple Benjamin Holloway. Among the several issues explained here, Woodward indicated that the Flood waters had not actually "dissolved" rocks; rather, God had suspended the normal operations of gravity, so that rocks had merely separated into their constituent particles. (Some readers had found this clear in the *Essay* of 1695.) Organisms, by contrast, had been held together by their interlocking fibers. He tried also to explain that, in the confusion attending the Flood and its immediate aftermath, one could not expect sediments and fossils always to subside precisely according to specific gravities. And finally, Woodward now abandoned any effort to present Moses as historian: Moses had known about the Flood not by observation or preserved traditions or conjecture, but by divine revelation.[50]

Among Woodward's earliest British disciples, John Harris, F.R.S., merits attention as one who both noticed and approved of the master's use of mir-

---

[47] The most extensive study of Scheuchzer is by Hans Fischer 1973, but more valuable is Jahn in Schneer 1969, 192–213. Converts on the origin of fossils and perhaps on the Flood included Elias Camerarius, K. N. Lang, and F. M. Misson; see Levine 1977, 45, 97, and De Beer 1948, 59–61. I have not seen Bianchi's *De Conchis* (1739); his adoption of Woodwardian theory is mentioned in *Novelle letterarie* 1 (1740), cols. 405–6, but *Trévoux* (April 1740), 603, says Bianchi allowed that at least some deposits could be "des sédimens anciens de la Mer qui couvroit ces terres."

[48] Woodward 1695, 165. For example, compare Tuveson 1950, 59, and Davies 1969, 77–79, with Levine 1977, esp. 40 for Arbuthnot, and 44–45.

[49] *Acta,* supplement, 3 (1702), 13, 15. *PT* 19 (1695), 118.

[50] Woodward 1726, 1:39–46, 157–62; 2:36–45, 97–100. As indicated above, n. 30, one topic of debate in this period was the invention of writing; more specifically, since Moses had not witnessed events antedating his own lifetime, how was such knowledge transmitted to him? By resorting to revelation, Woodward abandoned the kind of debate being conducted by some of his contemporaries. Discussion can be found in Rossi 1984, pt. 3, passim.

acles, his ferociously polemical *Remarks* (1697) presenting Woodwardian theory as a bulwark against all forms of materialism. If contemporaries shared such fears, Harris's tone and language led John Ray to describe his tract as "scurrilous" and Tancred Robinson, F.R.S., to label him an Anglican Inquisitor. Buried amid the name-calling, Harris insisted that to raise naturalistic objections to Woodward's theory was irrelevant. Several years later, in his *Lexicon Technicum* (1704), Harris had calmed down sufficiently to present a detailed analysis and defense of the theory; at the same time, however, he would recommend to his readers William Whiston's *New Theory of the Earth* (1696), a work not fully in accord with Woodward's.[51]

Whiston attempted both to add to Woodwardian theory and to patch up its deficiencies. Unlike Harris, he thought that resort to miracle should be avoided, and he would later admit that his own naturalism had been inspired in part by his reading of Burnet. His major addition to Woodward, therefore, was a natural cause of the Flood: a comet had passed near the earth, the vapors of its tail supplying water sufficient to cover the earth's mountains. That there was no historical warrant for the appearance of a comet at that time Whiston admitted, but he considered it permissible to propose a cause that would explain effects "perfectly unaccountable" by any other natural means.[52] Among the effects Whiston sought to explain were some in themselves hypothetical—such as a presumed change in the earth's orbit and in the length of the year—so that, in essence, Whiston manufactured effects as well as cause. As for the geological effects of the Flood, Whiston also perceived a significant flaw that he tried to rectify, namely, the earth's strata are not always laid down in order of their specific gravities. Revising Woodward, he claimed that such order had once existed but was scarcely to be observed any longer; instead, sedimentary strata gave evidence of having been the result of "a very Wild, Confus'd and *Chaotick* condition."[53]

By seizing on the issue of specific gravities, Whiston did not imagine himself to be using an empirical test to undermine Woodward's theory; on the contrary, the original theory had merely underestimated the amount of "confusion" obtaining both during and immediately after

---

[51] In Harris 1697, see p. 20 for an example of miracle. Ray to Sloane, n.d., in Ray 1848, 332. For Whiston, see Harris 1708, art. "Deluge." The dispute between Harris and Tancred Robinson is discussed by Rappaport 1997.

[52] Whiston 1696, 130; see also 127–29. For his admiration of Burnet, see Whiston 1698, preface. Additional information about the cometary hypothesis is in Force 1985, 53 and n. 99.

[53] Whiston 1696, 205; see also 200–205. According to Force 1985, 48, Whiston would later (1708) further modify Woodward by saying that it had taken one or two centuries for the waters of the Flood to subside.

the Flood.[54] Like Whiston, other British Woodwardians recognized the importance of this issue, and, like Whiston, they found ways to salvage the theory. Alternatively, some claimed that the empirical evidence either "generally" or in detail did conform to the pattern laid out by Woodward. Selected examples will illustrate how these naturalists proceeded.

In his study of Lancashire and other counties (1700), Charles Leigh showed acute discomfort about aspects of Woodward's theory: the biblical "abyss" baffled him, and he could not believe water capable of dissolving rocks. (Like so many, he interpreted the latter as a natural impossibility, ignoring Woodward's miraculous explanation.) On the subject of specific gravities, he sometimes follows Woodward without question, but on one noteworthy occasion Cambridge platonism comes to his aid. Here he observes that the Woodwardian order of strata could not be observed in coal mines, and he explains the anomaly by saying that "confusion" was great during the Flood, the deposition of strata then being "solely govern'd by an *Hylarchic Spirit.*"[55] In his peculiar fashion, Leigh did remain committed to the Flood as the only possible explanation for the ubiquity of marine fossils—and for the transport of large mammals—because he could see no better alternative. As he confessed, he wished Woodward's theory were correct, and, despite his many objections, he seems to have believed that Woodward had essentially been on the right track.[56]

Unreservedly Woodwardian, John Morton in 1706 communicated to the Royal Society some preliminary observations later incorporated into his volume on the natural history of Northamptonshire (1712). In the earlier study, Morton asked why freshwater and terrestrial forms were so much rarer in the strata than marine shells, and he claimed to know the answer: "these Bodies having been shown to be all Remains of the Universal Deluge," marine shells, being heavier than other forms, had merely sunk deeper into diluvial sediments and so had more often been preserved. In the detailed account of 1712, Morton analyzed samples from six different pits or quarries, indicating in each case that the strata succeeded each other in the correct Woodwardian pattern. Where exceptions existed, he argued that certain strata had been heavier at the time of their deposition, containing mineral matter that had later been leached out.[57]

---

[54] That this empirical test made easy the refutation of diluvialism has been asserted from time to time, for example, by Eyles 1958, 184, and Rudwick 1972, 90.

[55] Leigh 1700, bk. 1, pp. 29, 64, 66 (passage quoted), 98, 119.

[56] Ibid., 115, 117. Considering the jumble of Leigh's text, it is worth remarking that the reviewer for *Acta* (November 1701), 513 did notice his denial that the earth's crust had been dissolved.

[57] Morton in *PT* 25 (1706), 2210. Morton 1712, 130–33. Cf. Woodward 1726, 2:44–45.

If Morton can be said to have explained away discomfitting observations, William Derham scarcely seems to have noticed the contradictory evidence he and others were finding. Derham's Woodwardian sentiments first appeared briefly in the *Philosophical Transactions*, where he examined "the Subterraneous Trees in Dagenham, and other Marshes bordering upon the River of Thames, in the County of Essex." These deposits, he argued, were postdiluvial, because in each case the enclosing stratum was relatively "undisturb'd" and "of much less Specifick Gravity than the *Stratum* above it is."[58] If Derham here assumes the truth of the Woodwardian pattern and has to explain away an anomaly, in his *Physico-Theology* (1713) he undertook to test the pattern, being aware of Charles Leigh's assertion that strata in coal mines did not conform to Woodward's requirements. Resolved to "try the Experiment," Derham took samples from "divers Places" on his own property. In six cases, these bores confirmed Woodward, while in the other three, where he had not dug very deep, he found only a thick layer of sand of the same specific gravity as the superficial layer of soil. Then he proceeded to "some deep Chalk-Pits . . . but the [outcome] was not so uniform as before." What this failure means he does not explain, but remarks only that the Royal Society "ordered their Operator to experiment the *Strata* of a Coal-Pit." Nor does he tell his readers that the operator, Francis Hauksbee, had obtained a non-Woodwardian result: "the Gravities of the several *Strata* are in no manner of Order; but purely casual, as if mixt by chance."[59] Derham had understood the importance of such tests for diluvial theory, but one can only wonder what conclusions he drew from his own results and Hauksbee's.

This small array of British Woodwardians—selected because they gave the theory detailed attention—cannot convey the extent to which diluvialism was accepted or the degree to which diluvialists should be dubbed Woodwardian. As Robert Hooke had claimed as early as 1688, Fellows of the Royal Society increasingly believed that the Flood solved the transport problem for marine fossils. After Woodward's *Essay*, some observers considered diluvialism to be firmly established.[60] But disciples, as we have seen, questioned parts of the theory and introduced modifications they thought better in accord with nature. One of Woodward's Scottish readers, Sir John

---

[58] Derham in *PT* 27 (1712), 482.

[59] Derham [1713] 1727, 66, note f. Hauksbee's list and his conclusion in *PT* 27 (1712), 543–44. Where I have inserted "outcome" in the quotation, this is one of two places where Derham refers to the "success" of his and Hauksbee's trials; the latter word was then often used in a neutral sense to which qualifiers would be added ("good success," "bad success").

[60] For example, this is my impression from the lack of discussion by Stukeley 1724, 91 and index, s.v. "Antidiluvian [*sic*]." Hooke 1705, 412.

Clerk, put the matter succinctly: the gravitational sorting out of fossils and sediments was clearly wrong, and yet it was obvious that the Flood did account for the transport of fossils.[61] By tinkering with the theory, diluvialists expressed their instinct that a solution more or less like Woodward's would prove acceptable.

The same attitude characterized Continental diluvialists. Although a few could read English, most had to await the Latin translation (1704) of Woodward by J. J. Scheuchzer. Some were already familiar with earlier efforts to deal with the geological effects of the Flood—witness Ramazzini's response to Burnet and Jacopo Grandi—or, like Sir John Clerk, thought it obvious that the Flood and fossils should be causally related. Thus, when the Scheuchzer translation arrived at the Academy of Bologna, J. B. Beccari had already been arguing that fossils should be considered "monuments" of the Flood.[62]

Continentals seem not to have examined with any care the issue of specific gravities, Scheuchzer, for one, merely remarking that the strata "for the most part" do follow the Woodwardian pattern. Similarly, Giuseppe Monti in Bologna claimed that his own and Scheuchzer's observations showed "quite well that all these sediments were deposited in layers according to the law of gravity." Chemist J. F. Henckel noted that some strata do observe the Woodwardian order, whereas others are found either "confusedly or in layers, one atop the other, and it is easy to see that the Flood is the cause of these arrangements."[63] If examination was in short supply, these rapid conclusions resemble John Morton's: the pattern held good in general, and exceptions could be explained.

Some Continental Woodwardians nonetheless did query less readily verifiable parts of the theory, as did even the Scheuchzer brothers. In agreement with Woodward, the elder Scheuchzer thought it permissible to resort to miracle when "the explicit Word of God requires it, or when . . . the magnitude of the event surpasses the abilities of Nature." But for one of Woodward's miracles, the opening of the abyss, the younger Scheuchzer proposed a simpler alternative: God might have stopped the earth's rotation for a moment, causing the sea "violently" to erupt over the continents.[64] Again in contrast to Woodward, who argued that minimal geological changes had occurred since the Flood, the elder Scheuchzer seems to have planned a treatise on postdiluvial fossil plants "which are forming

---

[61] In Levine 1977, 265–66.

[62] Tega 1986, 1:70.

[63] Scheuchzer 1732, 1:62. Monti, 1719, in Sarti 1988, 54. Henckel [1725] 1760, 1:123.

[64] For the use of miracles, Scheuchzer 1732, 1:63. His younger brother's proposal for an alternative cause of the Flood is in *HARS* [1710] 1712, 21.

daily."[65] Indeed, if Edward Lhwyd is to be believed, the elder Scheuchzer for a time had been unable or unwilling to commit himself publicly to Woodward's notion of "an Atomical Dissolution of all Things the Terrestrial Globe consisted of at the Deluge." If true, then Scheuchzer eventually did support this essential feature of the theory.[66]

Clearly, Lhwyd thought "an Atomical Dissolution" so unlikely as to be absurd, but some of Woodward's disciples also had trouble that led them to propose modifications. For Giuseppe Monti, natural causes at the time of the Flood had lacked sufficient "force" to dissolve all strata, so some had merely been disturbed or ruptured, but not destroyed. For fossilist Dezallier d'Argenville, a universal dissolution was necessary neither for the punishment of mankind nor for the deposition of fossils. The Flood merely had to soak the earth in order for fossils to lodge at varying depths within the crust.[67] In both instances, these writers either misunderstood or rejected Woodward's use of miracle, substituting processes they considered to be more naturalistic and plausible.

The most famous moment in early diluvialism came when the elder Scheuchzer announced in 1725 his discovery of what he called *Homo diluvii testis*, or the man who had witnessed the Flood. To everyone concerned with natural evidence for the Flood, it had long been worrisome that so few remains of land animals could be found in the strata. Some collectors of curiosities had claimed to possess fossil human remains, but Scheuchzer thought these identifications dubious. His own discovery he considered decisive, and he quickly sent descriptions to the Royal Society, the *Journal des savants*, and two other periodicals.[68] If one editor expressed his approval, the others remained neutral, and it is difficult to judge whether Scheuchzer's specimen was widely accepted either as human or as proof of the Flood. One of his friends, Antonio Vallisneri, examined the published engraving of *Homo* and confided his conclusions to Louis Bourguet: the identification was doubtful, as was the assumption that the animal had died during the Flood rather than at some other time. Bourguet himself apparently disagreed with Vallisneri, for he would collaborate on a volume in which Scheuchzer's specimen featured among the illustrations.[69]

---

[65] *Trévoux* (January 1713), 75, in a review of Scheuchzer's *Herbarium diluvianum* (1709). I can find no evidence that Scheuchzer published what the reviewer calls a "Herbarium post-diluvianum."

[66] Lhwyd in *PT* 26 (1708), 157. But see Scheuchzer 1732, 1:63.

[67] Monti, 1719, in Sarti 1988, 48. Dezallier 1742, 161.

[68] Jahn in Schneer 1969, 206–10, lists the journals and translates two of the four accounts.

[69] Vallisneri to Bourguet, 16 September 1727, in Neuchâtel, MS 1282, fols. 293–94. Bourguet 1742, 2:90–91 and plate 60; discussion in Bork 1974, 56. Assessment can be found in Guettard 1768, 6:314–30, who concluded that *Homo* was either a fish or a crocodile.

Probably the most thoughtful of Continental diluvialists—if he should be called that—was Louis Bourguet, who for years had debated the issues with Woodward and Scheuchzer on the one hand and with Vallisneri on the other. Eventually, in 1729, he published a sketch of his own theory of the earth, designed to incorporate the Flood in such fashion that the most pressing objections would be resolved. In this brief outline, Bourguet proposed that there had been two periods during which fossiliferous strata had been laid down. First had come the stage when the earth's motions of rotation and revolution had somewhat diminished the solidity of the existing crust; in this period, mountains had been elevated and organisms were being deposited in sediments arranged in order of their specific gravities. Then, at the time of the Flood, there occurred a "successive dissolution" of the crust, followed by the gradual accumulation of strata; that these processes occurred successively and gradually is, according to Bourguet, the true explanation of why strata do not follow the Woodwardian pattern of deposition.[70]

Late in his career, Bourguet admitted that his theory needed considerable development if he were to convince doubters of the geological importance of the Flood. At the same time he did reiterate and develop slightly his earlier views in a treatise that reveals conflict between Bourguet and his co-author, Genevan pastor Pierre Cartier. Here Bourguet again insisted on successive stages in the earth's history, while Cartier, a faithful Woodwardian, preferred supernatural causes and proclaimed the accuracy of Woodward's statements about specific gravities.[71] The two collaborators agreed chiefly on the importance of the Flood, for reasons that Bourguet explained on more than one occasion: as a matter of historical record, the Flood could not be ignored.[72] With this general judgment, Woodward himself—and most other naturalists—would have agreed, but, as early as 1720, Woodward had heard that Bourguet intended to publish something on fossils and the Flood, and he warned the elder Scheuchzer: "He [Bourguet] will not do well to publish any Thing different from what I have published without consulting me, and knowing whether I cannot resolve his

---

[70] Bourguet 1729, 212–14. The full text of Bourguet's "propositions" about his theory of the earth is given in M. Carozzi 1986, 29–31. Paradoxically, one proposition also asserts that the earth took its present form all at once, except for small subsequent alterations caused by such events as earthquakes. My analysis is based on those propositions for which a little detail is provided. Nowhere in this text does Bourguet explicitly mention the Flood. Review in *BR* 4 (April–June 1730), 243–84; for the review in *Trévoux*, see below, at n. 116.

[71] Bourguet 1742, 1:43–44, 81–82. This volume is in the form of "letters," the former passage from an unsigned letter (I assume by Bourguet), the latter from a letter signed "C" (Cartier).

[72] Tucci 1983, 21, and above, Chapter 4, for Bourguet on the Flood and fossil bones.

doubts."[73] Woodward's suspicions were justified, in that Bourguet ultimately departed very far from the Englishman's theory.

Any effort to estimate the proportion of naturalists who adopted diluvial theory is fraught with difficulties, some of them taxonomic. On which side of the ledger should one place chemist J. F. Henckel and antiquarian William Stukeley, who, preoccupied with their own researches, hardly paused to examine issues or evidence? Did a writer such as Dr. J. F. Leopold, who admired Woodward sufficiently to address to him a compendium on the natural history of Sweden, adopt diluvial theory?[74] What shall we do with antiquarian Jacob von Melle, who in 1718 asked Woodward if particular fossils he could describe had been deposited by the Flood, and two years later was still arguing with himself about how to combine aqueous and igneous causes in the earth's history. And then there were the two Montpelliérains who examined fossils of the Mediterranean littoral but dismissed, without examination or discussion, all noncoastal sediments as diluvial. Perhaps easier to classify are J. J. Baier and the Berlin academician named Sack, who both promised to offer diluvial theories that would resolve the difficulties of Woodward's—but neither writer seems to have done so.[75]

Woodward himself eventually tried to provide two of his correspondents with some evidence of his success in making converts, but he tended to imply that admiration of his learning meant conversion to his ideas. Treatises were indeed dedicated to him, his international correspondence increased, and he doubtless did persuade some readers to tackle for themselves the study of fossils and the Flood.[76]

One may, in fact, track the Flood through treatises, pamphlets, and articles in learned journals, for the subject almost always does make its appearance, whether briefly or at length, and whatever the views of particular authors. In this morass, sorting out the Woodwardians is virtually impossi-

[73] In De Beer 1948, 57–58. Woodward died shortly before Bourguet's sketch of 1729 was published.

[74] Leopold 1720. I have not seen another such *epistola* addressed to Woodward by J. H. Linck in 1718. Both works are reviewed in *JS* (25 May 1722), 334–35, and (February 1725), 90–91. Discussion in Levine 1977, 98.

[75] Melle 1718 and 1720, 29–32. For Montpellier, see Jean Astruc, reported in *Trévoux* (March 1708), 516, and *JS* (supplement for 1708), 121; Guillaume Rivière, in *Trévoux* (April 1709), 607–8. The texts of these two memoirs were later published in Montpellier, *Mémoires* 1 (1766), 48–74, 75–84. Baier 1708, 67–68. Berlin Academy, in *Collection académique*, 8:xix–xx; for this useful set of volumes, see Kronick 1976, 212–15. Ellenberger 1980 analyzes the importance for geological theory of northern vs. southern French terrains.

[76] Letter to the Earl of Pembroke, 11 December 1713, in Woodward 1726, 2:7–8. Letter to unidentified correspondent, n.d. (but 1726 or later), in Woodward 1735, 359–62. Among admirers, he mentions Baier and Melle who did question aspects of his theory.

ble, for the examples of Baier and Sack remind us that not all diluvialists were Woodwardians. From Woodward's own point of view, any modification of his theory was a departure from truth, but we clearly cannot adopt such draconian measures in identifying his disciples. Nor do contemporary assessments provide much aid, the writers being insufficiently informed and sometimes partisan. Robert Hooke, for example, in a passage already cited, thought that by 1688 Fellows of the Royal Society were increasingly prone to adopt diluvial explanations of fossils; but the accuracy of this statement may be questioned, since Hooke for years was trying to persuade his colleagues of the validity of his own nondiluvial theory, and his writings on the subject often reveal a bitterness of spirit.

If Hooke perchance was correct, and if diluvial geology gained additional adherents—as it surely did—in the decades after Woodward's *Essay,* then we are left with inexplicable entries in John Harris's famous *Lexicon.* A Woodwardian himself, he nonetheless tells us that fossils were deposited by the Flood "or by some other means." In another entry he insists that fossils are genuine remains of organisms, even if "we cannot account presently how they came thus into the Earth, or solve all the Objections or Difficulties about them."[77] Had Harris reconsidered his own allegiance, or is he merely telling us that Woodward's views had not triumphed by about 1710?

These statements about Britain, inconclusive in themselves, do nothing to clarify the situation on the Continent, where a reviewer for the *Journal des savants* could remark in 1722 that diluvialism had been adopted by "some modern naturalists."[78] The difficulty here, as with Harris, is that historians retrospectively tend to look at diluvialism as constituting a solidly baked brick of accepted opinion. To Harris and the Paris journalist, opinion clearly had not solidified, as the preceding pages have attempted to show by recounting some of the debates taking place among Flood geologists.

CRITICS OF DILUVIALISM

If some naturalists welcomed Woodward's *Essay,* Robert Wodrow in Glasgow was aware of so much criticism that in 1700 he could say to one of his correspondents: "I am glade to hear Mr. Woodward has not given over his researches, for all the bad enterteanment his prodromus has met

---

[77] Harris 1708, s.v. "Fossils" and "Petrefication."
[78] *JS* (13 April 1722), 239, in a review of Melle 1718.

with."[79] If Wodrow remained hopeful, others were less so, as they tackled fundamental questions ranging from serious weaknesses in the theory itself, to the use of miracles in natural philosophy, to the relevance of the Bible to science. Rarely did critics conclude that the Flood should be omitted from geology, for they shared Louis Bourguet's view that one could not ignore the historical event. And only a few had alternative syntheses to propose. In the next pages, I focus on critiques, reserving for later discussion the several nondiluvial theories in existence before and after Woodward's *Essay*.

Certain British naturalists had known of Woodward's intentions before 1695 and had, among themselves, been sporadically concerned for some years about the natural effects of the Flood. John Ray had not only queried, in a tentative way, the universality of the Flood, but had also wondered about the effects of flooding: surely floodwaters would wash bodies downward and scatter them, rather than carrying them to high elevations or burying them deep within the earth. His friend Lhwyd wrote to Ray in 1692 about the possibility of attributing to the Flood the thousands of erratic boulders to be seen in Welsh valleys.[80] When Ray and Lhwyd learned of Woodward's planned treatise, both doubted that the enterprise would succeed. Lhwyd had come to dislike Woodward's arrogance, remarking in a letter to Martin Lister: "let him have never so much to say [in his book], he'll scarce outdoe his promises." More charitably, Ray hoped Woodward would solve the knotty problems he had long considered intractable. As he wrote to Lhwyd, "I know not but Divine providence may favour him with a peculiar illumination to penetrate further into this matter then other men have done."[81] In the event, both men judged that Woodward had failed.

Like some Woodwardians, Ray and Lhwyd recognized the issue of specific gravities as crucial to the theory. Lhwyd's more immediate reaction, however, was revulsion, expressed in a letter to Lister: "when we consider how far it [Woodward's book] may agree with reason and common sense; we find so many absurdities in it, that to me it seems scarce worth our consideration." If there was one saving feature of the book, Lhwyd hoped it would persuade people of the importance of studying fossils. As for specific gravities, Lhwyd thought Woodward's "a confus'd notion, of no real use, & scarce intelligible to himself or others . . . for when he names ye spec. [species] of shell yt should be lowest or highest; in chalk or stone &c.

[79] Wodrow to Nicholson, 18 October 1700, in Wodrow 1937, 118.
[80] Ray 1693, 122, 144–45. Lhwyd to Ray, 30 February 1691/92, in Gunther 1945, 157–58; this proposal will be explained in Chapter 6, at n. 59.
[81] Lhwyd to Lister, 1693, in Gunther 1945, 211. Ray to Lhwyd, 7 September 1694, in Ray 1928, 253–54; the next sentence in Ray's letter was quoted above, at n. 41.

our observations contradict him."[82] In a similar vein, Ray reported to Lhwyd that Samuel Dale, Ray's long-time friend and disciple, had also made observations that did not bear out Woodward's claims.[83] Comparable tests, with comparable results, would also be made by Robert Wodrow and, at the request of the Royal Society, by Francis Hauksbee. If Wodrow continued to hope that a theory "a little reformed" might obviate these difficulties, other critics were less optimistic.[84]

In contrast to the modesty of Wodrow, one of Woodward's first British critics was most acerbic, raising fundamental issues in a manner that makes Lhwyd's comments seem mild. In his *Two Essays*, the anonymous "L.P." charged that Woodward, by staying so close to the text of Genesis, had harmed both religion and philosophy. The language of the Bible having been adapted to the understanding of ancient and ordinary people, modern philosophers from Copernicus onwards "are not esteem'd the worse Christians, because they contradict the Scriptures in Physical or Mathematical Problems."[85] As for the specific phenomena investigated by Woodward, these should be examined

> by the Laws of *Gravity*, or by the *Hydrostaticks;* because the Controversie is about the descent of *Solids* in *Fluids;* in the managing of which there is no need of any extraordinary or Miraculous intervention of the first *Infinite Cause,* the principles and rules being originally established by it.

In other words, God had given nature its laws, and philosophers should not gratuitously suppose that "the *Machine* we call the *Universe*" stands in constant need of divine reform.[86]

L.P. then proceeded to discuss natural phenomena on the one hand and biblical interpretation on the other, displaying a mixture of caution and daring. Woodward's theory required that the earth's rocks be dissolved, even while delicate organisms remained intact; to L.P., "this may pass with *Romantick* Readers, but scarce with any sound or thinking *Philosophers.*" Equally clear to L.P. was that fossiliferous deposits are not arranged in order of their specific gravities. Well aware of the various interpretations of fossil

---

[82] Lhwyd to Lister, 28 March 1695, in Gunther 1945, 268–69, and notes appended to a letter from Ray to Lhwyd, 1697, in Ray 1928, 272.

[83] Ray to Lhwyd, 1699, in Ray 1928, 277. Dale would later summarize divergent views on the nature of fossils and the role of the Flood, giving no indication of his own sentiments about the latter. Dale in S. Taylor 1730, esp. 322–24.

[84] Wodrow to Sibbald, 28 October 1699, and Wodrow to Campbell, 9 November 1702, in Wodrow 1937, 27, 237. For Hauksbee, above, at n. 59.

[85] L.P. 1695, ii.

[86] Ibid., 3–4.

shells in particular, L.P. proposed that different specimens be interpreted in different ways: some might be marine organisms, and others might have grown in situ, as Martin Lister and Robert Plot had long been arguing.[87] If L.P. did not pretend to resolve such matters, he had less difficulty with the Bible. His sympathies here lay with some of the more daring expositors and historians, as he asserted that the ancient Hebrews had been as much given to myth and allegory as any other people. Latitude of interpretation was thus essential, notably in connection with the Flood which, in L.P.'s view, probably had not been universal.[88]

From a rectory in remote Cumberland (where he had trouble getting the latest books) came another critique by Thomas Robinson, who had already, in 1694, outlined a theory of the earth suffused by Neoplatonic and other traditional modes of thought. In 1696, he would reject both L.P.'s suggestion that the Flood had not been universal and Woodward's version of the causes and effects of the Flood. On the one hand, displaying the normal failure to read Woodward with care, he argued that Woodward had paid insufficient attention to miracle because surely the dissolution of the earth's crust could not have been accomplished without one. On the other hand, he declared that reason, his own observations, and Moses all tell us that the Flood had had only superficial effects: elevated strata had been tumbled down, river channels enlarged, large trees uprooted, and so on.[89]

In a fierce defense of Woodward, John Harris in 1697 attacked both L.P. and Robinson—he thought Robinson probably *was* L.P.—arguing, inter alia, for the place of miracles in natural philosophy. To critics who doubted that water could dissolve rocks, Harris replied that Woodward had never claimed for water any such natural power; instead, he had merely shown de facto that dissolution had occurred because fossils had become enclosed in rocks. To critics who noted that the theory required sediments to settle in order of their specific gravities, Harris replied that such would have happened naturally "had God Almighty thought fit."[90]

---

[87] Ibid., 4–7, 9–10, 40–47.

[88] Ibid., 8–9, 28–35, 45. For a discussion of this tract and its authorship (including the candidacy of Thomas Robinson, treated in the next paragraph), see Rappaport 1997.

[89] Robinson 1696, 87–90. Robinson caught up belatedly with some recent publications and mentions them, including L.P.'s tract, in a prefatory "Advertisement with Additional Remarks." His views were further developed in a volume published in 1709; main points here were summarized in *JS* (7 April 1710), 209–12, and the review ends with a very Gallic shrug about the oddities of the book. Robinson prided himself on his observations, notably in mines, but his idiosyncratic interpretations of such information may be seen, for example, in Robinson 1709, 30–31, 77–79.

[90] Harris 1697, 20, 26–27.

This method of argument came under severe scrutiny later in 1697 when Dr. John Arbuthnot produced a tract claiming in essence that Woodward had "explained" nothing. How had water from a central abyss been brought to the earth's surface? Why had the supposedly dissolved crust not descended into and filled up the abyss? Was there sufficient water for the Flood? Such questions prompted Arbuthnot to conclude that the several geological changes described by Woodward "appear to be all of them above the Power, and contrary to the Laws of Nature, and consequently exclude the Philosophy of Second Causes." Not only had Woodward produced a tissue of miracles, but he had also had the temerity to claim for his theory "an absolute and demonstrative Certainty."[91] One might even compare with Woodward the modesty of Nicolaus Steno, who had never made aggressive claims to certainty and whose analysis of the formation of strata was "more conformable to the known Laws of Nature."[92]

From the high ground of methodology, Arbuthnot descended to the most serious of empirical issues: specific gravity. Woodward, he argued, had misunderstood the physics of falling bodies. An oyster shell, for example, will descend faster in water than the dissolved particles of even a heavy metal, so that shells "could never be buried in Matter of the same Specifick Gravity with themselves." To confirm this analysis, Arbuthnot cited the well-known *Geographia generalis* of Bernhard Varenius for a long list of strata in the neighborhood of Amsterdam—this list was standard fare to naturalists both before and after 1695—which clearly contradicted the Woodwardian pattern of deposition. To be sure, Arbuthnot concluded, strata do appear to be "the Sediment of a Fluid," but laid down "by little and little, and at different times."[93]

British criticism did not end with Arbuthnot, although decades later the learned Thomas Birch seems to have considered Arbuthnot the most cogent of the critics.[94] To some contemporaries, including both Ray and

---

[91] Arbuthnot 1697, 8, 17–8, 44. Arbuthnot nowhere refers explicitly to Harris. See Beattie 1935, 194.

[92] Arbuthnot 1697, 18; see also 33–62 for a long comparison with Steno. Appended to Arbuthnot is a "vindication" of Scilla's 1670 treatise on fossils, by William Wotton. The impression conveyed by Arbuthnot and Wotton is that Woodward had borrowed, without acknowledgment, from both Steno and Scilla, and had made poor use of his borrowings. See Levine 1977, 40.

[93] Arbuthnot 1697, 21–24. The Varenius list, first published in 1650, was available in the many editions of this classic text. As a small sample of those who cite it: Birch 1756, 1:265; Ray 1673, 7–8; Hartsoeker 1706, 29–30; Leibniz [1749] 1993, chap. 48.

[94] Birch in Bayle [1697] 1734, 10:192–97. Conforming to the pattern of Bayle, the article is a biography, and details, with ample quotation, are in the long footnotes. Writing in 1739, Birch admits his debt to Ward (1740, 283–301), then in press; but Birch singles out Arbuthnot who is one of the many critics discussed by Ward.

J. J. Baier in Germany, Edward Lhwyd's objections seemed decisive, even if Baier continued to hope that he might find a way to salvage diluvialism. Lhwyd published in 1699 a long letter to Ray rehearsing some dozen points, including one that Ray himself had long since pondered: would not floodwaters merely scatter organisms on the earth's surface? Why, if fossils are diluvial remains, do we find so few nonmarine specimens? Why, if marine shells all date from the Flood, do we find some of them in a crystalline state and others not so transformed? Why are fossil shells in particular found in greater abundance than the living populations in modern seas? These and other questions Lhwyd did not pretend to have originated, but Ray considered his presentation so impressive and so expert that he induced Lhwyd to translate the letter into English for reprinting in one of Ray's own works.[95]

By 1705, then, Charles King had no difficulty in finding and summarizing an array of objections to Woodward's theory. He nonetheless considered the theory worth detailed examination because it seemed to be the only one to make a strong case for both the origins and the deposition of fossil shells. If Woodward failed to convince, then those who thought such shells to be the remains of organisms would have to look elsewhere for an explanatory theory.[96] In this prediction, King was both right and wrong: many naturalists did look elsewhere, but diluvial theory also survived.

On the Continent, as in Britain, different naturalists seized on aspects of diluvialism that struck them as peculiarly vulnerable. Not especially analytical, Nicolaas Hartsoeker's popular textbooks repeatedly cited the Varenius list of strata that Arbuthnot had used so effectively. By no means denying that the Flood had occurred, Hartsoeker did deny the Woodwardian pattern of deposition, and he concluded that sediments had been accumulated "during a very long time."[97] Luigi Ferdinando Marsili chose a completely different approach, stemming in part from his own research. In his widely known *Histoire physique de la mer* (1725) and in other writings, Marsili argued that the earth's basic structure had not changed since the Creation. Repeatedly he contrasted with the massive, unchanging, "essential" features of the earth (mountains, sea basins) what he called the "accidental" and relatively thin accumulations "of gravels, sand, testaceans, and many other diverse bodies" to be found in low-lying areas, chiefly on coasts

[95] I have used Lhwyd's English version in Beringer 1963, 142–53. See also Ray to Lhwyd, 13 February 1703, in Ray 1928, 284, and Baier 1708, 68.
[96] King 1705, 16.
[97] Hartsoeker 1706, 28–30; 1710, 169, 177–78.

and on the sea floor.[98] There was no place for the Flood in this general scheme. On one occasion, at the urging of Vallisneri, Marsili did enumerate his objections to diluvial geology. Included in this list are some of the familiar problems: could water dissolve the earth's crust? would not the retreat of the waters leave behind confused assemblages rather than orderly strata? He could only conclude that marine sediments were the work of the sea "not only at the time of the Flood, but since then as well."[99]

Among discussions of diluvialism in Louis Bourguet's correspondence, more than one friend would express doubts. Genevan Jean Jallabert in 1738 wrote: "I confess to you that I have always had trouble understanding how Woodward's system could suffice." In particular, how could a chaotic diluvial fluid assemble and deposit in specific locations groups of marine creatures of the same species and originating in foreign climates? Such deposits, often in parallel strata, surely looked as if "waves (*flots*) had laid them successively and gradually one atop the other."[100] To Réaumur in Paris, "the dissolution [of the earth's crust] occurring because of the Flood, as Woodward conceived it, has always seemed to me a very bad explanation, and not at all scientific." If Bourguet's own version of this process seemed a bit better, Réaumur still protested: "Stalactites clearly prove that certain rocks can be dissolved by water, which apparently cannot have much effect on flints. But the Flood would have had to last far too long for water to have dissolved [even] the softest rocks."[101]

Considerably more analytical and greater in range were the criticisms of Antonio Vallisneri, who for many years, in publications and in correspondence, attempted to undermine diluvialism. As early as 1701, Vallisneri began a long and intimate correspondence with J. J. Scheuchzer. Rightly regarding the Swiss as the more experienced naturalist, Vallisneri himself had barely begun to question diluvialism before he read Scheuchzer's Latin translation of Woodward. That same summer, 1704, he explored the Garfagnana region, writing a long description to be sent to the Royal Society of which he was a new Fellow. By the first months of 1705, he was describing fossils not as diluvial but as "antediluvial," a view he eventually tried to explain to Scheuchzer. The earth, he argued, is older than commonly thought, and successive fossiliferous strata ought to be attributed to

---

[98] The quotation is from Marsili 1711, 24. See also Marsili 1725, 14, 38–39.
[99] Marsili to Vallisneri, 24 October 1725, in Vallisneri 1728, 144–49, quoted passage on 144. This text is in the appendices intended as replies to critics of Vallisneri's first ed. (1721).
[100] Jallabert to Bourguet, March 1738, in Neuchâtel, MS 1273. See Bork 1974, 56.
[101] Réaumur to Bourguet, 25 March 1741 and 29 July 1741, in Neuchâtel, MS 1278, fols. 48–51. Persistent questions about solubility, plaguing Woodward's disciples and critics alike, would not be tackled systematically until after about 1750; see Laudan 1987, 61–63.

the era before the Flood when natural marine movements had deposited such sediments.[102] To this suggestion Scheuchzer responded by publishing and sending to Vallisneri more of his own books on the effects of the Flood. The correspondent who eventually proved more willing to argue about these matters was Louis Bourguet.

As his first published statement of his views, Vallisneri in 1708 wrote a review of the Scheuchzer translation for *La Galleria di Minerva*. Moving seriatim through the six parts of Woodward's treatise, he quickly dismissed certain topics, such as the several upheavals taking place at the conclusion of the Flood. To Vallisneri, "many are more imagined than proven." He preferred instead to concentrate on two issues: the supposed ubiquity of fossiliferous sediments, and naturalistic explanations. As a "really weighty argument," he cited his own observations in the Apennines where "one finds marine products only up to a fixed altitude in these mountains, and always in those parts that now still face the sea." Such strata clearly had been laid down when "the sea naturally occupied those places, and then retreated little by little, as we see in many places in Italy." Moving on to subjects which offended his commitment to naturalism, Vallisneri declared:

> When we can explain things without recourse to the omnipotent hand of God, this is more philosophical and does not diminish but increases the glory of the great Master who constructed the great machine in such fashion that what often appears miraculous to us really is subject to laws, even if these may be beyond our understanding.[103]

These sentiments are hardly surprising in a man who, in his youth, had translated some of Descartes and who came to admire Malebranche.[104] Similar issues and sentiments would later permeate his correspondence with Bourguet and the geological portion of his *De' corpi marini* (1721).

At least some of Vallisneri's writings provoked debate in the Bolognese Academy, which had been receiving books and letters from the Scheuchzer brothers. Among those few members concerned with natural history, diluvialism appealed to two, Giuseppe Monti and his son, who defended it for years, and to two others who used the Flood on occasion.[105] If the

---

[102] See letters to Scheuchzer, 1703–7, and one to Marsili, 20 February 1705, in Vallisneri 1991, 253, 260, 267, 291 (item no. 4 of list), 296–97, 370.

[103] *Galleria* 6 (1708), 17. The review occupies only one page, double columns.

[104] Vallisneri 1991, 17, 50, 184. Cf. Fontenelle's comment that unusual natural phenomena may be "extraordinary" but not "irregular," *HARS* [1703] 1705, 28.

[105] Sarti 1993, 451, insists that all the Bolognese academicians, "without exception," were Woodwardians. Compare the scholarly assessments by Tega 1986, 2:43–58; Cavazza 1981, 907–8; Soppelsa in Rossetti 1988, 349–50.

Academy's secretary is to be believed, Vallisneri's "academic lecture" (1715) on the origin of springs aroused interest in part because it challenged diluvialism. Here Vallisneri had offered evidence that subterranean waters could not rise to all altitudes above the earth's surface, the contrary position being an integral part of Woodwardianism. Not until after mid-century, however, did one of the academicians explicitly reject diluvialism, although the *Commentarii* for earlier years reveal that doubts and questions had occasionally been expressed.[106]

In France, local researches—on the Mediterranean and Norman coasts, and in Touraine—were regularly interpreted in nondiluvial terms either by the naturalists themselves or by Fontenelle. For about two decades after 1700, the Paris Academy received a number of pertinent communications from the Scheuchzer brothers, Leibniz, and Marsili, and from a few of its own resident members. Expressing deference toward the learned Scheuchzers, Fontenelle repeatedly queried diluvial geology in scattered phrases, allusions, and proposals. On one occasion, he alluded briefly to the problem of specific gravities, and he would elsewhere suggest the need for the study of the motion of bodies in fluids. Doffing his hat to the Flood, he declared that diluvialism did not exclude the possibility of local floods; elsewhere, he proposed using the Flood only for local deposits for which no other explanation could be found, and, on yet another occasion, he called the Flood a "hypothesis" that was "not absolutely required."[107] Eventually, in 1720, he welcomed Réaumur's study of the faluns of Touraine as clear evidence of successive, nondiluvial deposition. In the south of France, physician Pierre Barrère, relying heavily on the publications of the Academy, argued that the Flood had been too turbulent to produce an orderly succession of strata. Another southerner, engineer Henri Gautier, was willing to calculate the amount of water needed for the Flood, but he could find little use for that event in his analysis of how strata had been deposited and altered in the course of time.[108] Insofar as these and other

---

[106] For diluvial memoirs by the Bolognese, see the examples in Tega 1986, 1:67–70, 119; some queries are raised in ibid., 1:78, 271. Tega gives references to the pertinent pages in the *Commentarii*. Zanotti's own doubts are alluded to in the *Commentarii*, 2 (1745), pt. 1, 92–93. Vallisneri's *Lezione accademica* (1715, with later reprints and editions) was first delivered as a speech to the Paduan Accademia dei Ricovrati; further discussion below, in Chapter 6.

[107] The five examples here cited are in *HARS* [1708] 1709, 33; [1707] 1708, 6; [1708] 1709, 31, 34; [1710] 1712, 21. Discussion of the Academy as having something like a theory of the earth, and Fontenelle's role therein, will be found in Chapter 7.

[108] For Réaumur, *HARS* [1720] 1722, 5–9. Barrère 1746, 47–48. Gautier's treatise of 1721 has been analyzed in detail by Ellenberger 1975. This is probably the same Gautier who wrote to [Antoine de Jussieu?], 8 May 1722, discussing a fossil find and wanting to know what the Academy thought of the treatise he had sent to his correspondent; MHN, MS 3501, fol. 3. See also Ellenberger 1980.

writers treated the Flood at all, it became one of a series of events. In contrast to the Woodwardian claim that the earth's landforms had not changed appreciably since the Flood, local studies showed that sedimentation, erosion, and changing shorelines all argued for a more continuous history of the earth's crust.

In a different vein, some few critics declared or implied that diluvialism had failed and that one should abandon efforts to integrate a miracle into the earth's history. Unlike the anti-Burnet writers who had insisted that the Flood was miraculous in its cause but natural in its effects, these later critics considered the Flood to be wholly miraculous and thus completely outside the realm of philosophy. This position would ultimately be adopted by Buffon (1749), but it can be found earlier in the circle of Vallisneri's friends and admirers. And, indeed, in Vallisneri's own geological treatise of 1721. One friend, Antonio Conti, cleric and cosmopolitan, would exchange jokes with Vallisneri about the impossibilities of a literal interpretation of Noah's Ark, and he elsewhere criticized at length the use of mystery—in particular, Neoplatonic "plastic powers"—in natural philosophy. Another, Scipione Maffei, may well have been influenced by Vallisneri's critique of diluvialism; his published comments on the subject, however, date from late in his career when he expressed admiration of Anton Lazzaro Moro's nondiluvial geology. Moro himself, never acquainted with Vallisneri but a student of his writings, firmly excluded the miracle of the Flood from his own geological theory.[109]

In their desire to be impartial, editors of book-review journals may well have been hesitant to allow too much judgment to intrude on a matter as delicate as the Flood. Vallisneri's hostile review of Woodward thus seems to have been unusual. In the *Journal des savants*, by contrast, the issues dividing Woodward and one of his critics were fairly laid out and judgment explicitly left to the readers. A few years later, the *Journal* would use a single word, "system," to describe Woodwardianism; since this was commonly not a compliment, one may infer a degree of uneasiness and even disapproval on the part of the reviewer.[110] By 1727, even the staid *Acta eruditorum* displayed a little impatience with Woodward's efforts to explain how rocks could be dissolved while organisms remained undamaged.[111]

More explicitly judgmental—and over a longer period of time—than any other journal, the *Mémoires de Trévoux* reviewed many diluvial and antidiluvial treatises and also published a small number of original articles on

[109] Conti 1972, 434–35, and in *GLI* 12 (1712), 239–330. Maffei 1747 and Moro 1740 will be discussed in Chapter 7.
[110] *JS*, Amsterdam 12mo (April 1715), 383–400, review of Woodward, *Naturalis historia telluris* (1714). *JS* (13 April 1722), 239, at end of review of Melle 1718.
[111] *Acta* (July 1727), 315, concluding the review of Woodward 1726.

the subject. How the journal handled this matter reveals not only disarray among naturalists but also startling vacillations among the editors and contributors.

In 1706 the Trévoux journalists encountered a new edition of Abraham van der Myle's examination of postdiluvial migrations; so problematical were animal migrations to the Americas that the author concluded the Flood had not been universal, affecting only that part of the earth then inhabited by humans. Having considered diverse interpretations of various biblical passages, Myle reached a conclusion acceptable to the journalists: "This reasoning is not the least solid of his book." Later that year, the Jesuit editor apparently remembered that the Church had already, in 1686, condemned a work arguing for the non-universality of the Flood, and advocates of this position were now said to contradict physics, the Bible, and ancient traditions.[112]

Universality now being a closed issue, later numbers of the periodical began to take note of explications of the Flood. In 1713, the journalists, a little belatedly, turned to two of Scheuchzer's books, duly noting the author's wish to combat impiety and admiring his "singular observations" and "solid reflections." This Woodwardianism was more acceptable than Oratorian Bernard Lamy's adoption of the Burnetian position that the Flood had been produced by natural causes, "merely foreseen and arranged by Providence." To the reviewer, Lamy's was a deplorable attempt to abolish a biblical mystery. Reviewers did not become wholeheartedly Woodwardian, however, one of them chiding L. F. Marsili for calling the Woodwardian dissolution of the earth's crust *une sottise:* "Not every mistake is foolishness." When the journal took note of a French translation of Derham, a reviewer again reported one of Woodward's errors, namely, the claim that sedimentary strata had been deposited according to the laws of gravity.[113]

That there was no consensus among the editors is further suggested by the publication in the 1720s of two original articles, both by Jesuits. Louis-Bertrand Castel, a regular contributor to the journal, in 1722 advocated a return to an older theory of the earth as a sort of organism with an internal circulatory system. Such a theory, he argued, would explain the earth's structure and the transport of fossils better than diluvialism. A few years

---

[112] *Trévoux* (January 1706), 103; (October 1706), 1685–96. Myle's name was commonly Latinized as Milius; in modern works, he is sometimes given as Mijl.

[113] Ibid. (January 1713), 66–76; the works by Scheuchzer had been published in 1708 and 1709. Ibid. (July 1721), 1221–22, for Lamy. Ibid. (February 1727), 233–34, for Marsili. Ibid. (February 1728), 324–25 for Derham. The article "fossile" in the *Dictionnaire de Trévoux* (1721) gives considerable attention to Woodward's explication of the Flood.

later, Etienne-Augustin Souciet would defend diluvialism at some length. To interpret fossils as relics of the Flood confounded unbelief by providing concrete proof of "the most terrible and incredible event ever to take place." The event itself, Souciet implied, seemed incredible because it was not natural in its cause, but had had indubitable natural effects.[114]

Probably the first sustained attack on diluvial geology encountered by the journal was Vallisneri's, and the reviewer merely expressed irritation that Vallisneri had provided no alternative to the theories he had criticized. When the journal soon afterward tackled the belated French translation of Woodward, the reviewer at last understood what Vallisneri had had in mind: Woodward's theory was bad science and thus a bad support for the truths of faith—which had no need of such support in any case.[115] For the next decade or so, the journal seems to have adopted a consistent position, rejecting naturalistic explications and giving preference to abandoning efforts to find the causes or effects of a miracle. In reviewing a new edition of Bourguet, for example, the journal insisted: "The Flood should not and cannot be explained naturalistically." And it belatedly realized that Woodward himself had not been wholly naturalistic but had properly, "like a Christian philosopher," come to a halt at some delicate points.[116] Three years later, in 1743, Dezallier's diluvialism again provoked a warning: diluvial geology was pious and even plausible to some degree, "but one must be careful not to make dependent on [such a theory] matters of faith and belief in the supernatural."[117]

The publication of *Telliamed* in 1748 sent the journal into momentary retreat, alarmed by a book that seemed foolish and incomprehensible, but nonetheless dangerous. In reaction, the reviewer pronounced diluvialism to be well established and supported not only by respected savants but also "by the acknowledged principles of history, science, and revelation." Within a few months, however, equilibrium was restored. In a review of Buffon's theory of the earth, the journal happily discovered that Buffon considered the Flood a miracle and that he had criticized William Whiston for apply-

[114] Castel in *Trévoux* (June 1722), 1089–1102; this article begins with a critique of the non-diluvial memoir by Jussieu, in *MARS* [1718] 1719, 287–97. Castel's career is examined by Schier 1941. Quasi-animistic theories like his can still be found in this period, e.g., Robinson 1694, 1696, 1709, and Hobbs [ca. 1715] 1981. Souciet in *Trévoux* (March 1729), 480–81; this is in the second of a three-part article, the others published in February and April.

[115] *Trévoux* (December 1734), 2131; since this review dealt with Vallisneri's collected works (1733), little space was given to geology. (February 1736), esp. 246, 252, 254–55, in review of Woodward 1735.

[116] *Trévoux* (August 1740), 1660, 1662. The earlier review of Bourguet 1729 (October 1730), 1739–50, raised no such questions.

[117] Ibid. (March 1743), 433, in review of Dezallier 1742.

ing science and human reason to religious verities. Quoting Buffon's critique of Whiston, the reviewer declared that "these words [are] worthy of being inscribed in letters of gold."[118]

Had the journalists stopped there, one might conclude that *Trévoux* had at last found a satisfactory combination of fideism and the study of nature. Indeed, that position was later maintained when, in 1760, a reviewer regretted that J. G. Lehmann had tried to detect the natural causes that God had used to produce the Flood. But the Jesuit journalists were also encountering too much modern philosophy, so that the reviewer, in defense of the Bible, applauded Lehmann's integration of the Flood into the earth's history.[119]

The *Trévoux* reviews illustrate well the problems and choices confronting naturalists during the first half of the eighteenth century. The Flood was undeniable history, but its causes and effects had provoked literature ranging from the pious to the seemingly atheistic. Despite very substantial criticism, diluvialism did not vanish, although its advocates quarreled among themselves. More significantly, the critics, too, generally did not deny that the Flood had occurred. At stake, then, was not only the geological information and its interpretation, but also the status of Genesis as a work of history. For all its peculiarities, the story of the Flood possessed the attributes of historical narrative: it was factual, detailed, and precise, from the dimensions of the Ark to the number of days of rain. Unusual, therefore, were the Vallisneris and the Buffons, who relegated the Flood entirely to the realm of miracle, ignoring a "fact" of history even while they claimed to be reconstructing the history of the earth.

---

[118] Ibid. (April 1749), 635, 642, in review of Maillet 1748. (October 1749), 2232, 2244–45, in review of Buffon 1749.

[119] Ibid. (July 1760), 1696–98, in review of Lehmann 1759. For an excellent examination of how *Trévoux* dealt with dangerous books, mainly in the period 1745–62, see Pappas 1957; for how the journal dealt with Rousseau in particular, see Garagnon 1976.

CHAPTER 6

Alternatives to Diluvialism: Some Ingredients

Critics of diluvialism usually acknowledged that the Flood had occurred, there being historical warrant in texts both sacred and profane. They reached no consensus, however, about a geological role for the disaster or about whether one should even seek such a role. Nor did most critics propose an alternative theory of the earth. Rather, their chief point of agreement may be characterized as an emphasis on *sequence,* the earth's sedimentary strata having been laid down at more than one time and in more than one fluid.

If many commented on alluvial deposits forming plains and deltas, the ubiquity of marine fossils necessarily focused attention more often on the history of the sea: how had the sea successively covered and retreated from the continents? Without acceptable mechanisms to elevate the land, the sea obviously seeming more readily mobile, naturalists nonetheless could do little with this subject except to insist that marine transgressions had taken place. Behind the latter assertion lay not only the fossil evidence but also the observations recorded as early as 1650 in the often reprinted *Geographia* of Varenius. Here the author listed and described the strata found in the digging of a 232-foot well in Amsterdam. Well known in the decades before the publication of Woodward's *Essay,* this list would then provide ammunition for critics like Arbuthnot and Hartsoeker, even while

Woodward's disciples produced their own lists in efforts to validate his theory.[1]

The evidence for marine sedimentation proved hard to reconcile with another classic text, Robert Boyle's pamphlet about the sea floor, published in 1671 in both English and Latin. To the many readers of this tract, one baffling conclusion seemed obvious: the sea floor was so tranquil that any substantial accumulation of sediment could not take place. Woodward himself had noted the problem, arguing that only the Flood could have disturbed the sea bed sufficiently to wrest organisms from these depths and eventually deposit them at elevations visible to man.[2] Boyle's tract in hand, naturalists found themselves with a limited array of options. They could posit local floods or marine currents turbulent enough to deposit sediments. Or they could extend their knowledge of changing shorelines to argue that marine bays or gulfs had once existed much farther inland. Or they could accept such lists as Varenius's without providing any causal explanation for marine movements. In fact, as will become apparent in the next chapter, naturalists could not agree about what the sea was doing in their own day: was its level rising or falling?

Just as they disagreed about the sea, so, too, did naturalists fail to persuade each other about possible mechanisms for the elevation of land. Clear to everyone was that erosion had been destroying the earth's relief for a long time, and some undertook special studies of how rivers in particular had been altering the landscape. The study of rivers, especially by Italian writers, resulted in enhanced understanding of erosion, the building up of alluvial plains, and the formation of at least some valleys. But such analyses also led to the conclusion that the earth's topography was being progressively leveled and was here and there being further excavated. No answer was being found to the question that would eventually preoccupy James Hutton: was there any natural way for the earth to repair or restore what was being washed into the rivers and seas?[3] If some thought the earth's "decay" a proper prelude to the second coming of Christ, others concen-

----

[1] Varenius 1736, bk. 1, chap. 7, prop. 7. The text is the same in the various editions, but that of 1736 adds a pro-Woodward footnote to the above proposition. For some of those who cited Varenius, see Chapter 5, n. 93.

[2] In Boyle 1671. Woodward 1695, 26–27; 1726, 2:13. Among the many references to Boyle's tract, see Ray 1693, 28; Marsili 1711, 28 and 1725, 1, 48; Moro 1740, 179. For other efforts to study the sea floor, see Deacon 1965.

[3] Davies 1966 and 1969. The study of rivers was of especial concern in Italy for the better control of waterways, the building of canals, and the maintenance of ports. Many of these writings are assembled in *Raccolta* 1723; cf. Ciriacono in *Storia* (1986), 5/II, pp. 347–78. Manfredi 1732 summarized coastal studies in particular.

trated on whether volcanoes and earthquakes offered clues to how tracts of land might be elevated. Were such phenomena common in all times and places, or were they merely local "accidents" peculiar to certain terrains? Indeed, the associations of volcanoes and earthquakes with heat prompted naturalists to ask such basic questions as whether the earth possessed its own heat, whether heat was "central" or superficial to the earth, and whether "heat" should be equated with common fire.

Geological research could be approached in several ways, by the use of observation, experiment, and testimony. Observation supplied preliminary information without which further study would have been impossible. For at least some phenomena, causal inferences remained inconclusive until or unless they could be tested in the laboratory. As some naturalists found, however, laboratory experiments revealed what nature *could have done* but not necessarily what nature *actually had done,* it being hard to know if the experimenter had accurately replicated natural processes. Human witnesses might here be of some help, but all too often they had recorded particular events with insufficiently detailed attention to physical conditions.

Here as in earlier chapters, the use of human witnesses will again suggest to the reader a short geological timescale, and it is true that the critics of diluvialism generally did not extend—and saw little reason to extend—the commonly accepted scale. Indeed, observation and experiment usually reinforced this tradition. Experiments are "timeless," condensing into laboratory time what may have taken nature much longer; naturalists did recognize this process of contraction, but they usually assumed that some small extension—a year, a decade, a century—would be sufficient for nature to accomplish what they themselves tried to achieve in their laboratories. Similarly, the most readily observable geological processes (earthquakes, erosion, floods) were rapid; so, too, were changing shorelines, documented by ancient, medieval, and recent geographers. If modern historians have insisted that "catastrophic" geological mechanisms were required by a short timescale, the argument here will be quite different: rapid changes were observed and considered compatible with such a scale.[4] Some naturalists did exhibit that sense of time constraints so much exaggerated by modern writers, but, on the whole, most thought the biblical scale *to be long.*

---

[4] The conventional interpretation may be found in Toulmin and Goodfield 1965, 90, Davies 1969, 44–45, and Gohau 1990, 112–13. By contrast, see Piggott 1989, 43, on biblical time *seeming long.* The most analytical study of these issues is by Dean 1981.

SOME GEOLOGICAL EXPERIMENTS

Experiments did not figure prominently in early geological writings, although the ideal of replication was invoked from time to time. The contrast between ideal and reality may be illustrated by the propaganda of Fontenelle on the one hand and by the remarks and practices of several naturalists on the other.

In more than one of his influential texts, Fontenelle insisted that the laboratory allows us to imitate and thus to understand nature. Results might now and then raise doubts—did "our artificial Chemistry" accurately reproduce digestive processes taking place in the stomach?—but the experimental route remained the most promising recourse. Not being naive in these matters, Fontenelle added that the process of experiment was more revealing when one could manipulate known causes for the production of familiar effects.[5] The latter viewpoint had already been briefly queried, as early as 1667, by Nicolaus Steno when he remarked on the success of his teacher Borrichius in dissolving "a very hard limestone in ordinary water; why then should not we grant to Nature what we cannot refuse to art?"[6] Steno's rhetorical question hardly disguises the problem: did art really succeed in showing us not what nature can do but what she actually does? A few years later, a reviewer for the *Journal des savants* would be less cautious than Steno in pointing out the inadequacies of J. J. Becher's attempt to reconstruct the internal structure of the earth. In addition to unexceptionable processes like distillation and fermentation, Becher had used a tissue of suppositions, ranging from central caverns to a property called "gravity," to explain how water could rise to the earth's surface and then retreat again into its interior. More succinctly, Vallisneri would later ask whether experimenters could be sure they had reproduced the conditions and causes working in the unobservable bowels of the earth.[7]

Perhaps most illuminating are Robert Hooke's texts, in which he records

---

[5] The several texts by Fontenelle include *HARS* [1700] 1703, 51; eulogies of Homberg, 1715, and Poli, 1714, in Fontenelle 1709, 1720, 2:191, 155; and letter quoted in Marsili 1711, 22. Texts on digestion are in *HARS depuis 1666*, 1 (1733), 167, 253; cf. Salomon-Bayet 1978, 123–24.

[6] Steno [1667] 1958, Conj. 4; also, Conj. 5.

[7] Review of Becher in *JS* (3 February 1681), 29–32. Vallisneri, 1715, in M. Baldini 1975, 59–60. Although his text was hardly known outside of Germany before 1750, Henckel's difficulties in determining the origin of pebbles and gravels are worth noting. The translation in Adams [1938] 1954, 129, is not entirely accurate; see Henckel 1744, 341, and the further remark by Zimmermann, in Henckel 1760, 2:411: "When all the conditions are known, one can draw conclusions by analogy, but when they are lacking, this is a weak way of arguing."

experiments but does not seem fully to have understood or recognized their inconclusive nature. In one series Hooke soaked sticks in water with "petrifying" properties; after varying the time allowed, Hooke admitted that he had "never yet been able to petrify a Stick throughout," remarking that nature had more time available than did the human experimenter. If he thought these results promising, Hooke also acknowledged that nature might have used many means to petrify organic material, namely, by "a very long continuation of the Bodies under a great degree of Cold and Compression," by subterranean heat, by the mere passage of time (bodies may have grown harder with time"), by the encrusting action of water, or by water itself, within bodies, turning to rock.[8] Nowhere does Hooke suggest further testing, and he obviously selected for his experiments the only causal agent available in the laboratory and usable in a limited time. Nor, presumably, did Hooke imagine his list of processes necessarily operating concurrently, as is evident from his examples of heat and cold. One may suppose that he believed the different agencies might come into play depending on the conditions surrounding particular buried bodies. Elsewhere, Hooke did realize that at least some of his experiments could provide only rather dubious analogies to operations performed by nature. In the *Micrographia* he suggested that lunar topography might be reproduced by dropping a heavy body into a soft mixture of pipe clay and water, the result being an indentation with earthworks thrown up around it. But he rejected this bombardment model, noting that the moon might not be made of pipe clay. Instead, the more likely physical cause of all craters seemed to him analogous to what he had observed "in a pot of boyling Alabaster," where heat had caused bubbles to form on the surface, and where gradual cooling had left "craters" when the bubbles burst. Here he suggested, as "not improbable," that the earth and the moon, being so similar in topography, resemble one another in substance and in geological operations.[9]

Among thoughtful analysts, Locke and Leibniz both admitted that analogical reasoning "is the great rule of probability." Or, as Locke put it, one must at times employ "a wary Reasoning from Analogy." Leibniz himself occasionally used this method imaginatively when, for example, he proposed an experiment to show how moulds or impressions of fossils might

[8] The experiments with sticks, dated ca. 1668, are in Hooke 1705, 294; the rest of this discussion occurs on pp. 290, 293–95. In connection with "transmuting" water into "earth," he cites Boyle's *Origin of Forms*, 1666. Another experiment to petrify wood, starting with immersing specimens in boiling hops, is recounted by J. J. Baier, according to *HWL* (February 1709), 33–34.

[9] Hooke [1665] 1961, Obs. 60. See also Davies 1969, 42.

have been left in the rocks.[10] A few examples, however, will suffice to show that the difficulties lay in selecting analogies.

In trying to explain the formation of rocks, naturalists adopted one or more of the several available analogies or models: chemical crystallization, the formation of stalactites, the growth of plants. Some wrote of the mechanical agglomeration of particles, and some argued for the existence of a hypothetical "petrifying juice" that could penetrate the "pores" of both organisms and sediments. For particular rocks, pebbles and gravels, some rather warily proposed their origin as debris detached mechanically from larger masses; it proved easier to explain rounded shapes as formed by attrition in running water, but did the same explanation apply to deposits deep in the earth's crust? or to rounded specimens not yet completely indurated?[11]

After years of struggling with such problems, German chemist J. F. Henckel brought together his observations, experiments, and reflections in his *Idea generalis de lapidum origine* (1734), a work severely handled the next year in the *Bibliothèque germanique*. The reviewer noted that Henckel detected five ways in which rocks could be formed: by congelation, coalescence, germination, crystallization, and petrification.[12] He had failed to explain these clearly, in part because the subject itself was so difficult, but also because his method and style were faulty. In the reviewer's eyes, Henckel should have sought help in the publications of the Royal Society of London and the Academy of Sciences in Paris. As the preceding discussion suggests, however, the British and the French had as much trouble as Henckel himself. Later in the eighteenth century, naturalists would increasingly adopt chemical crystallization as the suitable model for the formation of certain large classes of rocks, but such a consensus was not possible in earlier decades when crystallization had not yet been extensively examined and when naturalists were not yet sure that chemistry could be of more than minimal aid in their own studies.[13]

[10] Locke, *Essay*, IV, xvi, 12, and Leibniz 1981, IV, xvi, 12. See also Davillé 1909, 605, 610. Summary of Leibniz in *HARS* [1706] 1707, 9–11.

[11] Useful as a survey is Adams [1938] 1954, chap. 4. A range of examples might include also: Plot 1677, 33; Bourguet 1729, in *DSB* 15:54; Geoffroy in *HARS* [1716] 1718, 8–13; Marsili 1744, 3:11, 21; Dezallier 1742, 86–87. For pebbles and gravels in particular, summary of Saulmon in *HARS* [1707] 1708, 5–7; Fontenelle in *HARS* [1721] 1723, 15, compared with Réaumur in *MARS* [1721] 1723, 255–76; and a later effort to resolve questions raised by earlier writers, by Frisi 1774, bk. 1, chap. 2.

[12] *Bibliothèque germanique*, 33 (1735), 158–61. The terms used by Henckel are clarified in Laudan 1987, 31–32. I am unable to discover reviews of Henckel's writings in such publications as *JS* and *Trévoux*, and it is unlikely that his works had Europe-wide distribution before being translated into French after mid-century. His treatise of 1734 is mentioned in Chambers, *Suppl.* (1753), s.v. Lithogenesia.

[13] This argument is developed to some extent for France in Rappaport 1994.

Apart from petrifaction and the origin of rocks, where experiments were not numerous, the study of sedimentation was based wholly on field work and, in the Woodward controversy, on measurement of specific gravities. As a striking exception, Emanuel Swedenborg did produce an account of his own simple and elegant experiments on sedimentation. He first agitated in water a mixture of sand, fine scrapings of granite, and wood shavings, and found that they settled in layers according to weight. Then, using a turbid clayey solution, but allowing particles to settle before adding more of the mixture, he obtained clearly defined, superimposed layers. The former experiment, he concluded, shows how strata are sometimes formed; the latter represents sedimentation occurring successively, at different periods of the past.[14]

Elsewhere, Swedenborg attempted to test Robert Boyle's description of the sea floor as so tranquil that virtually no sediment could be accumulated there. Using a glass vessel containing water and powdery sediment, Swedenborg found that agitating the surface of the water produced ripple marks in the powder. If this miniature effect could not adequately reproduce conditions in the ocean, he argued that there the great weight of superincumbent water would add substantially to any force generated at the surface. At the bottom of the sea, therefore, sufficient force must be available even to shift great boulders.[15]

For all Swedenborg's ingenuity, his experiments attracted little attention outside Germany and Sweden; even there, reviewers were sometimes distracted by his rationalist presentations or by the curious phraseology that led some to conclude that he was a diluvial geologist.[16] Not only did subsequent writers continue to cite Boyle, but one may also wonder whether readers would have thought that experiments on sedimentation added materially to Steno's discussion of superposition or to Varenius's well-known list of strata. That field observation, not experiment, continued to play a decisive role in showing that sedimentation had occurred sequentially is also evident in the reception of Réaumur's memoir on the faluns of Touraine, published in the same year, 1722, as a Latin edition of Swedenborg. This memoir quickly became the *locus classicus* for writers concerned with sedimentation.

Réaumur's chief point was stated at the outset: these massive deposits of fossil shells would prove beyond doubt what naturalists had long known, namely, that at least parts of the present land masses had "for a long time"

---

[14] Swedenborg [1722] 1847, 18–20.

[15] Ibid., 8, 15, and 156–59. Focus here is on the principles of hydrostatics, not the experiment (15). Boyle's name is not mentioned.

[16] Reviews, published mainly in Leipzig, Stockholm, and Uppsala, are assembled in Acton 1929, 1930. For further discussion of Swedenborg, see Chapter 7.

been at the bottom of the sea. He went on to describe the deposits, indicating both their considerable distance from the seacoast and the density of the shell population when compared with sparser assemblages in modern sea beds. The latter observation led him to conclude that Touraine must once have been a gulf into which marine currents had swept creatures not already living there. When, how, and why the sea had retreated he could not say, but modern rates suggested to him that thirty or forty centuries would have sufficed for the sea to move from Touraine westward to its modern location.[17]

Fontenelle immediately pointed out that Réaumur's memoir undermined diluvialism, and this surely is why the memoir quickly attracted attention. To Louis Bourguet in Neuchâtel, Réaumur belonged to a Parisian school of nondiluvialists. To the younger Vallisneri, Réaumur was one of the several who had added evidence to his father's nondiluvialism. For a diluvialist like Dezallier, on the other hand, marine gulfs had to be rejected.[18] In the different atmosphere of the 1830s, Charles Lyell would select the Touraine site, much studied by that date, as one of those "containing monuments of the era [Miocene] under consideration."[19]

SUBTERRANEAN HEAT

When Huttonian theory aroused debate about the nature of subterranean heat, the subject had already had a long history. Evidence of various kinds had been assembled to show that heat did indeed exist within the earth, but naturalists found it less easy to answer other questions. Did heat extend into the very bowels of the earth? (Was there what was commonly called "central" heat?) If the earth's heat were the same as common fire, how could it subsist without air and without exhausting a fuel supply? Did there exist caverns of water within the earth either capable of exciting heat or, conversely, capable of extinguishing fire? Investigation of such topics invited and virtually required experimental and analogical methods, there being obvious limits to how far beneath the earth's surface the observer could penetrate.

---

[17] Réaumur in *MARS* [1720] 1722, 400–416. A large part of the text treats also how local residents quarry and use the rock.

[18] *HARS* [1720] 1722, esp. 8. Bourguet 1729, 179; Vallisneri 1733, 1:xix; Dezallier 1742, 158–59. Other references include Ehrhart in *PT* 41 (1740), 549–50, as well as several texts by Voltaire, dated 1746, 1767, and 1768, in Voltaire 1877, 23:223–24; 26:407; 27:150–54.

[19] Lyell 1830–33, 3:203.

In the considerable literature on volcanoes, much of it stimulated by eruptions of Vesuvius and Etna, naturalists were virtually unanimous in identifying what Nicolaus Steno called "the material cause" of heat: sulfur. In Italy G. A. Borelli's study of Etna and Gaspare Paragallo's of Vesuvius; in France discussions at the "conferences" of Renaudot, Rohault, and Denis, as well as papers by members of the Academy of Sciences; in England articles in the *Philosophical Transactions* and Hooke's lectures on earthquakes— the common theme is that volcanic eruptions entail the burning of sulfur compounds or "bitumens." For the many who had never seen a volcano, and for the many more who had never witnessed an eruption, the decisive element in eyewitness reports was the sulfureous odor accompanying eruptions. Such observations were confirmed again by accounts in 1707 and the next few years of the appearance of the new volcanic island near Santorini in the Aegean Sea.[20]

In the Aristotelian language of his youthful *disputatio* (1660), Steno found the "material" cause of subterranean heat easier to identify than the "efficient" cause. His chief topic being thermal springs, he argued that

> every fire needs fuel, and there scarcely seems to be any cause that could provide a limitless supply to these subterranean fires, since bitumen, sulphur, and other similar materials are consumed by burning.

If, he continued, one imagined the heat to be produced by the contact of water with quicklime, one still could not explain the "lasting heat," which implied an unlimited supply of quicklime.[21]

Steno's text may conveniently serve as a model for the content and arguments of subsequent treatises: the identification of sulfur and bitumen, the question of fuel (others would add the problem of an air supply), and the proposal of a chemical model for ignition. Borelli, for example, would repeat the reference to quicklime, and an anonymous correspondent of the Royal Society tentatively asked whether ignition might be explained

> by Fire struck out of falling and breaking stones, whose sparks meet with Nitro-sulphureous or other inflamable substances heap'd together in the bowels of the Earth.[22]

---

[20] See especially the detailed account of Santorini, with interpretative remarks, in *PT* 27 (1711), 354–75. See also *HARS* [1707] 1708, 11–12, and [1708] 1709, 23–26.

[21] Steno [1660] 1969, theses 8, 9, pp. 55–57. Thesis 5, p. 55, also considers kinetic heat.

[22] *PT* 4 (1669), 967–69, quotation from 969. Borelli's treatise of 1669, reviewed in *PT* 6 (1671), 2264–68, at p. 2267 for quicklime.

In 1674 chemist John Mayow would propose his own theory of ignition, taking as his starting point the need for both air and fuel. Air, he argued, "is possessed of a highly fermentative nature," so that the effervescence of "something aërial with saline-sulphureous particles" produces almost all forms of heat, including the heat of the blood. The particles of "something" in the air enter the earth's crust, carried there by rain, and they encounter the saline-sulfureous particles already present in minerals. The latter particles, constituting the fuel supply, never become exhausted because the earth itself contains a "mineral seed" that, "like vegetable seed," constantly grows and replenishes itself.[23]

Parts of Mayow's theory incorporated a fairly long chemical and alchemical tradition, based on analysis of the ingredients of gunpowder and readily available to contemporaries in Varenius's *Geographia*. The explosive noises of gunpowder, thunder, earthquakes, and volcanic eruptions might have seemed merely coincidental had it not been for the sulfur component of gunpowder and the sulfureous odor associated with thunderstorms and eruptions. For Mayow and others, the nitre or saltpetre in gunpowder was essential for ignition and thus must also be present in these other violent phenomena.[24] If not all scientists could adopt the details in a theory like Mayow's, the Royal Society on at least one occasion was presented (by Hooke?) with an experiment showing that the nitre—described as "air contracted"—in gunpowder could produce the fire and explosive force characteristic of volcanic eruptions.[25] At other times, however, different suggestions occurred to Fellows about how spontaneous combustion could begin. Hooke, for example, claimed that fire could start when rain fell on "a sort of coals, called brass lumps; as also on quick-lime lying against dealtimber," to which his colleagues responded by citing instances of the spontaneous combustion of "hay wet, green grass heaped up, malt, cotton-wool, rose-flower leaves," and other plant material.[26]

Among Continental writers, Nicolas Lemery, the Paris academician and author of a most successful textbook of chemistry, ultimately persuaded naturalists that he had come close to a convincing replication of volcanic eruptions. Like Mayow and others, Lemery actually believed that he had found related causal explanations for several phenomena—including

---

[23] Mayow [1674] 1926, 177–78, 180–82. Detailed discussion in Guerlac 1954.

[24] Varenius and others are examined by Guerlac 1954.

[25] Details are in a letter from Evelyn to Tenison, 15 October 1692, in Evelyn 1883, 3:325–30. Evelyn says the experiment was "long since made at Gresham College." Studies of gunpowder, in seventeenth-century London and Paris, are analyzed by A. R. Hall 1952, 62–64.

[26] Birch 1756, 3:435, dated 7 November 1678.

earthquakes, violent storms, lightning, and thunder—but his interpreter, Fontenelle, drew the reader's attention almost exclusively to Lemery's volcanic experiment. To a mixture of sulfur and iron filings, Lemery added sufficient water to produce the consistency of a paste; when the mixture had been buried for some hours under a thin layer of soil (with a few cracks provided for air), fermentation began, the soil swelled, hot vapors began to emerge, and flames eventually appeared.[27] Eighteenth-century geologists commonly cited Lemery's experiment as wholly compatible with their observations that volcanoes are associated with sulfureous odors. Furthermore, Lemery had made no problematical assumptions about subterranean conditions and fuels—although he did say that air was necessary to promote fermentation—so that these matters could still be discussed, even while geologists believed that the basic chemical reactions had been plausibly reproduced.[28]

At the same time, it also seemed clear that volcanoes provided no clue to the internal structure of the earth. For one thing, sulfur and bitumens were by no means near the earth's center, but only as deep as what would later be called the Coal Measures. For another, volcanoes were uncommon, and they were still judged to be so after 1750, when many hitherto unsuspected volcanic sites were being discovered in Europe. Often associated with volcanic heat in the minds of naturalists, earthquakes, too, were uncommon, even if quite a few came to European attention in the last decades of the seventeenth century.[29]

With volcanoes and earthquakes relegated to the realm of "accidents" or epiphenomena, naturalists lost two ways of explaining the elevation of tracts of land. Some did wonder whether such an explanation might be found in nonvolcanic heat, if it could be shown that such existed within the earth. More generally, however, interest in this subject stemmed from quite different considerations, as some naturalists wanted to test Cartesian cosmogony in which the earth had an igneous origin. Many saw the question of central heat as also linked to the ancient belief that the earth pos-

---

[27] *HARS* [1700] 1703, 51–52, and *MARS*, for the same year, pp. 101–10.

[28] Lemery's experiment was still being cited by Dezallier 1755, ix–x; Bertrand 1757, 10–11; and the *Encyclopaedia Britannica* 1771, s.v. Vulcano. Although not mentioned by name, Lemery was clearly the source for a passage in Query 31 of Newton's *Opticks* 1952, 379–80. In the aftermath of the Lisbon earthquake, Lemery's text was one of those included in Bevis 1757. For the common, well-established view of earthquakes before the 1750s, Aldridge 1950 is convenient.

[29] See, for example, Kenneth Taylor's articles on two vulcanologists, Desmarest and Dolomieu, in the *DSB*. As for the rarity of earthquakes, Hooke extracted all he could from reports dating chiefly from the 1690s, but see the comments by Kennedy and Sarjeant 1982, 214–16.

sessed internal reservoirs of water, elevated by heat to the earth's surface to form rivers and springs. If volcanic fire had an identifiable fuel, and if one could imagine ways for air to reach these relatively shallow depths, the same solutions were harder to apply to the earth's core. In debating central heat, therefore, chemists and physicists played an important role, for they had some familiarity with varieties of heat generated in laboratories.

In his popular *Traité de physique* (1671), Rohault added to the Cartesian hypothesis the comment that "the Heat which is in the Bowels of the Earth . . . is found by Experience to be the greater, the deeper we go." This did not seem obvious to critic Pierre Perrault, who complained that neither Descartes nor Rohault had explained the cause of such heat, "as if [they] would have us believe that the earth's interior is naturally hot, which no philosopher that I know of has yet claimed."[30]

Perrault seems not to have known one of Robert Boyle's tracts, "Of the Temperature Of the Subterraneal Regions" (1671, also published in Latin), the first methodical attempt to examine a subject that had long been part of the lore of miners. His dispatch of measuring instruments to a mine having "miscarried," Boyle relied on the testimony of informants who led him to conclude that there is within the earth, first, an area relatively warm, then an area (as in deep cellars) of cold, and finally a deepest zone generally warm and sometimes hot. That the latter phenomenon most puzzled Boyle is evident in his long series of conjectures, where he struggled with the possibility that mineral "exhalations" produce heat. In fact, Boyle could not decide whether he was dealing with heat or fire or both, and he could only say that we know too little about "these yet inpenetrated Bowells of the Earth" to judge whether these consist of "a continued solidity, or great Tracts of Fluid matter."[31]

In the next decades, several members of the Paris Academy tackled the nature of subterranean heat, without committing themselves about its location within the earth's crust. Moïse Charas, for example, found to his surprise that adding a little water to sulfuric acid produced enough heat to shatter the container; this induced him to suggest that comparable chemical reactions might explain the heat of thermal springs. A colleague would later undertake similar experiments, mixing heat-producing liquids. Since both Charas and Wilhelm Homberg used sulfur compounds, they pre-

---

[30] P. Perrault [1674] 1721, 2:783–84. Perrault's work, published anonymously, was reviewed in *PT* 10 (1675), 447–50. Rohault [1671] 1728, pt. 3, chap. 10, sec. 5. Rohault also treats (3.9.23) volcanic fire, mentioning sulfur and bitumens, but it is not clear if he regarded this as a manifestation of *central* fire.

[31] Boyle 1671, esp. 23–24, 28–29, 42–43 of this tract. It is not clear if Boyle thought that solidity or fluidity would be preferable as the better container for the "great store-houses of either actuall Fires, or places considerably Hot," mentioned on p. 28.

sumably had in mind the relatively superficial heat of volcanoes and, as Charas said, thermal springs.[32] Another academician, Guillaume Amontons, took a different route. Interested in the effects of the compression of air, Amontons wondered whether the earth's interior might not be filled with air so compressed that it formed an almost solid body, heavier than any bodies we are familiar with. If this were the case, then air, not Cartesian central fire, would have sufficient force to shatter and uplift the earth's crust.[33] Unlike Charas, Homberg, and Nicolas Lemery, Amontons was moving in the direction of analyzing central heat as a phenomenon different from fire.

In 1719 Amontons's problem would be taken up by a new young academician, J.-J. Dortous de Mairan, who pointed out that Amontons had not properly differentiated between "local and accidental causes"—such as the seasonal variations in heat detectable at the earth's surface, thanks to the changing position of the planet—and the "general cause" of the earth's heat. That the earth possesses internal heat Mairan did not doubt, but he confessed his failure to determine its cause, leaving the reader with four causal options: the motions of Cartesian subtle matter within the earth's crust, varieties of fermentation, emanations from a central fire, or solar heat stored by the earth over a period of some centuries.[34]

More clearly than Boyle and Mairan, chemist Hermann Boerhaave, in his much-reprinted textbook, undertook to clarify the distinctions one ought to make between subterranean heat and fire. Boerhaave was familiar with the research of the Paris academicians who had shown that the earth's heat increases "in proportion to the depth, till at length it becomes so suffocating, that unless it be attempered . . . it overcomes the Miners." If some had argued that subterranean fire "was all a fiction" because it required both air and fuel, Boerhaave replied with what may be an allusion to the theory of Amontons: any air at all in the bowels of the earth would be so highly compressed that "the smallest attrition [friction of the particles] must produce the greatest Heat." Kinetic heat of this kind would require no fuel to sustain it, nor any additional air to promote combustion.[35]

---

[32] Charas in *MARS* 1692, 155–58. Homberg in *MARS* [1701] 1704, 95–99. For Leibniz's great interest in such experiments, see his correspondence with Fontenelle, in Birembaut 1966, esp. 125, 130. Notable in the Charas text is his admitted preference for chemical explanations, because ordinary fire requires fuel.

[33] Amontons in *MARS* [1703] 1705, 107–8. In summarizing Amontons, Fontenelle (*HARS* for the same year, esp. pp. 8–9) reintroduces the role of heat: compressed and *heated* air would expand in explosive fashion.

[34] Mairan in *MARS* [1719] 1721, 104–5, 132–35.

[35] Boerhaave 1735, 1:279. This discussion does not occur in all editions of Boerhaave, some being "unauthorized" versions of his teachings. I have used one of the authentic versions, as identified by Lindeboom 1959, 80–84.

If the studies by Boyle and Mairan went far toward establishing the existence of nonvolcanic heat within the earth, and if Boerhaave offered yet another solution to the fuel problem, none of these authors could begin to suggest how far beneath the earth's surface such heat might extend. When Mairan considerably expanded his analysis, three decades after his first memoir, he could only say:

> whether this is a fire truly central or very deep, innate to the earth or acquired, . . . I will not here discuss; although many reasons lead me to believe that it has to do with the internal structure of the earth and the planets in general. For me it is enough [to show] that its existence is not doubtful.[36]

In other words, "a fire truly central" remained in the realm of speculative cosmogony.

Central heat was also crucial, in the decades before and after 1700, to those naturalists who debated the origin of springs. For students of this subject, the controversy revolved not around seasonal springs—dry and wet seasons could readily explain their variations in abundance of flow—but around those which seemed invariant or perennial. The latter required a steady and limitless supply of water, available only, or so it seemed, from the oceans. It was thus commonly said that sea water either penetrated the earth's crust, emerging as lakes, rivers, and springs, or was collected in caverns within the earth and then elevated to the earth's surface. In both cases (and the two hypotheses were not mutually exclusive), the water had to be desalinated within the earth. And in both cases, but more crucially for advocates of deep caverns, one needed some means of elevating the water. That encyclopedic author, Athanasius Kircher, S.J., examined these issues at length in his *Mundus subterraneus* (1664), concluding that water could be raised in two ways namely, by mechanical forces of pumping and suction, or by heat. The existence of subterranean heat seemed obvious to him, as he was acquainted with Italian volcanoes and thermal springs. If heat was responsible for elevating the waters of thermal springs, then it could readily be imagined to work on a larger scale, vaporizing the water collected in subterranean caverns; when the vapors rose to the earth's surface, they were condensed to form rivers and springs.[37]

---

[36] Mairan 1749, 59–60. On p. 61, Mairan cites both Boyle and Boerhaave.

[37] For Kircher, see Adams [1938] 1954, 434–36. Adams also gives some indication of ancient sources for the ideas of subterranean waterways and caverns. Greater detail is in Plot 1686, chap. 2. The *locus classicus* for underground communication as providing an outlet for landlocked seas is Aristotle's discussion of the Caspian, in *Meteorology*, I, xiii.

Although far from being a Cartesian, Kircher's views were compatible with those of Jacques Rohault, who expounded on the role of heat in elevating water stored in the earth's bowels. For Rohault, however, the idea of a mysterious "suction" performed within the earth was too non-mechanical to be accepted.[38] John Mayow disagreed, regarding the existence of caverns as a gratuitous assumption. Nor did he see any mechanism available to propel subterranean water upward, for his own conception of subterranean heat was limited and localized; that is, Mayow's heat was the product of fermentations associated only with certain mineral deposits and capable only (or mainly) of supplying the warmth of thermal springs.[39]

Considerably more impressive than these writings for range and detail was a work that appeared, coincidentally, in the same year as Mayow's, Pierre Perrault's *Origine des fontaines* (1674). This treatise and the subsequent work by Edme Mariotte (published posthumously, in 1686) are usually described as the first quantitative examinations of the adequacy of rainfall to supply rivers. Both writers and some of their successors found springs to be more problematical.[40]

Arguing that rain supplied rivers and that rivers in turn supplied springs, Perrault had to explain how river water somehow rose to the tops of hills and mountains where some springs could be found. Perrault disliked the Cartesian concept of subterranean heat not only because it had no acceptable causal explanation, but also because he considered the so-called heat of the earth to be a mere sensory illusion produced by the contrast with the coolness of the atmosphere. Without heat as a causal mechanism, Perrault resorted to the analogy of the pump. To be sure, pumps could raise water only to a height of 32 feet, but would not a larger pump do better? (Christiaan Huygens, to whom Perrault had presented this suggestion before 1674, carefully explained why the answer had to be no.) On the other hand, how could pumping actions take place within the earth where there existed no air pressure? Here Perrault answered that nature might use a variety of causes, including attraction and the *horror vacui*, to raise water—and he added that such concepts should not be interpreted as the attribution to nature of sentiments and desires.[41]

Mariotte presumably did not care for Perrault's explanation, for he extended his calculations and applied the rainfall theory to springs as well

[38] Rohault [1671] 1728, pt. 3, chap. 10, secs. 3–4.
[39] Mayow, as cited above, n. 23.
[40] Useful discussions are in Biswas 1970; Delorme 1947/8; and Broc 1974, 208–9.
[41] The debate with Huygens is in Huygens 1888, 7:287–301. For his discussions of heat, see P. Perrault [1674] 1721, 2:763, 776–78, 783–84.

as rivers.[42] In the next decades, Robert Plot, Edmond Halley, Bernardino Ramazzini, Philippe de La Hire, and others all more or less agreed that springs remained more problematical than rivers. (Plot wrote too early to be aware of Mariotte.) Halley therefore proposed a further explanation applicable to springs: in addition to rainfall, the cycle of evaporation and condensation must also supply especially those perennial springs found at high elevations. For Plot, Ramazzini, and La Hire, some kind of internal circulation, within the earth, was still essential, as was the agency of heat to raise vapors to the surface.[43]

One of the more controversial writers on this subject, as on others, was Antonio Vallisneri. His interest in petrifying waters and medicinal (especially thermal) springs eventually led him to deliver a lecture on the general topic at the Paduan Accademia dei Ricovrati; the first edition of this lecture (1715) contained bulky annotations longer than the text itself, and later editions would be further expanded. By 1715, he was fully familiar with such writers as Perrault; he had come to dislike those who revived ancient "systems" like the existence of central caverns of water; and he had already expressed his opposition to the use of the Bible in science. In his lecture he charged other authors—and here he did them less than justice—with basing the theory of the subterranean circulation and ascent of sea water on interpretations of biblical passages.[44]

For Vallisneri, the abyss (or caverns), the earth's circulatory system, and the role of heat were all unproven suppositions, and these in turn required the further assumption that sea water was being desalinated within the earth. If, he argued, we examine the geological structure of hills and mountains, we find (and he described in detail) sedimentary strata no longer in their originally horizontal position; not only are the strata curved and flexed in various ways, but they also contain substances, such as clayey materials, more or less impenetrable by water. One could not reasonably suppose that underground water followed tortuous paths, insinuating itself between strata and finding ways to circumvent the clays, in order to

[42] Mariotte 1686, pt. 1, sec. 2. Mariotte's calculations would be further extended, the author reaching the same conclusions as Mariotte, by Sédileau in *MARS* 1693, 81–93.

[43] Plot 1686, chap. 2. Ramazzini in St. Clair 1697, 132–40. La Hire in *MARS* [1703] 1705, 56–69. Halley in *PT* 16 (1691), 468–73; see also Biswas 1970. As baffled as his contemporaries was Guglielmini, in *Raccolta* 1723, 2:253–56. See also the discourses on springs by Neapolitan academician Lucantonio Porzio, discussed by Dini 1985, 32–34.

[44] Vallisneri 1715, vol. 2, and the analysis in M. Baldini 1975, chap. 2. Pertinent also is Tucci 1983, 9, as well as early (1705) letters to Marsili, in Vallisneri 1991, 1:281–83, 292–98. For Vallisneri's rejection of Johann (I) Bernoulli's arguments favoring the ascent of water, see Riccati 1985, 20–22, 125–28. The key biblical passage was Eccles. 1:7.

ascend a hill or mountain. *Descent* along such routes was perfectly natural and possible.

Vallisneri's arguments added substantially new evidence to the dispute but did not end it, and a new edition of his *Lezione,* with additional material, appeared in 1726. Irritated by at least one review that continued to support the opposition, he planned an expanded third edition, which would be published posthumously.[45] To understand why the debate continued for so long, one must lay aside Vallisneri's insinuations that his opponents were unthinking bibliolaters or revivers of ancient systems—they actually included one of the Bernoullis who produced hydrostatic arguments in favor of the ascent of water within the earth. Perhaps the strongest part of the opposition's arguments was its advocacy of subterranean heat as the motor force for raising water vapors. No one could deny the existence of such heat, but nor could anyone be certain whether heat extended far into the earth's interior—a necessity for Vallisneri's opponents—or was merely a more superficial phenomenon entailing fermentations or burning in the vicinity of coal deposits.

In summary, discussions of heat produced agreement on two points: volcanic phenomena did not provide clues to the basic structure of the earth, and subterranean heat did exist in nonvolcanic regions. The former position survived until at least the end of the eighteenth century. As for the latter, Boyle's collection of testimony and Mairan's calculations took the subject about as far as it could go. Left on the agenda was the apparently insoluble problem of how far heat extended into the earth's interior. Similarly, despite the ingenuity of chemists and physicists, efforts to distinguish heat from fire resulted in a number of competing suggestions and models that could not be confirmed.

GEOLOGICAL TIME

To most modern historians, the short geological timescale used by almost all naturalists of the period 1665–1750 evokes sentiments of regret

[45] The unfriendly review was in *Acta* (November 1726), 488–95. A long and favorable review was in the periodical Vallisneri helped to edit: *GLI,* 38, pt. 1 (1726–27), 190–243. About preparing a third edition, details are in Riccati 1761, 4:xxix–xxx, xxxiv–xxxv; Riccati, an old friend and patient of Vallisneri, was asked to contribute to this edition. Contents of the several editions are indicated in Riccati 1985, 171–72. See also the discussion in Vallisneri 1991, 1:54–56 of the editor's introduction. An excellent treatment of the whole debate, concluding in favor of the rainfall theory, can be found in Desmarest, "Fontaine," in Diderot, *Encyclopédie,* vol. 7 (1757), 81–101.

and even pity. The poor fellows (so the argument runs) were perceptive in some ways, but they failed to recognize the obvious evidence in favor of lengthened time. Alternatively, we are told that when evidence of the latter kind was noticed, religious constraints forced naturalists to adopt a "catastrophic" dynamics in order to compress long processes into a short time.[46]

For a variety of reasons, these interpretations will not do. The supposedly obvious evidence for long time included changes that must be inferred but cannot be directly observed; knowing the difference between the two, early naturalists preferred observation (either personal or by historical witnesses) to what they deemed speculation. Furthermore, volcanic eruptions, earthquakes, flooding, rapid erosion, landslides, and changing shorelines constitute genuine clues to how nature has worked in the past. Such observable phenomena did not have to be invented by men in search of speedy mechanisms. In fact, in both private correspondence and their published works, naturalists rarely reveal any sense of constraint. Accustomed to think of time as measured by human history, many concluded that a few thousand years were very long indeed. Some did flirt with a longer timescale; how they did so, in what terms, and in what framework deserve careful examination.

In the decades around 1700, there existed only two common options: biblical time or ancient philosophies of eternalism. To John Wilkins, as to Lorenzo Magalotti, the latter ideas merited discussion and refutation, and they and their contemporaries assembled a range of evidence to show that the world had a beginning and that it was not very old. In 1745, one of the Rouen academicians was still repeating some of the standard arguments against Epicureanism: the uniformities and regularities of nature confute any idea of the mere chance combination of atoms. Similarly, the existence of laws implies a lawgiver and hence a definite moment (or period) when the world was given those laws.[47]

Some of the same writers, joined by others, also gave attention to whether the world was older than the Bible seemed to allow. As we have seen (in Chapter 2), evidence about the antiquity of Egypt, Chaldea, and especially China provoked a large scholarly literature, either in rebuttal or in accommodation to biblical time. In a nonscholarly vein, some argued against both

---

[46] In addition to works cited above, n. 4, see Whitrow in Yourgrau 1977.

[47] Guérin, 1745, in Gosseaume 1814, 1:168–69. Wilkins [1693] 1969, 62–77. Magalotti 1741, letters 26–28; dated 1681, these letters were first published in 1719. Other examples include Burnet 1684, 34–44; Nicholls 1696; Cheyne 1715, 1:149–51. See Rudwick 1972, 91, 93.

eternalism and a lengthened timescale on grounds superbly summarized by Don Cameron Allen:

> imagine the state of a world in which all creatures had been breeding forever. The seas would be solid fish; the air dense with wings; and man would find no purchase on the earth for his feet.[48]

Others took up the relatively recent development of the mechanical arts and indeed of knowledge, arguing that mankind could not have existed for a long time in a primitive state and without the conveniences of life. Even to the more daring writers, evidence of a naturalistic and historical kind led only to a conviction that the age of the world could not be determined with any precision.[49]

Biblical chronology had the great virtue of providing a framework, even if learned chronologists could not agree among themselves about the age of the world. That they assigned dates to Creation ranging from about 5870 B.C. to 3945 B.C. merits a moment's attention, for a difference of almost two millennia in a scale so short implies a degree of elasticity as well as the comforting feeling that sufficient time exists to accommodate any new discoveries. This feeling manifested itself among the scholarly Jesuits of Peking (Beijing), as it did in Leibniz, who told one of his correspondents that the seemingly great antiquity of China need not distress us because it is readily compatible with the longer chronology of the Greek Septuagint.[50]

Most naturalists not only found the biblical scale comfortable, but some also explicitly defended it as long. Steno, Molyneux, and Dezallier all present one argument of this kind: How could buried organisms be preserved in the rocks and soils for *as long as* four thousand years? Molyneux, in fact, thought that the antlers of the Irish elk could not have lasted that long; when he emphasized that the animal had left Ireland "so many Ages past," he actually meant that migration to North America had occurred sometime before the first voyage of Columbus.[51] Other writers expressed

---

[48] Allen 1964, 137.

[49] Virtually the whole range of arguments can be found in Berkeley's *Alciphron*, 1732, dialogue 6, esp. sections 21–27, in Berkeley 1901, vol. 2, and in Goguet [1758] 1775. For the daring, see Bolingbroke 1809, 4:395–483, for letters written in about 1720, and Tyssot 1722. Additional examples are in McKee 1941, 30–33. On Bolingbroke and Lévesque de Pouilly, see Fletcher 1966, 211–14, and 1967, 36–40.

[50] For Leibniz, above, Chapter 2, n. 116. Chronological tables and discussion are in *Universal History* (1740), 1:l–lii. A different range of dates was published by Benjamin Franklin on the title-page of *Poor Richard*, 1733, in Franklin 1959, 1:287.

[51] Molyneux in *PT* 19 (1697), 499, 506. Steno [1669] 1916, esp. 258; see also Greene 1961, 59; Rudwick 1972, 72–73; and Ellenberger 1988, 1:294. Dezallier 1742, 156. See also Souciet in *Trévoux* (March 1729), 467.

their conviction that traditional chronology allowed room for gradual changes, as when Leeuwenhoek asserted that there had been ample time since the Flood for organisms to petrify. Comparably, to an anonymous Italian writer, the major part of the earth's sedimentary crust had accumulated "little by little" since the era of the Flood. And Swedenborg explained to a correspondent that "lands, especially the boreal, lay hidden under a deep ocean for a long period after the time of the flood, and, with the subsidence of the sea toward the [equator], came to view little by little."[52]

Perhaps most interesting is Robert Hooke's sense of time because it can so easily be misconstrued. When Hooke failed "to petrify a Stick throughout" in his laboratory, he asked: "what may not Nature do that can take her own time, and knows best how to make use of her own Principles?" The answer was forthcoming a few years later when Hooke informed his colleagues of "a ground in Bedfordshire, which would in a twelvemonth's time turn wood and other matter, that was not stony, into stone, without vitiating the figure."[53] Elsewhere, Hooke's declaration that fossils can provide a "natural chronology" of the earth should be placed in the context of other remarks, such as his assertions that most geological changes took place in "fabulous" times and that changes in England antedated the use of writing.[54] In the latter instance, he is precise: this means pre-Roman Britain. "Fabulous" also had a precise meaning: this referred to the age of myth, when human traditions were transmitted orally. It being difficult to construct a chronology from such materials left by man, fossils presumably (albeit not easily) might be used to better effect. Certainly, Hooke on a number of occasions rejected long chronologies for Egypt and China.[55] If he did not lengthen the human past, nor did he do so for nature.

Hooke's effort to petrify a wooden stick provides another key to the acceptability of biblical time. From experiment one could only conclude that some slight extension of laboratory time would suffice for nature. As

[52] Leeuwenhoek in *PT* 24 (1704), 1774. *Raccolta* 1723, 1:xvi. Swedenborg to Melle, 21 May 1721, in Acton 1948, 252.

[53] Hooke 1705, 295, text of ca. 1668; Birch, under date 5 February 1672/3, as given in Gunther 1930, 7:408. Among those who see Hooke as adopting a non-biblical time scale are Davies 1969, 90; Morello 1979a, 73; Drake 1981, 964. Rudwick 1972, 73–75, identifies Hooke's timescale as short but argues that his use of earthquakes and large-scale changes in the early history of the earth enabled him to make geology and chronology compatible. Compare the different argument below, at n. 70.

[54] Hooke 1705, 389, 418–19, texts dated 1693 and 1690, respectively. The famous passages on fossils providing a "natural chronology" are in ibid., 335, 411, texts dated 1686–87 and 1688.

[55] Ibid., 328, 372–73, texts dated 1668 and 1687.

Réaumur informed Bourguet, "I do not see why two or three hundred years would not be sufficient [to form] crystalline rocks."[56] The rapidity of chemical crystallization seemed to some writers to be in accord with other processes in nature, such as the almost visible growth of stalactites. To antiquarian William Stukeley, as to chemist J. F. Henckel, the two examples, mentioned in virtually the same breath, both showed that nature works quickly.[57]

Those naturalists who began to contemplate some extension of the timescale usually stayed within the biblical framework by posing a traditional question: were the six days of Creation to be interpreted as twenty-four-hour days? When Newton and Burnet discussed the matter, they agreed that days need not be taken literally, but neither saw much to be gained by extending "days" to "years." As Burnet put it, could land and sea separate in less than a year? "I thinke not, in a much longer [time]." If Burnet found a modest extension legitimate, albeit not very valuable, in his theory of the earth, he explicitly rejected any further extension that would be necessary if naturalists were to insist on the importance of such minor geological events as volcanic eruptions: "how many thousand Ages must be allow'd to them to do their work, more than the Chronology of our Earth will bear?"[58] In this passage, one recognizes a "catastrophist" of the kind familiar to modern historians, for Burnet deliberately chose to use a cataclysm, the Flood, that could fit into both traditional chronology and biblical narrative.

If Burnet exhibits constraint, remarkably few comparable cases have come to light. One such occurs in the pages of the *Philosophical Transactions* as part of a discussion of whether the earth's poles have shifted in the course of time; as the writer indicated, such changes (if they occurred at all) "would require a prodigious number of Ages," unless one supposed a cataclysmic event like the shock of a comet. Quite as tentatively, Edward Lhwyd once remarked to Ray that the accumulation of so many erratic boulders in Welsh valleys would require more time than history allowed,

---

[56] Réaumur to Bourguet, 25 March 1741, in Neuchâtel, MS 1278, fols. 48–49.

[57] Stukeley in *PT* 30 (1719), 966. Henckel [1734] 1760, 2:398, 411. Although Laudan (1987, 10) suggests that those who had a short timescale could more readily use laboratory analogies than those with a longer scale in mind, I argue only that such analogies might confirm short time but were also widely accepted as the best way to understand nature when direct observation was impossible, or, more generally, when causal explanations demanded investigation.

[58] Burnet in Newton 1959, 2:325; Burnet 1684, 161. Burnet also pertinently remarked (ibid., 146) that if mountains had been produced by earthquakes, "In what Age of the World was this done, and why not continued?" The search for witnesses is here combined with the question of uniformity in nature.

and so perhaps one should attribute their transport to the Flood.[59] This suggestion apart, Lhwyd never became a diluvialist.

John Ray, too, wondered about the age of the earth, working his way from unsettling observations to biblical reinterpretation to a noteworthy solution. In an early discussion of the subject, Ray found it "a strange thing, considering the novity of the World," that Varenius should have reported more than two hundred feet of accumulated sediments and that fossil shells should be found at high elevations. And he summarized the alternative implications of these facts:

> if the Mountains were not from the beginning, either the World is a great deal older than is imagined or believed, there being an incredible space of time required to work such changes as raising all the Mountains, according to the leisurely proceedings of Nature in mutations of that kind since the first Records of History: or . . . in the primitive times and soon after the Creation the earth suffered far more concussions and mutations in its superficial part than afterward.[60]

Ray did not decide between these solutions, but later in his career allowed that a literal interpretation of the six days required reexamination. He then briefly hoped that John Woodward could find a way to reconcile geological observations with "the novity of the World," but eventually confessed to Lhwyd that, "whatever may be said for ye Antiquity of the Earth it self & bodies lodged in it, yet that *ye race of mankind is new* upon ye [earth], & not older then ye Scripture makes it, may I think by many argumts be almost demonstratively proved."[61]

Ray's tentative solution entails an expansion of pre-Adamic time, and other naturalists discussed this possibility without exhibiting the distress one detects in Ray. With what amounts to a shrug, J. J. Scheuchzer, Woodward's Swiss translator, admitted that "days" could mean anything from years to seconds; having no way to be sure of the meaning, he judged it best "to accept the common hypothesis."[62] A different conclusion was proposed in the correspondence of two prominent Quakers, Thomas Story and James Logan. Having examined a sizable accumulation of sediments during a trip to Yorkshire, Story informed Logan that "days" were a Scriptural locution "suited to the comon Capacitys of human kind" and should be interpreted to mean "certain long & competent periods of time, & not

---

[59] *PT* 16 (1687), 406. Lhwyd to Ray, 30 February 1691/92, in Gunther 1945, 157–58.
[60] Ray 1673, 7–8, 126–27.
[61] Ray 1693, after p. 162, and 1928, 252, 253–54, 259–60. Italics added.
[62] Scheuchzer 1732, 1:11.

natural days." If he never completed what amounts to a theory of the earth, he nonetheless told Logan that he wanted to trace how nature worked from the present "bacckword, . . . the better [to] perceive the manner of her procedure from her fountain & origin."[63] In comparable fashion, but with added clarity and acute self-consciousness, Anton Lazzaro Moro in 1740 published his own geological theory in which expansion of the biblical days occupied a crucial place. The theory required more time than allowed by conventional chronologies, and it also required many volcanic eruptions that no human witnesses had recorded. Since Moro could find no substantial gaps in written history, he placed his eruptions in pre-Adamic time, reinterpreting the meaning of the days of Creation.[64]

Among Continental naturalists troubled by matters of time were Montpellier physician Jean Astruc and Swiss polymath Louis Bourguet. For the Mediterranean littoral, which he examined with some care, Astruc thought historical time adequate, using ancient and medieval documents to supplement his study of nature. About marine fossils far from modern seas, he wrote to Tournefort:

> A philosopher who does not want to part company from his duty has trouble finding a solid explanation of this phenomenon which will not shake the foundations of his beliefs. The earth seems too new for all the changes one must suppose it has undergone in order to explain the distribution of these shells.

His solution, expressed briefly and never analyzed, was to attribute non-coastal marine formations to the Flood.[65] That Astruc fully recognized the textual and interpretative problems posed by Genesis in particular became apparent much later in his career when he published a scholarly work in the tradition of Richard Simon.[66]

Louis Bourguet was urged by at least two of his correspondents, Leibniz and Vallisneri, and by his reading as well, to consider biblical time as too

[63] Story to Logan in Penney 1926, 70–73. If I read correctly Story's way of dating, the letter was written 8 December 1738. See also Thomas 1983, 168, and Raistrick 1950, 261. I am indebted to Linda Stanley of The Historical Society of Pennsylvania for the information that the theoretical sketch Story seemingly sent to Logan cannot be found among the Logan papers.

[64] Moro's theory is discussed below, Chapter 7.

[65] Astruc to Tournefort, 9 July 1708, MHN, MS 1391.

[66] I refer here to his anonymously published *Conjectures sur les memoires originaux dont il paroit que Moyse s'est servi pour composer le Livre de la Genese* (Brussels, 1753). The title alone shows kinship with Richard Simon. Astruc separated the passages in Genesis and Exodus using the two different Hebrew words for God; the result was two narratives, each fairly complete. See Lods 1924.

restricted. In effect, Leibniz disregarded time in any measurable sense when he wrote about the pre-human history of the earth. As he eventually told Bourguet, time is the *order* of events, and no system of absolute dating could alter that order.[67] From his Paduan friend, Bourguet received even more pointed comments on geology and the Bible. Vallisneri insisted that the earth is older than commonly supposed and that Bourguet's efforts to explain what happened during the Flood required more time than Bourguet seemed to realize.[68] The immediate responses to these letters are unknown, and it was more than a decade later that Bourguet published an outline of his theory of the earth. Here he sketched the order of events, indicating that the Flood, too, should be considered a series of "dissolutions." In his only allusion to time, he declared that the originally soft crust of the earth acquired its basic structure (mountains, valleys, sea basins) during "at least the time of a revolution around the sun."[69]

Bourguet's solution was not uncommon, but if his seems to have been dictated by some concern for (short) time, the same cannot always be said for his contemporaries. As it happens, not only Genesis, but also Cartesian cosmogony and Newtonian physics either specified or implied that the earth's crust had once been more malleable than it is now. (In the Newtonian case, only thus could the earth's rotation have produced the shape of an oblate spheroid.) For some naturalists, the Bible and physics agreed: geological changes could have been larger and could have taken less time during the earth's early history. Since Fontenelle also adopted this reasoning, we cannot assume that a desire to save biblical chronology necessarily lay behind this argument.[70]

One result of the physical argument was to undermine any hope that naturalists would be able to find some "geological constant," invariant in time and place, to serve as the basis for a chronometer. Indeed, one did not have to go far back into the earth's early history to notice the difficulty, for, as J. J. Scheuchzer indicated, it would be hard, perhaps impossible, "to make any Computation of the Depth a Valley sinks in a Century, by marking

[67] Leibniz's geology will be discussed in Chapter 7. Leibniz to Bourguet, 2 July 1716, in Leibniz 1875, 3:595. Although the letter clearly reflects issues peculiar to the controversy with Samuel Clarke, he elsewhere indicated his impatience with efforts to establish chronological precision. For example, Leibniz 1875, 3:221; Davillé 1909, 372. Whether Leibniz also lengthened *human* history is an open question. Aarsleff 1982, 48, 89, argues that he did do so, but I find Leibnizian locutions far from clear. (I am grateful to Professor Aarsleff for corresponding with me on this subject; neither of us can clarify the matter any further.) See also Heinekamp in Parret 1976, esp. 537–39.

[68] Notably in the letter from Vallisneri to Bourguet, 23 November 1717, in Neuchâtel, MS 1282, fols. 225–26. See also Ellenberger 1980, 65 and n. 53.

[69] Bourguet, 1729, as translated in M. Carozzi 1986, 30. Bourguet's theory was outlined above, Chapter 5.

[70] *HARS* [1716] 1718, 14–16.

Yearly the Height of the Water on the Stones in the narrow Passages of the Mountains." Any such effort would be frustrated by the yearly and even daily variations in the amount and force of water and by the equally variable resistance of the different rock types.[71] Under the circumstances, one may appreciate how unusual was Edmond Halley's thinking that he had perhaps discovered a physical constant—the rate of increase of the ocean's salt content—that would enable future naturalists to calculate the age of the earth. Like others, Halley began with the biblical days:

> But whereas we are there told that the Formation of *Man* was the last Act of the *Creator*, 'tis no where revealed in Scripture how long the *Earth* had existed before this last Creation, nor how long those five Days that preceeded it may be to be accounted; since we are elsewhere told, that in respect of the Almighty a thousand Years is as one Day.[72]

Proceeding then to cite such examples as the Caspian Sea, Halley argued that the world's oceans had one thing in common with inland bodies of water: all received water from rivers but had no "outlets" other than evaporation. (He here discarded the traditional option of an outlet into the earth's subterranean circulatory system.) Constant evaporation left these bodies of water with a constantly increasing salt content. If one were to measure the rate of increase, one might extrapolate backward in time to when the process had begun. By this calculation, one would "refute the ancient Notion . . . of the Eternity of all Things; though perhaps by it the World may be found much older than many have hitherto imagined."[73]

This proposal seems to have inspired no one, and, on the whole, those who sought to extend biblical time hardly knew where to begin, how to proceed, and what to conclude. Antonio Conti, for example, vaguely wondered whether astronomical cycles—changes in the angle of the ecliptic and in the location of the earth's magnetic poles—implied similarly long periods for the motion of the earth's seas. Vallisneri used historical records to show that even small changes in shorelines had taken several centuries to become detectable, but he could add nothing more to his early conviction that the earth is "of nearly incomprehensible antiquity."[74] Nor could Antoine de Jussieu, in his several studies of fossils, move beyond his first

---

[71] Lhwyd's summary of Scheuchzer, in *PT* 26 (1708), 144–45. If Guglielmini tried to cast in mathematical form his studies of rivers and streams, he nonetheless found he had too many variables and had to introduce qualitative modifications; see Rouse 1957, 71.

[72] Halley in *PT* 29 (1715), 296.

[73] Ibid., 300. Analyses in Biswas 1970 and Gould 1993, chap. 11. See also Schaffer 1977.

[74] Conti 1972, 390, text written after June 1722. The Vallisneri quotation is from a letter to Scheuchzer, 12 June 1707, in Vallisneri 1991, 1:370. His use of incremental geological changes will be discussed in Chapter 7.

announcement that fossils constitute "the oldest library in the world." If this statement does conjure up visions of pre-human "libraries," the more typical locutions during these decades were such phrases as, "a long time," "an extremely long time," and "a long succession of centuries."[75]

"Libraries" and "centuries" serve as useful indications that those who wished to lengthen the age of the earth continued to think of periods well within the order of magnitude of historical time. (It is no accident that Conti's reference to astronomical cycles was coupled with an allusion to ancient Egyptian traditions that he admitted might be mythical.) Similarly, those who extended time by reinterpreting the biblical days reveal their need for a comprehensible framework of some kind, rather than an inconceivable and formless eternalism. Finally, those who employed the biblical timescale often insisted, although with some exceptions, that biblical time was either sufficient or indeed long.

To the various kinds of evidence and argument just presented, it is appropriate to add the reflections of Aristotle that geological changes take "immense" periods of time when compared with the brevity of human life. Transformed into a comment on human psychology, suitable in a text decrying anthropocentrism, similar sentiments would be expressed in Fontenelle's *Plurality of Worlds* (1686).

> If roses, which last only a day, wrote histories and left memoirs for one another to read, the first roses would have depicted their gardener in a particular way; and after more than fifteen thousand generations of roses, who would also have left records for those who followed, nothing would have changed. And the roses would say: "We have always seen the same gardener; he is the only one within rose-memory; he has always been the same as now; surely he does not die as we do; he does not change." Would such rose-reasoning be good?[76]

The human penchant to measure everything by human standards did not go wholly unperceived by Fontenelle's contemporaries, as when Lord Bolingbroke remarked: "Philosophers reason often, and the vulgar always, like the roses in Fontenelle." Nowhere, however, did Fontenelle suggest that it would be easy for a rose to acquire the point of view of the gardener. One

---

[75] Jussieu in *MARS* [1718] 1719, 289. The phrases quoted can be found in *HARS* [1720] 1722, 8; *MARS* for same year, p. 400; *JS* (24 July 1713), 469, contrasting diluvialism with "a long succession of centuries." Short times could also be described by such phrases, as Molyneux did (above, at n. 51).

[76] Fontenelle [1686] 1973, 118. This passage occurs near the end of part 5. Aristotle's *Meteorology* is quoted in Haber 1959, 39–40.

rose who tried to do so was Nicolas Desmarest, in his prize-winning "dissertation on the former junction of England with France" (1753). In the event, he failed, for, in addition to the physical evidence he cited, Desmarest remained as attached to written histories as had his predecessors.[77]

[77] Bolingbroke 1809, 4:480. Desmarest 1753, 143 for his reference to Fontenelle's roses. Desmarest's views were to change by 1757; see Chapter 8.

CHAPTER 7

Alternatives to Diluvialism: Some Syntheses

To attempt a reconstruction of all or most of the earth's past was, by 1700, considered to be philosophically rash, for a synthesizer could be labeled a mere "system-builder" or "world-maker" after the fashion of Descartes or Burnet. Not surprisingly, relatively few—compared with the many who wrote on fossils or volcanoes—did make the attempt, and these few displayed acute awareness of the limits of their knowledge and the probabilistic nature of their conclusions. After 1695, a main impulse behind the new theories was Woodwardianism, as naturalists sought to explain the sequence of events that had produced strata not laid down in the order of their specific gravities. In addition, all theorists, before and after 1695, knew that marine deposits could be found far above the level of modern seas. Had the land somehow been elevated? As shown in the earlier discussion of volcanoes, earthquakes, and subterranean heat, to discover an acceptable causal explanation of elevation proved almost impossible. Had the more obviously mobile sea changed in level? That this had in fact occurred seemed obvious to most writers, but here, too, possible mechanisms were hard to discern.

The several theories to be treated here fall into few detectable patterns if one seeks to discover national styles in early geology. On the whole, the number of writers is so small that to speak of "groups" becomes a matter of assembling clusters of perhaps three or four. One apparent cluster, for example, consists of several Swedes who gave considerable weight to the evidence that the Baltic Sea appeared to be subsiding. An incongruous addition to this little group was Benoît de Maillet, whose *Telliamed* (1748) contained a comparable theory, but developed and argued in a way utterly different from the Swedes. As another group, naturalists in the Paris Academy were said by some contemporaries to constitute a "school" of geology. Looked at more carefully, this group almost dissolves into the person of Fontenelle, who produced the larger synthesis that his colleagues avoided, and who, moreover, was inspired somewhat by the work of Leibniz.

The following examination of geological syntheses will be conducted in a modified chronological order. The well-known writings of Steno, Hooke, and Leibniz will first be presented as pre-Woodwardian theories, formulated before diluvialism had attained a degree of respectability. For these philosophers and their successors, I will attempt to identify what evidence they relied on, how or to what extent they used human records, how theories changed as new evidence either came to light or assumed new importance, and how contemporaries commented on the several proposed versions of the earth's history.

Nicolaus Steno

To put Steno's *Prodromus* in context, it is worth recalling that in 1669 there existed no developed theory of the earth and no diluvial geology, although the possibilities of each had been sketched, respectively, by Descartes and by Fabio Colonna. What Steno did confront were competing explanations of the nature and origin of fossils. His cautious conclusions on this subject in 1667 gave way to more settled convictions by 1669, so that the main object of his *Prodromus* became to explain the origin of various "solid bodies"—crystals as well as fossils—enclosed in the earth's strata. More than this, the text offered a reconstruction of the series of events that had produced the strata themselves.[1]

The structure of Steno's argument having been examined earlier, it is appropriate here to focus on the famous diagrams depicting the six stages of the history of the Tuscan landscape (and of the whole earth). These diagrams, with the accompanying text, present an initial period of the deposition of sediments in horizontal layers. For reasons left unclear, a

---

[1] Among the many analyses of Steno, I have found most valuable those by Morello in *Niccolò Stenone* 1986, 67–111; Rossi in *Niccolò Stenone* 1988, 13–21; and Rudwick 1972, chap. 2. For Fabio Colonna (his work unknown to Steno?), see Morello 1981.

hollowness developed in the foundations of this structure, so that sediments collapsed to form valleys. Further sedimentation then took place in these valleys, after which there followed another cycle of weakening, collapse, and sedimentation.[2]

In this abstract analysis of mechanics and hydraulics, various ingredients are missing, must be sought in other parts of the text, or may be inferred from small clues. The basement upon which the first layers were deposited is not indicated; that this core seems to have been smoothly spherical was either a supposition useful for what followed or a possibility derived from a combination of Genesis with Descartes. Missing, too, is any reference to the composition of the different strata, except that the oldest were unfossiliferous. Elsewhere, Steno does analyze the various strata by arranging them in order of substance and contents (228–29); arrangement by order of deposition would become a major desideratum only after Woodward. Why horizontal sediments collapsed was a matter of conjecture, Steno proposing both "the violent thrusting up" due to subterranean fires and "the spontaneous slipping or downfall" due to the (unexplained) "withdrawal of the underlying substance" (231). Comparison of the text with one of the diagrams reveals another puzzle: if the Flood was universal, covering existing mountains, why, then, does the corresponding diagram (stage 4) show diluvial sediments being deposited only in valleys? If Steno seems to have attributed to the Flood all deposits of marine fossils, he also refers without explanation to the "sediments of the turbid sea" and "how often the sea had been turbid in each place" (206). Is this an allusion to postdiluvial marine sedimentation? Although he thought that the four thousand years since the Flood constituted a long and presumably eventful time, Steno left undeveloped the postdiluvial history of the earth, except for noticing some recent fossils such as the remains of Hannibal's elephants; his final paragraphs seem to promise attention to this subject in the longer dissertation that he never completed.

From this complex and odd little book, historians of geology have traditionally extracted chiefly the clearer parts, such as the fossil question and the principle of superposition. Steno's contemporaries, however, read the book in diverse ways. Most evenhanded was the Roman journalist Nazari, who reviewed the main purpose of the text, its handling of fossils and sediments, its attention to the Flood and other aspects of history, and the six diagrams. Nazari also thought it appropriate to assure readers, both at the beginning and the end of his review, that nature and Scripture were here

---

[2] Steno [1669] 1916, 262–70. In the next paragraph, references to this edition are incorporated in the text.

in accord.[3] For Paolo Boccone, the fossil question had already been on his mind, and he produced a lucid account of Steno's arguments; but he gave little attention to the Flood, merely pointing out that marine deposits could easily be explained as the effects of "floods or storms" and perhaps also of sea water being vomited by volcanoes.[4] What Henry Oldenburg thought can be inferred from his preface to his English translation of Steno. Here was an "ingenious" text, certain to be of interest to the "curious," for it had much to say about the structure of the earth, the changes it had undergone, and the "productions" to be found in it. Much of the preface then singled out fossils, crystals, and various stones, Oldenburg informing readers that two Fellows, Boyle and Hooke, had been examining the very same topics.[5]

To these examples of what the first readers saw in the *Prodromus,* one may add other fragments, such as Malpighi's preference for Steno over Martin Lister on the origin of fossils; Jacopo Grandi's approving rehearsal of Steno on fossils, sedimentation, and the Flood; and the admiration expressed by Leibniz. Wherever *glossopetrae* entered into later writings, Steno's views usually received attention, and, in fact, as modern scholars have shown, Steno's name became especially associated with the interpretation of marine fossils.[6] What has long baffled historians, however, is the absence of reference to Steno in so much of the geological literature of the eighteenth century. It will hardly do, as an "explanation," to suggest that later writers were less empirical than Steno and more concerned to reconcile Genesis with geology.[7] Other factors deserve consideration, even if they entail a degree of conjecture. For one thing, as the fossil question became less controverted, his writings would have lost their immediate relevance. Heresy this may be, but it is likely, too, that the principle of superposition did not require the elaborate proofs Steno devoted to it; certainly, his first readers did not find the idea startling, and once attention had been drawn to the process of sedimentation, to say that the lower strata antedate the upper ones became commonsensical. (The latter way of thinking typified Woodward's critics.) Finally, issues did change, as did styles. Woodward succeeded in calling attention to the need to examine strata in the kind of

---

[3] Nazari 1669, 111–14. Also evenhanded was *PT* 6 (1671), 2186–90.

[4] Boccone 1671, 38–66, esp. 44–45, 65–66.

[5] Steno 1671, preface; reprinted in Steno [1669] 1916, 199–201.

[6] Rome 1956; Eyles 1958; and refs. in Chapter 4, n. 65. See also Adelmann 1966, 1:470; review of Grandi in Nazari 1676, 8–12; Roger in *Leibniz* 1968, 137.

[7] This is the suggestion of Victor Eyles in Schneer 1969, 163. During the eighteenth century, there was one Latin reprint (1763) and one French abridgment (1757) of the *Prodromus;* see Steno [1669] 1916, 195–96, 201.

detail absent from the *Prodromus*.[8] As for style, the demonstrative mode of the *Prodromus,* so familiar in 1669 that readers did not comment on it, gradually lost favor during the first half of the eighteenth century.

ROBERT HOOKE

Independently of Steno, and as early as 1667–68, Hooke began to develop his theory of the earth in a long series of lectures delivered at Gresham College. Discussion of their content was initially confined to the internal debates of the Royal Society and to the correspondence of some Fellows. When the lectures were finally assembled and published by Richard Waller in 1705, Hooke's theory had already been rejected by most Fellows.[9]

Hooke now and then remarked in his lectures that fossils could provide a "natural chronology" of the earth, but there is no evidence that he himself undertook such a sequential ordering or had the conceptual tools and the information to do so. As he admitted on one occasion, it would be difficult to detect from fossils "the intervalls of the Times wherein such, or such Catastrophies and Mutations have happened," except by combining fossil evidence with "other means and assistances of Information." Like Steno, for whom nature alone did not reveal the "how and when," Hooke's "other means" included historical accounts of earthquakes.[10] Indeed, Hooke focused consistently not on fossils but on his thesis that the earth's history could be explained by the recurrent phenomenon of earthquakes. Many tremors were actually reported in the late seventeenth century, and Hooke seized on such reports, combining them with ancient accounts and with the imaginative reinterpretation of ancient myths. All this evidence, he argued, showed that earthquakes had been common in all times and places, elevating and depressing great tracts of land—and, in particular, elevating above sea level those terrains containing marine fossils.

In his earliest lectures, Hooke had in fact made little use of historical testimony, because, as he admitted, few pertinent narratives existed, and these

---

[8] The catalogue of Steno's collection of specimens, published in Scherz 1958, 202–75, shows that he sometimes identified where particular objects came from, but never the strata in which they were found. Compare Woodward 1696, 5–7. Woodward's own classification of his collection did not consistently include the details he called for in 1696, and the peculiarities of his taxonomy are visible in Woodward 1728 and Price 1989.

[9] The next paragraphs are based on Rappaport 1986. For a different approach, emphasizing physical more than historical evidence used by Hooke, see Ito 1988.

[10] Quotation from Hooke 1705, 411. This text dates from 1688 when Hooke had just begun his systematic search for written testimony.

few were limited to some classical texts. Instead, he preferred to rely on modern accounts and to argue that nature had worked in the same way in the past. At one stage in the lecture series, he also hoped to enlist the data of astronomy in order to devise a physical explanation of the periodic recurrence of earthquakes. If one could show that the earth's poles had shifted in the course of time, the result would have been changes in the earth's center of gravity; the latter in turn would have provoked those seismic disturbances that had repeatedly altered the landscape. This hypothesis of polar shift had its attractions in Hooke's eyes for two crucial reasons. First, a physical law, if such could be established, would obviate the necessity to find human witnesses to individual earthquakes. Second, a law of this kind would circumvent one of the problems inherent in the earthquake theory: since Hooke, like so many of his contemporaries, associated earthquakes with eruptions of volcanic fire, his initial theory entailed a secular decline in frequency and intensity of disturbances as the earth gradually lost its supply of subterranean fuel. To say that the earth was losing its early vigor did not distress Hooke, but he came to prefer the greater uniformity and law-like behavior implicit in the astronomical hypothesis. Although he never abandoned hope that the latter idea would prove valid, Hooke himself failed to find reliable evidence in its support. That failure, coupled with the criticism voiced by Fellows and by members of the Oxford Philosophical Society, drove Hooke to search more systematically for human witnesses to past earthquakes.

Oscillating between caution and recklessness, Hooke tried to supplement the small stock of ancient narratives, by such authors as Pliny and Strabo, with a quasi-original interpretation of ancient myths. During Hooke's lifetime, the common moralistic readings of myth were being superseded by a revival of ancient Euhemerism, that is, by an effort to see in ancient gods, demigods, and heroes real historical personages whose deeds had been magnified after their deaths. From equating gods and heroes with real men, it was perhaps inevitable to begin equating them with events of a natural kind. If Fontenelle was willing to transform one of the exploits of Hercules into a naturalistic opening of the Straits of Gibraltar, Hooke went much further, interpreting many myths as disguised allusions to earthquakes. Now and then, Hooke did allow that some myths might simply have been works of imagination and that "Geometrical Cogency" had not yet been applied to the analysis of such tales. These moments of caution alternated with what may be called episodes of license. When Fellows of the Royal Society doubted that Hooke had properly understood ancient texts and that his evidence was adequate to prove his theory of earthquakes, he ultimately acknowledged defeat and returned to his earliest argument: nature worked uniformly, and modern earthquakes gave evidence that the same events had occurred in all places and at all times in the past.

From Hooke's long struggle with his audience, one may extract two significant points. First, Hooke himself believed that the true origin of marine fossils would not be accepted unless coupled with an explanation of transport and deposition—and, above all, an explanation of how fossiliferous strata were no longer at sea level. The latter difficulty, he believed, would be resolved by the theory of earthquakes. Toward the end of his life, Hooke admitted that the two questions had been separated: his colleagues could accept the organic origin of fossil shells without subscribing to his theory, for they had come to prefer the "easy" solution of diluvialism.[11]

More revealing about the common expectations and mentality of this period, however, is the second point faced by Hooke: what constitutes proof? Hooke's colleagues expected scientific conjectures about the past to be confirmed by what Robert Plot called "the records of time." If Hooke could not produce sufficient witnesses in all times and places, then his theory would have to remain an example of mere speculation. As John Wallis and the Oxford *virtuosi* put it, the lack of historical confirmation removed Hooke's earthquakes from historic time to the period before Adam, and there was no evidence at all in ancient writings about that part of the past. None of the disputants doubted that nature worked in uniform, lawful ways, but Hooke had claimed a *frequency* of earthquakes that was not within modern experience.[12] To Wallis (and Plot and others), even the most dimwitted of chroniclers could not have failed to notice an event as startling as an earthquake; if the records were silent, then the events had not occurred. Ironically, as we have seen in an earlier chapter, Wallis himself would later encounter just such a silence about the hypothesized land bridge linking England with France, and its presumed destruction. Like Hooke, he assembled such fragmentary historical evidence as he could find, even as he wondered why the chroniclers were mute.

Hooke's arguments with his colleagues centered on matters of evidence and proof. Viewed in this way, it becomes less surprising that he had only one known disciple, John Aubrey, for what Hooke tried to do was, in eighteenth-century parlance, to elevate an "accidental" phenomenon into a normal or "general" one. In other words, earthquakes were as uncom-

---

[11] Hooke began to make this complaint about diluvialism as early as 1688, and he would repeat it in the years after Woodward's *Essay*. Hooke 1705, 412, 437.

[12] Details about the Hooke-Wallis dispute, using hitherto unexploited manuscripts, are in Oldroyd, in Hunter and Schaffer 1989, 207–33.

mon as volcanoes, and one could not legitimately build a theory on such occasional events.[13]

When Hooke's so-called "Discourse of Earthquakes"—actually consisting of lectures delivered over a period of more than thirty years—was published in 1705, it seems to have aroused no interest in Britain; nor did it attract attention on the Continent until more than half a century later.[14] At the same time, earthquakes continued to provoke discussion, so that John Ray, in one of his much reprinted treatises, wrote at length about their frequency and their power to elevate land. Although he cautiously concluded, "I cannot positively assert the Mountains thus to have been raised," some subsequent writers would associate an earthquake theory—or at least undue emphasis on earthquakes—with Ray, not with Hooke.[15] And they continued, as had Hooke's audience, to protest the magnification of accidents into general causes.

The possibility of polar shift had a more continuous lifetime than the theory of earthquakes, having entered into academic discussion probably by way of Burnet's *Sacred Theory*. There the Flood was said to have caused a change in the orientation of the earth's axis, bringing to an end the antediluvian paradise, initiating the cycle of the seasons, and shortening the lifespan of man. In William Whiston's *New Theory of the Earth* (1696), a comparable change was proposed, this time caused by the passing comet that had produced the Flood. Hooke, by contrast, had argued not for such a singular event, but for progressive small shifts that might be detected by a comparison of ancient and modern astronomical data. If Hooke could find no confirmation of this theory, and if Halley had no better success, Halley and other Fellows nonetheless remained willing to entertain Hooke's proposal as likely, although they rejected the geological consequences that Hooke thought followed from this theory.[16] Independently of Hooke, members of the Paris Academy debated the same theory, one of the

[13] For Aubrey, see Hunter 1975, 58. The concept of the "accident" is treated in Rappaport 1982, 38–44.

[14] Hooke's theory was revived by Raspe [1763] 1970. Compare Davies 1969, 44, who remarks that reports of earthquakes would have impelled naturalists to recognize their importance.

[15] Ray 1693, quotation from p. 24, discussion on pp. 9–18. Ray's views, first published in 1692, aroused Hooke's displeasure, for Ray seemed to have appropriated the theory that Hooke had not published; reported in a letter from Aubrey to Anthony à Wood, 13 February 1691/2, in Williams 1969, 517. For the association of Ray with both earthquake theory and vulcanism, see discussion of Moro later in this chapter and Buffon's critique in Chapter 8. Abraham de la Pryme, in *PT* 22 (1700), 677–87, would transform the Flood into a giant earthquake, but there is no reason to suppose he was influenced by Hooke. Cf. Porter 1977, 81; also Pryme 1870, 236–37, for a reference to Hooke in about 1700.

[16] Rappaport 1986, 135–36.

astronomers arguing at length that he had found historical data showing a gradual change in the angle of the ecliptic. As with Halley and others, however, the Parisians did not link the astronomical problem with the history of the earth's crust.[17]

Before the Lisbon earthquake of 1755 renewed interest in the general subject, the only known traces of Hooke's theory occur in articles by two Fellows, exchanging views in the pages of the *Philosophical Transactions.* Both Henry Baker and William Arderon allowed that even a small shift in the earth's center of gravity could plausibly account for changes in the location of sea basins and land masses. When Arderon specifically mentioned Hooke, he rejected the theory of earthquakes for reasons Hooke himself had heard from his colleagues: "why don't such Things happen now? And why is all History silent upon this Head?"[18] For Arderon, displacements of land and sea had occurred slowly enough to be unremarked by chroniclers and unobservable by moderns. Dramatic events like earthquakes were another matter: the silence of past records and the infrequency of upheavals in modern times both meant that Hooke had been wrong.

GOTTFRIED WILHELM LEIBNIZ

That Leibniz's chief geological text was not published until 1749 should not be taken to mean that his views remained unknown until that date or that the brief, sometimes cryptic sketch published in 1693 constituted the only clue to his theory of the earth. In addition to the outline printed in the *Acta eruditorum,* more detail appeared in 1706 in the *Histoire* of the Paris Academy (a volume actually published in 1707), and Leibniz reinforced his theory to some extent in 1710 in both the *Miscellanea Berolinensia* and the popular *Théodicée.* Furthermore, he now and then discussed aspects of his theory with his correspondents, on one occasion also transmitting to the Royal Society his reflections on the elephant bones made famous by Tentzel.[19]

[17] That Hooke knew of, and welcomed, French research is indicated in Hooke 1705, 536–40, text dated 1695. The results of French investigations are in *HARS* [1716] 1718, 48–54, and in Fontenelle's eulogy of Louville, 1733, in Fontenelle 1825, 2:321–22.
[18] Quotation from Arderon in *PT* 44 (1746), 280 note (k). Baker's article, without explicit reference to Hooke, is in *PT* 43 (1745), 331–35.
[19] The 1693 text is also available in a translation by Oldroyd and Howes 1978, and that in *Miscellanea Berolinensia* in a translation by Ellenberger 1988, 2:145–46. For the correspondence with Tentzel and the Royal Society about elephant bones, see above, Chapter 4. The best synthesis of these and other fragments is by Sticker (1967) who has also (1971) examined Leibniz's correspondence with Louis Bourguet.

The Leibnizian theory was notable in going back, more explicitly than Steno or Hooke had done, to a pre-sedimentary and pre-human period in the formation of the earth's crust. In this respect Leibniz confessed his own penchant for what he called the Cartesian view that the earth had once been a star, that is to say, a molten body whose heat had diminished in the course of time. Pre-sedimentary cooling could be detected, he argued, in the oldest "vitrified" materials of the earth's crust, with fragmentary evidence visible in the sands that Leibniz regarded as detritus from earliest igneous deposits. (Although life presumably did not yet exist, Leibniz later wondered whether the effects of residual heat might be detected in fossils. He asked John Woodward if, in his great collection, any specimens showed the effects of fire. Woodward could find none.[20]) After this igneous period, waters gathered so as to cover most or all of the earth, depositing organisms now found as fossils. In the 1693 sketch, organisms are bafflingly described as being "brought in by sea or by land." In the 1706 text offered to the Paris Academy, it becomes clear that Leibniz distinguished between marine organisms and those found in slates that had once been the muddy waters of lakes. Further clarified in 1706 and 1710 was Leibniz's view that some fossil forms, not corresponding to living species in Europe, had been transported (in some unexplained way) from the Indies to the strata of Europe.[21]

Apart from his treatment of an igneous phase in the earth's history, Leibniz's theory contained a number of elements peculiar to him and unusual at that time. He did envision a pre-human past for which historical documents did not exist; at the same time, he also thought that Genesis, reinterpreted, provided clues to that past if one properly read the "separation of light from darkness" to mean—and Leibniz is here as obscure as Genesis—a distinction between the igneous earth and the dark (dead) planets.[22] Nowhere did he question that the Flood should be regarded as an event in history, but the several texts differ about whether floodwaters covered all or only most of the earth. From the earliest sketch, however, it seems that he also thought marine sedimentation to have had a more or less continuous history, beginning before the Flood. For such remote

---

[20] The correspondence with Woodward is mentioned by Levine 1977, 112, without indication of date.

[21] Transport from the Indies has been attributed to Fontenelle, adding to Leibniz, by M. Carozzi 1983, 26–27, because the topic is missing from the *Protogaea*. But such comment does occur in Leibniz 1710, 119. A transcription of Leibniz's 1706 text has now been located in AAS, P-V, t. 25, fols. 336r–338v (4 September 1706); reference to the Indies occurs on fol. 337r.

[22] Leibniz 1693, 40, and further elaboration in a letter to Bourguet, 22 March 1714, in Leibniz 1875, 3:564–66.

periods—and indeed also for more recent, postdiluvial ones—he encountered the common problem of the paucity in human records; when he could find such evidence, notably about shorelines and watercourses, he wove them into his arguments, but he elsewhere confessed to the vagueness or the absence of human documents.[23] Perhaps as a result of these gaps, Leibniz's texts are not especially clear about establishing a sequence of geological events, and the reader can only infer that the nonmarine forms in German slates dated mainly from postdiluvial times. If he was willing to suggest the transport of some organisms from the Indies, for the larger animals, like Tentzel's elephants and Spener's crocodile, he rejected the Flood as agent of transport and deposition; instead, he more than once proposed that these fossils be interpreted as creatures formerly native to German seas.

To track the *fortuna* of these texts is no easy matter, for books, not articles, received more attention in the journals of the period. Still, Leibniz's fame guaranteed that journals would not wholly neglect articles by him in the volumes of the Paris and Berlin academies. Reviewers, however, did not see here a theory of the earth, but rather a series of discrete observations and ingenious suggestions. The *Journal des savants,* for example, singled out from the Paris Academy's *Histoire* the Leibniz experiment designed to suggest how moulds of organisms might be preserved in the strata when the organisms themselves had vanished. Even more briefly, the Trévoux journalists reported the same experiment, without mentioning Leibniz's name.[24] When in 1710 Leibniz defended his theory as a way to explain Spener's crocodile, the *Journal des savants* gave much space to the fossil and little to Leibniz's arguments.[25] As we shall see, later readers, Vallisneri and Buffon among them, also used selectively particular elements in one or another of Leibniz's short texts.

In fact, the several Leibniz texts published during his lifetime are not as clear or coherent as Leibniz himself thought. Buried in his mind, but not obvious to readers, were implications he associated even with the brief outline carried in the *Acta eruditorum* of 1693. When he replied to Tentzel and Spener, he invoked that little sketch as if it would help to explain the large fossil finds of the next decades. Even Fontenelle, who corresponded with Leibniz and summarized his geology in 1706, somehow concluded, eventually and inexplicably, that Leibniz was a diluvialist.[26] Nor did the most

---

[23] Letter to Schmidt, 30 July/9 August 1691, in Leibniz [1749] 1993, 188–91. Leibniz 1710, 118–20. See also Leibniz [1749] 1993, chap. 19 of *Protogaea.*

[24] *JS* (suppl. for 1708), 24. *Trévoux* (April 1708), 585. *HWL* (January 1708), 23–24.

[25] *JS* (15 December 1710), 657–63.

[26] Fontenelle's comment occurs in his eulogy of Leibniz, 1717, in Fontenelle 1709, 1720, 2:286. I can find no explanation for this, except to wonder if Homer nodded.

popular and widely disseminated of Leibniz's works, the *Théodicée* (1710), clarify his theory of the earth, here presented in subordination to his controversial philosophy of pre-established harmony.[27]

In 1719, not long after the death of Leibniz, J. G. Eckhart proposed to publish the (unfinished) *Protogaea,* but first sought the judgment of the Paris Academy.[28] Although we do not know exactly what worried Eckhart, the Academy's response implies that he asked about the possible conflict between geology and Scripture. (What conflict: the apparently long prehuman history of the earth? the nonliteral reading of the narrative of the Flood?) As a formal declaration—a "certificat en forme"—the Academy announced that it did not consider matters of religion to be within its purview, but, being "composed of Christians knowledgeable in their faith," found nothing in the *Protogaea* that could be construed "as contrary in any way to what the Bible teaches us." True to its philosophy, however, the Academy could not endorse even this "ingenious hypothesis" about the earth's history. What it did endorse was the general enterprise of introducing a local history of mankind—the *Protogaea* having been intended as a preface to the history of the House of Brunswick—"by comparable expositions of the former [physical] condition of the region."

AT THE ACADEMY OF SCIENCES, PARIS

From almost the beginning of his long tenure as permanent secretary of the Academy (1697–1740), Fontenelle conceived of his function as making clear to readers both the content and the context of the technical writings of the members. As he put it in the preface to the first annual *Histoire et Mémoires,* he recommended that his own "clarifications" in the *Histoire* be read in conjunction with the specialized papers in the *Mémoires.* Thanks to his intelligence and his inimitable style—that successors tried in vain to emulate—the *Histoire* was judged by one contemporary to have achieved a clarity not visible and perhaps not attainable in the relatively abstruse *Mémoires.* This judgment can be confirmed in the writings of contemporaries who found Fontenelle more understandable and more quotable than the academicians whose researches he was ostensibly "summarizing."[29]

---

[27] Leibniz [1710] 1952, secs. 243–45.

[28] What follows is drawn from Scheel in *Leibniz* 1968, 56–57, where the full text of the Academy's response, written by Fontenelle, is printed. Added comment by Roger in *Leibniz* 1968, esp. 143.

[29] The next paragraphs are based on Rappaport 1991a, and footnotes here will indicate only additional material not in that article. See the judgment of Camusat 1734, 2:169, on the *HARS.*

Since Fontenelle inserted into these summaries his own version of implications, commentary, judgment, and possibilities for future research, his remarks in the *Histoire* became an integral part of what some readers perceived as a geological viewpoint they associated with the Academy at large.

When Fontenelle first encountered the problem of fossil shells, this subject had not yet aroused much interest in the Academy, despite some earlier presentations of this or that fossil deemed a curiosity worth reporting. Occasionally, one or another of the astronomers—the elder Maraldi, the La Hires father and son—showed more than passing interest in fossils, aroused by their familiarity with Italian terrains or with Italian treatises.[30] What seems first to have attracted Fontenelle's attention, however, was Tournefort's theory of "seeds," presented to the Academy in 1702. Here Fontenelle detected a promising analogy: if plants grow from seeds, then perhaps organic-like remains in the rocks have similar origins. In the next few years, the Academy received communications from the Scheuchzer brothers, Leibniz, and Marsili, and Fontenelle began to contemplate the earth's history in a fashion that had nothing to do with Tournefort's seeds. Without much explanation—but with a brief allusion to the problem of specific gravities—Fontenelle never considered diluvialism as a serious explanation of marine sedimentation. When, for example, treatises by the Scheuchzers were examined by academicians in 1708, one *rapporteur* pointed out that some evidence conflicted with diluvialism, the other *rapporteur* asserting that the elder Scheuchzer had made a plausible case for the role of the Flood in depositing fossils. Fontenelle modified both reports, especially the favorable one, to focus clearly on the inadequacy of Woodwardian theory.[31] His own sympathies inclined in the direction of Leibniz and the clues he found there to a nondiluvial, sequential history of the earth's crust.

Fontenelle's exposure to diluvialism seems to have come mainly from the Scheuchzers, rather than from any previous knowledge of the writings of Burnet or Woodward. That he would not move in this direction is not really surprising, in view of his "libertine" heritage. Libertinism is almost indefinable, as the word was used to label anyone who promoted unconventional views of history, morality, and religion. In Fontenelle's case, his own earlier writings indicated his daring allegiances. In his *Origine des fables*, probably written before he became an academician, Fontenelle had argued

---

[30]   See Chapter 4, at nn. 76 and 84, for these early texts.
[31]   Reports by Homberg and Maraldi, AAS, P-V, t. 27, fols. 289r–291v (4 August 1708), 337r–339r (17 November 1708). Cf. *HARS* [1708] 1709, 30–33.

for the view that ancient myths contained no historical information but should be seen as products of primitive ignorance and superstition. Early man had had little knowledge or understanding of nature, and had tried to explain the incomprehensible by attributing to gods what later studies showed to be merely the workings of nature. In this superb analysis of primitive mentalities, Fontenelle briefly exempted the Hebrews from the category of ignorant primitives, but there is every reason to think that this fleeting reference should be taken as a graceful bow to convention.[32]

With this background behind him, Fontenelle was clearly no candidate for conversion to diluvial geology. In the decade before 1716, he assimilated such information as he thought valuable in the works especially of the elder Scheuchzer—chiefly a knowledge of the great variety of fossils, marine, freshwater, and terrestrial—and had read with attention the theory of Leibniz. He also found significant the observations of an obscure academician named Saulmon and a sketch submitted to the Academy by L. F. Marsili. In Leibniz, Fontenelle found and accepted his friend's rejection of the concept of *lusus naturae*. He noted also Leibniz's distinction between marine and freshwater organisms found as fossils, and wished that Leibniz had proposed some means of transport for fossil forms discoverable in European strata but resembling species still living in the West Indies. In Saulmon's descriptions of terrains in Normandy and Picardy, Fontenelle found it notable that deposits of larger and smaller gravels were not mingled in any one place and were not to be found at the highest elevations explored by Saulmon. To Fontenelle, this meant that the higher ground had never been under the sea—not even in a universal Flood—and that further study of the motion of bodies in fluids ought to shed light on how marine currents had distributed the fragments observed by Saulmon. Details of that distribution should be mapped in order to clarify the locations and operations of marine currents.[33] In Marsili's *Brieve ristretto*, published in 1711 but presented to the Academy in 1710, Fontenelle found well-developed the distinction between what Marsili called the "essential" and the "accidental" parts of the earth's crust. The former, later to be called "primitive," Marsili dated from the Creation; the latter consisted of very thin sediments deposited gradually on this basic structure.[34]

[32] For dating of *Origine des fables*, see Chapter 2, n. 121. The best study of Fontenelle's early career, to about 1702, is by Niderst 1972. The more recent biography by Niderst (1991) says very little about Fontenelle and the Academy, except for comments about individual eulogies, each inserted into the strictly chronological framework of the book.

[33] See also the eulogy of Guglielmini, 1711, in Fontenelle 1709, 1720, vol. 2, esp. 73–80.

[34] The noncommittal reply of the Academy was published in Marsili 1711, 22–23. Marsili seems not to have noticed that approval of his research did not mean endorsement of his conclusions.

Before any Parisian member of the Academy paid much attention to the earth's history, Fontenelle produced a synthesis of the several studies he had been encountering.[35] The Flood was not a consideration here, as Fontenelle declared it might be useful to explain only those fossiliferous deposits "which are located in places where no other event (*accident*) could have carried them, and where one cannot believe that any water has been present since the time [of the Flood]." Having relegated the universal Flood to this trivial role, Fontenelle moved to his reconstruction of the earth's past. Here he ignored Leibniz's vitrified core and Marsili's "essential" part of the earth, and he concentrated instead on sedimentation. There had existed a universal ocean in which lived "the oldest inhabitants of the earth"—that is, fish and mollusks. Parts of the earth's crust had then collapsed here and there, allowing water to drain from the surface and producing a few mountains. On the exposed land there appeared the non-marine flora and fauna reported as fossils by Leibniz and the elder Scheuchzer. To explain how these creatures had in turn been buried and petrified, Fontenelle suggested that "local and less sizable changes (*révolutions*)" had occurred, parts of the exposed land sinking beneath the sea or to the bottom of lakes. These small crustal collapses might also have produced some mountains, whereas other elevations might have resulted from earthquakes and volcanic eruptions.

Fontenelle never claimed any originality for this synthesis, and some elements are indeed familiar in a general way in the writings of his contemporaries and predecessors. The universal ocean and crustal collapses, for example, are reminiscent of Descartes, Burnet, and Leibniz, but Fontenelle never tried to explain the formation of this ocean—it is simply there, stocked with marine life. By the time that Fontenelle produced his sketch of the earth's history, Woodwardians and their critics were examining in detail the sequence of sedimentary strata, but this subject Fontenelle ignored, perhaps because such minutiae did not enter readily into his general vision of his own role as interpreter of the larger meanings of specialized research. One of those larger meanings, however, was Fontenelle's suggestion that there might be some correlation between forms of life and past conditions in the earth's history. Nothing remotely evolutionary should be read into such statements, and Fontenelle never did comment on Leibniz's argument that fossil elephants and crocodiles could be marine

---

[35] That Fontenelle wrote the synthesis attached to his summary of Geoffroy, in *HARS* [1716] 1718, 8–16, was proposed by Rappaport 1991a, 291; that can now be confirmed by inspection of Geoffroy's text, AAS, P-V, t. 35, fols. 286v–304r (19 August 1716).

ancestors of modern animals.[36] What Fontenelle proposed was a very general scheme, starting with early marine (nondiluvial) deposits, succeeded by an era in which the nonmarine fossils reported by Scheuchzer and Leibniz had been buried.

After this sketch of 1716, Fontenelle began to encounter detailed works by resident members of the Academy. The first of these, produced in 1718, was Antoine de Jussieu's memoir on the fossil ferns of Lyonnais. Noting that these deposits included marine shells, Jussieu concluded that the fossil plants had been deposited in the same (nondiluvial) sea as the fossil mollusks. But how had "exotic" species of plants been transported to Europe, and could such fossils actually be identified with species still in existence? The tough-minded Fontenelle had some doubts about extinction, "a rather daring idea," but Jussieu had none. Jussieu also announced a schematic history of the earth in which "most of the lands which seem to have been inhabited [by man] from time immemorial were originally covered by the waters of the sea which then, gradually or suddenly, retreated." To a diluvialist, or indeed to a careless reader, this latter comment might seem either diluvial or reconcilable with diluvialism. But Jussieu was clear and explicit: the Flood had been too turbulent to account for fossil plants whose leaves were not curled or folded but laid out "as if they had been mounted." No flood could account for this observation.

Fontenelle integrated Jussieu's memoir into the theory of the earth he had sketched two years before, placing Jussieu's fossil ferns into their chronological position in his own synthesis. After the first Fontenellian "revolution," which uncovered some land masses, terrestrial plants had begun to thrive. This second period could have been a time when various floods occurred, thus transporting plants from the Indies to the strata of France. Somewhat cautiously, Fontenelle added that the problem of transport deserved further study, as did the identification of "exotic" species with those found fossilized in France. One would like to know, for example, about the geographical distribution of living species and about the species or genera to which fossil forms belonged. Such studies in hand, one could then began to ask whether the fossils in any one stratum (in Europe) had come from identifiable regions of the earth.

If Fontenelle never developed this sketch further, its nondiluvial message did receive dramatic confirmation in 1720 in Réaumur's study of the Touraine. Here Réaumur described in detail massive deposits of marine shells, the strata showing none of the turbulence of the Flood. The sea

[36] Leibniz did present his ideas about elephant fossils to the Academy in 1706, but Fontenelle omitted the topic from his summary in *HARS*. Above, Chapter 4, n. 30.

therefore had covered the region "for a long time." Fontenelle seized on this memoir to launch another attack on diluvial geology. As he put it, the faluns of Touraine could not have been the remains of the Flood because they "could only have been deposited successively." With his usual graceful bow in the direction of Genesis, Fontenelle added that there did exist "many remains of the universal Flood," but he never explained just what those "remains" were.

Although Réaumur did not return to the subject of his memoir, Jussieu would produce in rapid succession several studies of fossils, in which the problem of transport puzzled both the author and Fontenelle. Had there been floods in the past? Had the sea changed its basin? Could any change of the latter kind be attributed to floods, to "extraordinary tides," or to "revolutions" of some sort? Notably absent from these discussions, as from Réaumur, too, was any reference to human history, human witnesses, and, more generally, the scheme of ancient history deducible from Genesis. When, for example, Jussieu produced a study of primitive flint implements, he failed to ask those questions commonly raised by his contemporaries— such as, how did primitive peoples lose that knowledge of iron that had existed in Noah's day? When Jussieu worried about transport and Réaumur about the retreat of the sea from Touraine, neither man seems to have searched in ancient Roman writings for clues or for testimony. (Whether they actually did so but found nothing is not outside the realm of possibility.) On one occasion, Jussieu even absorbed into his own prose both Fontenelle's language of successive revolutions and a clear commitment to nonwritten evidence.

> I have postulated several extraordinary revolutions antedating us by a long time, such as some flood by means of which the seas, having changed their bed, would have abandoned . . . our continent which they had formerly covered, . . . leaving deposits of marine plants and animals. . . . More such discoveries in the future will provide monuments to confirm the likelihood of these revolutions.[37]

Having only physical evidence, Jussieu carefully gave only one example of possible revolutions and affirmed the "likelihood" of his conclusions. In comparable fashion, Réaumur's analysis had also changed abruptly after his detailed description of the faluns of Touraine; faced with trying to explain how and why the sea had left the area, he briefly offered a choice of conjectures, admitting that it would be hard to decide among them.

[37] *MARS* [1723] 1725, 205.

For Réaumur, Jussieu, and Fontenelle, it seems that the earth had had a pre-human history of some length and eventfulness. Jussieu treated the subject allusively: French soil had an older history than its "immemorial" occupation by man, and fossils were an older "library" than man's books. Réaumur allowed thirty to forty centuries for one minor geological event, the departure of the sea from Touraine. In his role as interpreter, Fontenelle went further than his restrained friends, perhaps because early in his career he had already contemplated the vastness of space, a plurality of worlds, and relative perceptions of time. When, in 1727, Hans Sloane sent the Academy his long analysis of fossil elephants, Fontenelle gracefully avoided saying that Sloane attributed to the Flood the transport of these animals to Siberia and elsewhere. In Fontenelle's language, "It is easy to see what conclusion we wish to arrive at with Monsieur Sloane. There have been great upheavals (*bouleversements*) on our Earth, and especially great floods." This elliptical refusal to mention the Flood represents Fontenelle's consistent position that the earth's history is not one recorded in human texts; geologists, not historians, could decipher that remote past, using as their materials "kinds of histories written by the hand of nature itself."

The several texts in the *Histoire* and *Mémoires* attracted considerable attention, the Academy itself enjoying enormous prestige and Fontenelle's prose being powerfully seductive. Even the brief summary of Leibniz in 1706, for example, would be analyzed at length by Vallisneri fifteen years later, Vallisneri clearly finding this text more pertinent to his research than the earlier sketch by Leibniz in the *Acta eruditorum*. In Bordeaux, Fontenelle's prose was discussed at the local academy, and as late as 1746 Pierre Barrère would still be using (unacknowledged) quotations from the *Histoire* of more than twenty years earlier.[38] One of the Jesuit contributors to the *Mémoires de Trévoux* singled out Jussieu's memoir on fossil ferns as meriting refutation. On the one hand, Father Castel argued, it seems "the most natural" explanation to say that the sea for a long time covered what is now land; on the other hand, such a view "seems hardly to be based on History" and was espoused in ancient times especially by those who thought the world eternal. Antonio Conti, then living in Paris, thought the Jussieu-Castel conflict worth reporting in detail to Vallisneri in Padua.[39]

More significant than these examples is the fact that some writers began to see in the Academy a "school" of geologists characterized by the belief that marine fossils testified to "the long presence" of the sea. To the widely

[38] Barrère 1746, notably 50–51, and Barrière 1951, 167. Recognizing these phrases requires familiarity with the prose of *HARS,* for early writers supplied no footnotes.

[39] *Trévoux* (June 1722), esp. 1090–93. Conti's comments are in a letter to Vallisneri, n.d., in Conti 1972, 388–91.

read diluvialist Dezallier d'Argenville and the quasi-diluvialist Louis Bourguet, the Academy had become the center of what Dezallier called the revival of an ancient philosophy: "The continents used to be part of the sea floor, so that what now is land was formerly water."[40] Bourguet coupled members of the Academy with a number of other writers—including Scilla, Boccone, Leibniz, and Vallisneri—whose emphasis on the "long" submersion of land masses constituted a theory rivaling the two basic kinds of diluvialism, the Burnetian and the Woodwardian. If the Academic school actually consisted of two Paris naturalists and Fontenelle, the impression would linger in 1749 that the Academy had for decades been a hotbed of impiety and materialism and a breeding ground for Buffon's theory of the earth.[41]

VALLISNERI AND MORO

Vallisneri surely would have disliked Bouguet's coupling of his name with some Parisians, for he had a well-developed antipathy toward the Paris Academy, which he saw as epitomizing the worst in French cultural arrogance. That he nonetheless read attentively at least some parts of the Academy's *Histoire*, and perhaps the *Mémoires* as well, is evident in his chief geological publication, *De' corpi marini* (1721). For reasons unknown, he did not revise the geological parts of this text in its second edition (1728), by which time he had perhaps become familiar with the memoirs of Réaumur and Jussieu.[42] Among the villainous French, Vallisneri found only

[40] Dezallier 1742, 159. Bourguet 1729, 177–80. Cf. Guérin 1745 in Gosseaume 1814, 1:68–70, and Bertrand 1766, 32–33.

[41] This was the charge later leveled at the Academy in the aftermath of Buffon's publication; see below, Chapter 8. Remarkably little of geological interest appeared in the Academy's *Histoire et Mémoires* for about twenty years after the writings of Réaumur and Jussieu, but a kind of resurrection of their general viewpoint can be found in Sauvages, *MARS* [1746] 1751, 713–58. Sauvages's memoir was actually presented to the Academy in 1749 or 1750, as indicated in a note to the text (713n). When Sauvages mentioned (714) "un des plus beaux génies de [ce] siècle" who predicted that scientists would be able to "deviner l'histoire, quoique si ancienne," of the earth's "revolutions," he was quoting Fontenelle, *HARS* [1721] 1723, 3–4. A detailed analysis of two Sauvages memoirs can be found in Ellenberger 1980, 54–57.

[42] Revisions in 1728 consisted of various appendices, not modifications of the body of the work. There are several studies of Vallisneri as geologist, but see Rappaport 1991b, 83–87, for an attempt to trace his views as developing during some years. References to Jussieu and Réaumur are in the sketch written by his son, in Vallisneri 1733, 1:xix. It is not clear if Vallisneri senior read these memoirs, or if Vallisneri junior merely mentioned such studies as harmonizing with what his father believed. In a letter to Bourguet, 8 June 1721 (Neuchâtel, MS 1282, fols. 247–48), Vallisneri expressed his desire to read Jussieu's memoir of 1718.

Jean Astruc, the provincial not associated with the Academy, deserving of some admiration.

De' corpi marini does not offer a theory of the earth, the text described by Vallisneri himself as an exercise in skeptical questioning: it was easier to criticize the theories of others than to provide answers to all the problems raised. After rapidly disposing of "plastic forces"—those "ancient occult faculties of generation"—and defending the organic origin of marine fossils, Vallisneri devoted most of his attention to a critique of Burnet and especially Woodward. In the course of his analysis, certain of his own views and preferences become visible.

Vallisneri's commitment to naturalistic explanation, to the "inviolable laws" of nature, is evident throughout his book and most notably in his assertion that one should seek explanations "without violence, without fictions, without assumptions, without miracles."[43] Easier said than done, this ideal led him to reject the few available attempts to explain changing levels of the sea, even while he was quite sure that such changes had occurred and were still occurring. Where had all the water come from, and where had it all gone? (With some reluctance, he rejected the traditional hypothesis, espoused by Leibniz, that there existed caverns within the earth into which the sea could retreat.) What mechanism could one propose to explain changing levels of the sea? Could the sea lift and transport bodies as heavy as shells? In fact—and here he was aware of Robert Boyle's tract about the sea floor—did not the tranquillity of the sea's depths militate against any accumulation of marine sediments? To answer the latter question, one would have to posit turbulence of the sort used by Woodward, but such fictions still could not account for the orderly assemblages of fossils in marine sediments.[44] What Vallisneri found so attractive about Jean Astruc's analysis of the Mediterranean littoral was that here, at least, one could document how the sea had behaved in the past. But could coastal fluctuations be extended to explain marine deposits at high altitudes and far from modern seas?

If Vallisneri could provide no answers, nor could contemporaries like Jussieu and Réaumur. Readers of *De' corpi marini* were thus left with a number of assertions, suggestions, implications, and lacunae. It seemed evident to Vallisneri that fossiliferous sediments had been left by the sea, and that orderly sequences of such strata implied a series of marine incur-

[43] The quotation is from Vallisneri 1728, 34–35. How to apply uniformity of laws to *biological*, not geological, phenomena preoccupied Vallisneri and his contemporaries far more than did the history of the earth. The classic analysis is by Roger 1993.

[44] Vallisneri 1728, esp. 25–27. Repeated references to Leibniz (pp. 34, 42, 46–47) reveal Vallisneri's reluctance to reject the views of a man he respected.

sions. In addition, it also seemed likely that certain low-lying areas, such as the whole plain between the Alps and the Apennines, had once been under the sea, the terrain then gradually transformed into a marsh and, ultimately, by successive deposits, into dry land. This transformation was implied not only by an examination of the strata, but also by consultation of ancient Roman texts that described the Modenese plain as marshy. Comparable retreats by the sea could also be inferred from Astruc's study of the coasts of Languedoc, but Vallisneri was also uneasily aware that the sea seemed to be *rising* in some localities, as at Venice.[45]

For none of these phenomena could Vallisneri produce a causal explanation. He could only conclude that the earth had had a history of local alterations that ought to be understood as natural and law-abiding. Nor did he find much help in historical texts, except for minor matters like the Modenese plain and the silting up of ports, but such texts confirmed his belief that the earth had had a long history of incremental changes. Indeed, even in his earliest correspondence, Vallisneri had proclaimed that ancient texts were silent on matters geological and that the earth was older than human records implied.[46] Intimate friends such as Muratori, Conti, and Porcia knew that Vallisneri was in trouble conceptually and historically, although none could appreciate the geological evidence that bothered Vallisneri. For Conti and Porcia, ancient human records, possibly older than Genesis, might be helpful, but how Vallisneri responded to such suggestions is not known.[47] Another intimate friend, Louis Bourguet, struggled for years with Vallisneri's insistence that there was no natural or rational way to explicate the Flood. To Bourguet, the Flood, an undeniable historical event, must have left traces of its occurrence. To Vallisneri, any naturalistic interpretation raised so many insuperable problems that the Flood was best left as a miracle, irrelevant to the deposition of "strati sopra strati." In setting the Flood aside, Vallisneri had in mind not only the geological evidence but also the fate of Galileo. What this meant to him was not an avoidance of natural investigation, but rather the need to refrain from mingling science with biblical exegesis.[48] In effect, he agreed with Galileo that science should be autonomous.

[45] Ibid., notably 35–41, 43–44.

[46] For example, the several letters in Vallisneri 1991, 1:253, 370, 441, and Vallisneri 1715, 2:25. Also, his son's preface in Vallisneri 1733, 1:xx.

[47] Conti and Porcia both responded to *De' corpi marini* by citing *human* records making the earth perhaps older than in Genesis. Conti 1972, 388–91, and letter from Porcia to Vallisneri, 24 July 1721, in Brunelli 1938, 230.

[48] Rappaport 1991b, 84. Vallisneri's views are amply expressed in letters to Bourguet, the first of these now published in Vallisneri 1991. Bourguet's few extant replies are analyzed by Tucci 1983.

Some of Vallisneri's readers were baffled, for the author had criticized existing syntheses but provided no alternative of his own. *De' corpi marini* attracted no attention from major book-review journals until the publication of Vallisneri's collected works by his son, and at that date (1733) reviewers complained that skepticism was all very well but hardly admirable. Even Bourguet, after years of frank correspondence, could not understand his friend's views and wondered why the Paduan refused to explicate a miracle.[49] From the Royal Society, however, Vallisneri would eventually (in 1726) elicit an appreciative response, and he then arranged to have the second edition of *De' corpi marini* sent to London. Perhaps the warmest tribute, albeit a posthumous one, came from Giacomo Spada, who in 1737 called his own book an effort to show that "the petrified marine bodies, found in the mountains near Verona, are neither tricks of nature nor diluvial, but antediluvial." In the latter word, Spada echoed the interpretation maintained for years by Vallisneri.[50]

Just when Anton Lazzaro Moro read Vallisneri's treatise is unknown, but the two heroes of his *De' crostacei* (1740) were Vallisneri and Newton. Moro shows familiarity with more than one of Vallisneri's publications, his allusions ranging from descriptions of strata in Vallisneri's *Lezione* of 1715 to approving citations of a variety of methodological maxims.[51] In addition, Moro did his own independent reading of Burnet and Woodward, but his long critique of diluvialism gives his treatise a distinct family resemblance to Vallisneri's *De' corpi marini*. The Newtonian ingredient, based on the "rules of reasoning" in Newton's *Principia*, Moro adapted to his own ends.

In his letters, sometimes more clearly than in his book, Moro explained what he conceived to be the chief shortcoming of existing geological theories: all assumed that the sea had somehow risen to cover the continents, rather than that land had been elevated out of the sea. That the latter possibility had been neglected he found startling, especially because just such an event had occurred recently with the appearance in 1707 of a new

---

[49] These reactions are documented by Rappaport 1991b, 87; not mentioned there is Niceron 1731, 16:73–88, where the commentary is neither critical nor querulous. The *silence* of one journal has been interpreted by Kurmann 1976, 45, to mean opposition; that is possible, since one of the editors was Scheuchzer, both a friend of Vallisneri and a vigorous diluvialist. Vallisneri is not listed in the long bibliography by Scheuchzer 1732, 1:x–xxxiv.

[50] For London, letters from Vallisneri to Bianchini, 3 November 1728, and from Dereham to James Jurin, 1723–26, in Royal Society, "Early Letters," s.v. Bianchini, Dereham. Spada is discussed in Pomian 1987, 259. Although Gimma 1730, 2:264–67, does not mention Vallisneri, he uses many of the same texts and arguments to show that local changes are preferable to the Flood.

[51] E.g., Moro 1740, 2–3; bk. 1, chap. 29; bk. 2, chap. 12.

Aegean island near Santorini. As Massimo Baldini has remarked, with only slight exaggeration, the whole of Moro's theory of the earth was based on this single observation.[52] Nature had supplied a clue to the past, and Newton was invoked to justify Moro's emphasis on that clue. As Moro put it, Newton's Rule III, that nature works uniformly and is always "consonant to itself," permitted one to argue analogically from the elevation of one new island to the emergence from the sea of all mountains. Recalling Newton's remark that a single experiment could invalidate a received theory, Moro concluded that the single example of the new island could undermine existing conceptions about the earth's past.[53] Furthermore, Newton's Rule I—as well as other maxims discoverable in Aristotle and Vallisneri—dictated that philosophers not multiply causes unnecessarily, for nature does not use many causes where few will do. To Moro this rule meant that it would be illegitimate to attribute the elevation of only some mountains, but not others, to volcanic forces.[54]

In addition to what Moro called "metaphysical" arguments—those stemming from Newton and other framers of maxims—*De' crostacei* made use of a variety of available observations while attempting to explain away some less convenient ones. Moro was aware, for example, of Robert Boyle's tract (or summaries of it) showing that the sea floor remains undisturbed even during storms; this was grist for Moro's mill, showing that sedimentary strata could not have been accumulated by the sea. Nor, according to Moro, could the sea have assembled marine fossils in curious, family-like groupings; on the contrary, the sea could not have been so selective, but would have left unsorted, miscellaneous populations of organisms in all strata. Less easily reconciled with his ideas were the difficulties for which the Venetian Republic called in expert engineers, namely, the level of the Adriatic seemed to be rising at a rate that demanded the attention of specialists in hydraulic science. Moro could not dispute the fact, but he struggled to find an explanation. Perhaps—and here he violated his own Newtonianism—the modern rise of the sea was not relevant to an analysis of the remote past. Alternatively, was the modern rise part of a long trend dating back to those early times when seas had been at a lower level than in recorded history? In proposing the latter, Moro knowingly contra-

---

[52] M. Baldini in *Moro* 1988, 124. Moro to Giovanni Bianchi, 3 October 1739, in *Moro* 1987, 88. One of the Bolognese academicians, Tiberio Codronchi, had earlier attempted to use new islands as a key to the earth's history; see Tega 1986, 1:78. Only a summary was published in *Commentarii* 1 (1731), 205–7, and I have encountered no contemporary references to these proposals.

[53] Moro 1740, 239.

[54] Ibid., 263. For Newton's Rule II, ibid., 265.

dicted the prevailing view that the seas had once been higher than at present. Furthermore, he could not explain the causes of the apparent rise of the Adriatic—where was the seemingly additional supply of water coming from?[55]

In short, Moro's treatise had built into it a number of methodological problems as well as conflicts with available evidence and contradictions of conventional wisdom. The theory itself can be summarized in few words. Moro began with the new Aegean island and moved quickly to the proposition that all land had once been under the sea; the volcanic island then provided the clue to how all land masses had emerged from the sea. Nothing in this theory had to do with worldwide "catastrophes," Moro's volcanic eruptions being local phenomena occurring at a variety of times and responsible for the elevation of single mountains and other tracts of land. All these eruptions had taken place locally and many times in succession, each such episode lifting marine organisms and sediments from the sea floor and depositing them on a mountain's flanks. (Eruptions had the power to do this, as Boyle's tranquil seas did not.) As for the level plains, Moro attributed such deposits to debris brought down from the nearest high (to him, volcanic) elevations.[56]

If Moro's theory is fraught with problems, only one of these aroused his own sustained attention: historical evidence. Like Robert Hooke, who looked for human testimony about the frequency of earthquakes, Moro wanted to find witnesses to the frequency of volcanic eruptions. Hooke failed, and he argued that such events, like other major episodes in human history, simply had not been written down. Human history contained gaps, and Hooke argued that fossils might fill in those gaps. For Moro, dealing with a more continuous written history of Italy than the sparse fragments for northern Europe, there were no perceptible gaps during which many volcanic eruptions, unreported by chroniclers, could have occurred. Moro's eruptions thus got moved to the only relatively vacant place in history, namely, the days before the creation of Adam. In fact, Moro had already tried some naturalistic exegesis of Genesis in the privacy of his correspondence; the unfavorable response may have been enough to move him away from too much reinterpretation of the Bible.[57]

[55] Ibid., 179 (no explicit mention of Boyle's name); bk. 1, chaps. 27–29. Questions about the sea level are clear especially in letters to Bianchi of 1739–40, in Moro 1987, 91–92, 93, 94–95, 96.

[56] Moro 1740, bk. 2, chaps. 10–12. For a modernized version of Moro's theory , see the *DSB*. A good analysis is in M. Baldini 1975, chap. 6.

[57] Piutti in *Moro* 1988, 96–97. Piutti does not quite see that the evidence she quotes shows the lack of historical gaps and thus the retreat to reinterpreting the days of Creation. For his earlier efforts to explicate Genesis, see Moro 1987, 62–63, 132.

Given these historical, scientific, religious, and prudential considerations, Moro emulated Vallisneri in insisting that the Flood was a miracle that no geologist could hope to explain. At the same time, he tried a rough correlation between his geological theory and the biblical days of Creation, pointing out that even early Church Fathers had been puzzled by the measurement of "days" before the sun was created on the fourth day. Abandoning such efforts to fathom mysteries, Moro explained that God had established "second causes" and normally did not interfere with their operation. The earth's crust had therefore evolved in a more or less gradual manner, beginning on the third "day," when a spherical earth, covered by water, had begun to suffer the first volcanic disturbances. Succeeding "days" had seen the appearance of new forms of life, and these creatures, now extinct, could be found in the materials emitted by volcanoes. God's administrative role in allowing natural laws to function unimpeded meant that each "day" lasted as long as necessary for natural materials and organisms to go through the normal cycles of growth, multiplication, accumulation, and so on. Although each day was thus an epoch, Moro could not estimate duration, and he used phrases like "hundreds of years" and "thousands of years."[58]

In addition to his vulcanism, Moro attempted a classification of the earth's rocks into two divisions, which he called primitive and secondary. The former, crystalline and unchanging, he believed corresponded to what L. F. Marsili had described as the "essential" part of the earth's crust. The secondary were the successive layers emitted by volcanoes; this stage in the earth's history had begun on the third "day" and had been in progress ever since.[59]

In the dozen years or so after 1740, Moro's book attracted limited notice, much of it unfavorable. It was quickly reviewed in the Florentine *Novelle letterarie*, the writer concentrating on Moro's critique of diluvialism; for reasons unknown, the journal abandoned any effort to explicate Moro's own theory.[60] If a degree of admiration or at least interest was expressed in the *Bibliothèque raisonnée* and the encyclopedia of Ephraim Chambers, these and other writers immediately noticed the obvious weakness in Moro's evidence: where in the sedimentary strata and their fossils could one detect traces of heat or fire? More generally, how could one reconcile the violence

[58] For God's role and the extension of time, see Lino Conti in *Moro* 1988, 49–74.

[59] Moro 1740, esp. 262–63, 284. In Altieri Biagi 1983, 448, n. 1, the editors suggest that Moro had read J. G. Lehmann's classification of mountains. Lehmann's study was not published until 1756.

[60] *Novelle letterarie*, 17 February 1741, cols. 102–7; 3 March 1741, cols. 133–37; 12 May 1741, cols. 296–99.

of eruptions with the obviously gradual and nonviolent process of marine deposition?[61] Such objections carried little weight with Scipione Maffei, who declared that Moro's theory, bizarre at first glance, became more attractive the more one thought about it. By arguing for the elevation of land, Moro had circumvented the difficulties inherent in all theories based on inexplicable changes in sea level. If Moro appreciated this tribute from the famous Maffei, he did not much care for the effort of another disciple, Carmelite G. C. Generelli, to summarize and clarify the theory. In Moro's view, Generelli had mingled too much sacred history with geology.[62]

Sometimes cited as evidence of Moro's international repute is the German summary published in 1745 and followed, six years later, by a German translation of *De' crostacei*. Moro's correspondence with the translator, however, reveals why Balthasar Ehrhart became interested in the theory. What aroused Ehrhart's enthusiasm can only be described as a confused identification and assimilation of Moro with other writers, including Leibniz, who had dealt with the igneous origins of primitive rocks. In short, the translation was undertaken because the theory was not quite understood.[63] Although the German summary did receive a noncommittal report in the *Philosophical Transactions*, other evidence suggests that Moro's work remained unknown to many naturalists. Maffei's laudatory analysis of 1747, written in the form of a letter to LaCondamine, seems to have elicited nothing from the Frenchman who would later claim that he was still unfamiliar with Moro's theory in 1754 when he made the ascent of Vesuvius. (Nor, he added, did he wish to revive Moro's "system.") Even the erudite Dezallier, who could read Italian and who liked to supply his readers with an extensive bibliography, remained ignorant of Moro until 1757.[64]

These several reactions and silences are hardly surprising. By 1740, Moro's critique of diluvialism would no longer have been novel. His emphasis on volcanoes, on the other hand, could not be taken seriously. Not only were volcanoes products of the burning of bitumens, but, like earthquakes, they were relatively uncommon. Buffon in 1749 would put volcanic eruptions and seismic shocks in the same category as "accidents" irrelevant to

---

[61] *BR* 36 (1746), 3–23; the concluding lines declare that "il a donné une étendue universelle à un phénomène particulier." Chambers, *Suppl.* (1753), s.v. "Marine Remains."

[62] Maffei 1747, 114–27. Generelli's essay, published in 1757, was criticized in a letter from Moro to Bertoli, 7 September 1757, in Moro 1987, 132–33. Other responses to Moro include the critique by V. M. Amico (1764) and the attack on Moro and Vallisneri by publicist Costantini (1747).

[63] Ehrhart to Moro, 9 September 1747, in Moro 1987, 113.

[64] *PT* 44 (1746), 163–66. La Condamine in *MARS* [1757] 1762, 377. Dezallier 1757, 72–74; references to Moro are lacking in the 1742 and 1755 editions. Dezallier did allude to Moro in a letter of 1748 (Nîmes, MS 136, fol. 134), but may not yet have read the book.

geological theories; apparently unfamiliar with Moro and Hooke, Buffon directed this criticism at John Ray. A few years later, Baron d'Holbach, who had read Moro, coupled his name with Ray's: both had produced monocausal "systems."[65]

THE DIMINUTION OF THE SEA

In 1743 Antonio Conti asked the Paris Academy whether it had any precise information about changing sea levels, and the Academy replied that such changes had been documented historically but without the detailed observation and analysis that modern science required.[66] When Conti posed the question, he had in mind "what some have detected as the Elevation of the Sea in relation to some buildings in Venice." To Italian writers, such evidence had long been familiar. North of the Alps, however, the more common perception was quite the opposite: the sea level had lowered. Diluvialists understandably argued for an exceptional surge of floodwater, which then subsided, usually into caverns within the earth. A comparable scheme can also be found in Leibniz, whereas Réaumur, without broaching the question of the absolute amount of water, had argued for the sea's long retreat from Touraine.

The theory of the "diminution of the sea" did not take its departure from these observations and theories, but rather from the research of Swedish scientists who claimed that the Baltic Sea had retreated from Swedish shores within living memory. From this historical testimony and other evidence, several writers concluded that a universal ocean, not necessarily identifiable with the Flood, had gradually subsided and was still doing so. Where all the water was going remained problematical, but the phenomenon itself seemed thoroughly documented.

In a series of publications during the period 1718–34, Emanuel Swedenborg attempted to integrate this new information into a theory of the earth. For the most remote times, he followed Descartes and especially Burnet in depicting the whole earth as a paradise and finding "confirmation" in Ovid and other ancient texts. When he moved on to his own investigation of the earth's crust, he argued that only some sedimentation could have occurred during the Flood but that more dated from submersion in

---

[65] Buffon, d'Holbach, and others are cited in Rappaport 1994, 238–39 and n. 37. See also *BR*, quoted above, n. 61. What may signal a desire to publish a French translation of Moro, in 1770, came to nothing; a manuscript fragment of such a translation exists among the Guettard papers, MHN, MS 175, folder "Traductions de Plusieurs ouvrages."

[66] *HARS* [1743] 1746, 40–41.

what he called a "universal diluvian ocean." The choice of phrase was unfortunate, leading some reviewers to call him a diluvialist—an error he tried to correct. In fact, it is not at all clear how Swedenborg's universal ocean originated, unless these were Flood waters that were still, in the eighteenth century, withdrawing from the continents. The visible evidence of such withdrawal he found in the retreat of the Baltic, which he generalized into the proposition that northern seas were subsiding toward the equator.[67]

In this scheme Swedenborg faced the classic problem posed by Robert Boyle's observations on the tranquillity of the sea floor. In reply he produced hydrodynamic arguments to show that sufficient force must be available at great depths even to shift great boulders.[68] At the same time, his theory had no need to account for any absolute diminution of water, since the northern ocean was merely moving southward. Nor, of course, did any question arise about the elevation of land: new land was not rising but was only being exposed by the ocean's retreat.

The Swedenborgian analysis had its own peculiar features that would not be adopted by two of his compatriots, Celsius and Linnaeus, when, in the 1740s, they considered the diminution of the sea. Celsius relied on his own observations over a period of almost twenty years, on measurements by colleagues, and on the testimony of living witnesses. Use of personal testimony did not provide the basic evidence but confirmed the rapidity of diminution, as old people recalled how formerly deep harbors could now admit only shallow-draft boats, or how former fishing grounds were now too shallow, or how areas of deep water now showed hazardous boulders where none had formerly been visible. Measurements confirmed these impressions, and Celsius calculated that two thousand years earlier the seas surrounding Sweden had been an astonishing ninety feet higher. If such projections were necessarily tentative—Celsius admitted that the rate of subsidence might have been different in the past—he could not doubt that an absolute loss in amount of water had occurred and required explanation. And he cautiously offered two traditional possibilities: some water had disappeared into caverns within the earth, and some had been lost when taken into plants. Absent from this scheme was the Flood, which Celsius refused to subject to naturalistic explication.[69]

---

[67] Swedenborg 1847, 5–38, passim, and p. 152 on the Flood as lasting only one year. See also Frängsmyr 1983, 11–13. A collection of reviews is in Acton 1929, 1930; another review is in Berengo 1962, 55–56.

[68] Above, Chapter 6, n. 15.

[69] This account is based on the translation of Celsius by d'Holbach, in *Recueil* 1764, 112–29. For Celsius and his students on whether the Flood was a miracle, see Frängsmyr 1983, 131.

The Linnaean theory did employ an obviously biblical framework, although Linnaeus refused to attribute marine fossils to the Flood, "which came suddenly and was as suddenly past."[70] Such fossils being ubiquitous, they were deposited in a nondiluvial universal ocean that was still in process of diminution. If one extrapolated from the calculations of Celsius, one would arrive at a date, six thousand years ago, when the whole earth, except for one island, had been under water. The island Linnaeus described as the biblical paradise, where Adam and all other living creatures had their birth. Then Linnaeus the botanist proceeded to develop in expert detail an essay on the geographical distribution of plants. As more land emerged from the sea, downy seeds, for example, could have been borne by the wind to the new locations, whereas other varieties would have been transported by migrating birds and other animals. Linnaeus would, a few years later, somewhat alter one aspect of his analysis, when he indicated that fossils seem to imply an earth older "than any historicus can follow."[71] The diminution theory, however, remained unchanged.

Except for some attention to Swedenborg, especially in German and Swedish periodicals, diminution theory went apparently unnoticed for several years. Nor in fact did Swedenborg's analysis receive much commentary from reviewers, as the most internationally known of the German journals, the *Acta eruditorum*, merely remarked that Swedenborg thought the origin of mountains "must be sought in inundations."[72] Although debate in the Academy of Stockholm continued in lively fashion, the pertinent tracts by Celsius and Linnaeus were ignored by such journals as the *Philosophical Transactions*, the *Mémoires de Trévoux*, and the *Journal des savants*. This silence ended when diminution theory appeared in the sensational *Telliamed* (1748), but the author, Benoît de Maillet, did not know of Swedish research, and a few more years elapsed before Europeans paid attention to the Swedish studies.

When *Telliamed* was published, its author had been dead for a decade, and his editor, abbé Le Mascrier, had not yet manifested publicly his willingness to dabble in heterodoxies.[73] Perhaps because the work had already been in circulation in manuscript copies, Le Mascrier remained generally

[70] Linnaeus [1744] 1781, 1:86. The indispensable study of Linnaeus as a geologist is by Frängsmyr 1983, 110–55.

[71] Quoted by Frängsmyr 1983, 155.

[72] In Acton 1929, 93.

[73] See the assessment of LeMascrier, written during the period 1748–53, and quoted by Darnton 1985, 158. More will be known about Maillet, the preparation of *Telliamed*, and the circulation of manuscript copies when Claudine Cohen publishes her doctoral dissertation. Meanwhile, see Cohen 1991; Rothschild 1964, 1965, 1968, 1977; and Benitez 1980, 1984.

faithful to the author's intent, even while he introduced some modifications. Among editorial changes, Le Mascrier deleted references to the importance of Epicurean "chance" in the physical world and added some conventional teleology to his own preface; nonetheless, the lack of teleology in *Telliamed* could hardly be disguised. Le Mascrier also softened Maillet's advocacy of eternalism, but he could not alter the text's cosmological cycles, obviously endless, of suns becoming planets and ultimately evolving again into suns. In the only place where Maillet offered a figure of two billion years as the minimal past duration of the process of the diminution of the sea, Le Mascrier changed this to the equally unorthodox figure of two million.[74]

Although *Telliamed* includes discussions of cosmic cycles and of the "transformations" of life on earth, its focus is on diminution theory, which occupies fully two-thirds of the text. Here Maillet presented a large array of observations, some of them his own, of the structure of mountains, valleys, shorelines, and the offshore sea floor. The chief evidence for diminution, as he realized, consisted of horizontal sediments and their enclosed fossils: these could only have been produced at a time when the sea covered now-exposed land masses. For other deposits (tilted, fractured, unstratified) and for nonmarine fossils, he devised explanations to show how these, too, could have been the work of the sea. In one of the relatively simpler examples, he summarized at length Antoine de Jussieu's memoir of 1718, using this description of fossil plants laid out "as if mounted" to show that the plants had floated peaceably from their place of origin to that of their deposition. When he went on to claim that Jussieu also supported diminution theory, he misinterpreted that author, who wrote only of marine currents and not of a universal ocean or of changing amounts of water. In some of the more complex passages, Maillet argued that certain tilted strata could have been laid down as we now find them, that others, initially horizontal, could have collapsed when marine currents undermined them, and that still others consisted of debris detached from mountains and tumbled or transported by rivers to the shores where they were redeposited. As for nonmarine organisms, these had first begun to appear on exposed shores, and they could easily be imagined as having

[74] My discussion of *Telliamed* is based chiefly on the English version of 1750, with comments sometimes also drawn from Carozzi's translation and edition of 1968. Carozzi focusses on particular manuscripts and the 1755 edition in order to distinguish what Maillet wrote from the editorial additions. Although Carozzi uses the admirable article by Neubert (1920), the latter is still worth consulting. See Maillet 1968, 381, n. 52, for billions and millions of years; neither figure appears in the 1750 edition, but one finds there locutions like "a long Time" (206) and "an Infinity of Years" (208). Maillet's geology is analyzed in Carozzi's Introduction, 33–50.

been washed into the sea to become incorporated into sediments being formed there.[75]

Maillet's universal ocean included in its depths marine currents responsible for building the landforms that gradually emerged as the waters retreated. Like other writers, he had to account for such a quantity of water and for its progressive disappearance. Asked by Fontenelle to include some cosmology in his book, Maillet decided to make his cosmological speculations serve important geological functions. The universal ocean, he argued, did not sink into hypothetical caverns but was gradually dispersing into space; this cosmic vapor, carrying with it seeds capable of germination, would land on now-lifeless heavenly bodies where new universal oceans would ultimately form. While new inhabited worlds were being created, our earth was heading for eventual desiccation and temporary death. Not only would the earth's water eventually vanish, but the "fats and oils" of organisms buried in the strata had already given rise to volcanic fires; with more such deposits accumulating, fires would become more widespread, until the earth became wholly enflamed and transformed into a sun. In time, fuels and fires of the new sun would be exhausted, a period of opacity would ensue, and cosmic vapors would then arrive to form a new universal ocean.[76]

For at least the earth's part in these eternal cycles, Maillet tried to provide a timescale. He claimed to have devised a means of measuring diminution rates, arriving at a figure (much smaller than that of Celsius) of three inches per century. Calculating on this basis the date at which the ocean had covered the whole earth, he estimated that the process of diminution had begun at least two billion years ago. As he knew, so vast a time span challenged not only the Bible but also any system of chronology based on human records. Like his contemporaries, however, Maillet did not want to abandon human testimony entirely, for such evidence could confirm physical conjectures. His strategy involved both a lengthening of the human past and the deployment of myths to bolster his interpretation of certain ancient events. These devices could not take him back to earliest times, but they could supplement his physical analysis for ages beyond the confines of conventional chronologies.[77]

In tackling biblical chronology, Maillet in some respects went further than even the *esprits forts* of his day. If some had pondered the evidence for the antiquity of Egypt, Chaldea, and China, Maillet asserted that India and

[75] Discussion drawn chiefly from Maillet 1750, part 1 and the initial pages of part 2.

[76] Most of the cosmological details are in ibid., pt. 5.

[77] Ibid., 175: "Let us not measure the past Duration of the World, by that of our own Years."

China both had histories older than the conventional date of the Flood. If it had become commonplace to say that non-Hebrew claims to antiquity were products of mere national vanity, Maillet transferred this logic to the Jews, "a People whom History and Experience prove to have been ... the vainest and most credulous Mortals in the Universe."[78] If some had wondered whether any or all of the ancient flood traditions should be identified with the universal Deluge, Maillet declared that non-Hebrew writings were astonishingly devoid of references to an event that, had it occurred, ought to have left many traces in pagan literature. In perhaps unconscious imitation of the dreadful LaPeyrère and preadamitism, Maillet further claimed that Moses had not written world history, but only the history of the Jews. Noah, therefore, had survived a local flood that, in his ignorance, he assumed to be universal; so, too, had the ancient Greeks elevated a local flood into the myth of Deucalion, who, with his sister, had survived the destruction of the "whole" earth.[79]

In analyzing the various kinds of human testimony, Maillet, like everyone else, had no difficulty with the relatively recent (ancient and medieval) descriptions of ports, deltas, and shorelines that could be compared with modern conditions. But any text as much as three or four thousand years old "is not only obscure, but even totally destitute of Facts,"[80] This judgment applied as much to the Bible as to the ancient dynastic lists of Egypt and China. Still, mankind had existed long before scribes were producing texts, as evidenced by some archaeological remains. In a strikingly imaginative passage, Maillet turned his attention to the ruins of cities located along a north-south axis from Alexandria to Nubia. Although now surrounded by inhospitable desert, each settlement must in the past have had access to waterborne trade; each, therefore, beginning with the southernmost, had been a port abandoned as the sea retreated northward.[81] Equally startling were the results of Maillet's efforts to employ the convention of seeking in ancient myths and traditions evidence of a pre-literate history of man and the earth. This he did even on behalf of his peculiar cosmol-

---

[78] Ibid., 74. If Carozzi is correct in identifying this passage and its context as added by LeMascrier (1968, 296, n. 155), then LeMascrier was not nearly so conventional as Carozzi believes. For the intellectual circles to which LeMascrier belonged, see Wade 1938, chap. 6.

[79] Maillet 1750, 72–77. The psychology behind thinking a local flood universal occurs also in BN, MS fr 9774, fols. 202r–v, a version of *Telliamed* which Neubert (1920, 207) dated about 1725–29. Works by J.-B. Mirabaud, circulating clandestinely, contained a similar analysis; see Bernard 1740, 1:121, and Mirabaud [1751] 1778, 95–96. Links between these works and *Telliamed* are explored by Neubert 1920, esp. 91–110.

[80] Maillet 1750, 112–13.

[81] Ibid., 79–82.

ogy, when he reported Ovid's account of Arcadian tradition: the ancient Arcadians had inhabited the earth before the appearance of our particular sun and moon. Grist for his mill, too, when he suggested the evolution of mankind from mermen and mermaids, was the ancient Chilean tradition "that their Ancestors, who were the first Men, proceeded from a certain Lake."[82]

One may assess *Telliamed* in any number of ways: "modern" in some respects and merely modernized pagan (Lucretian) in others, often meticulous in detail and yet bristling with contradictions and tendentious arguments, boldly imaginative and yet already outdated in 1748 in elements of its cosmology. Undeniably, the book was a sensation to its first readers, and it rapidly achieved a second edition (1749) and an English translation (1750). At least one reviewer found it charming, as he enjoyed the prospect that a future generation might walk dryshod from Calais to Dover. Others were horrified, and the *Mémoires de Trévoux* rightly pointed out that, for all its Fontenellian frills and fantasy, the author's tone was not at all lighthearted: this was a serious book and a dangerous one. Indeed, the Jesuit journal suddenly found it necessary to defend diluvial geology. No quarter could be allowed a book that "necessarily assumes the eternity of matter" and has mankind emerging from the sea. As for the diminution of the sea, the reviewer regarded the theory as unproven because there did exist evidence that the sea had been gaining on some shores.[83] Evidence of the latter kind would also be produced by one of the Swedish critics.

Johan Browallius had already shown his opposition to the Celsius theory of diminution, but the critique he published in 1755 indicated that *Telliamed* confirmed his belief that the theory could have dangerous implications. Bishop of Abo and an experienced naturalist, Browallius coupled Christian outrage with a remarkable range of evidence. If, he argued, diminution was universal—and it was bad physics to say, as Swedenborg had, that only northern seas were subsiding—how could one explain the unchanged elevation of a city as ancient as Alexandria? Moreover, several non-Swedish scientists, including Manfredi, Buffon, and Hartsoeker, had

[82] Ibid., 187–89 (Arcadians), 264–65 (Chileans). Impressive as a traditional exercise in assembling testimony, for non-traditional purposes, are pp. 230–44 on mermen; Roger (1993, 522–23) rightly finds this a noteworthy performance.

[83] The charmed reviewer was Clément, letter of 1749 (1755, 1:135–37). *Trévoux* (April 1749), 631–47, esp. 632–33, 635, 644–45. The *Monthly Review* 2 (November, 1749), 37–52, included long quotations, but indicated clearly that the book was non-Newtonian in both facts and principles, irreconcilable with sound philosophy and religion, and wildly improbable. I have not seen the unfavorable review in *Bibliothèque impartiale* (January–February 1750), mentioned by Neubert 1920, 14 n. 19; for the character of this journal, see Marx 1968.

produced credible evidence that the sea had gained on some shores and had neither advanced nor retreated on others. In any event, one could not possibly believe that mountains had been built up on the ocean floor. On the contrary, no sediments could have hardened under such conditions, however long the elapsed time. And where had all the water gone? To say, as *Telliamed* did, that it was dispersing into the cosmos was another piece of bad physics. Furthermore, diminution theory meant that the earth would increasingly lack for water, and this conflicted with the saner view that the hydrologic cycle was a powerful example of God's wisdom.[84]

Despite the solid support drawn from Manfredi and others, Browallius could not be wholly convincing because there still remained the accumulated data about the Baltic Sea. About the latter he could merely point out that Swedish calculations conflicted with those of Maillet, and he charged that the Swedes had relied too much on the memories of old fishermen and peasants. But the Baltic data refused to go away.

This is not the place to examine the post-1755 career of diminution theory, except to notice that, thanks in some measure to *Telliamed,* translations of the relevant Swedish tracts began to appear in the 1750s and especially the 1760s.[85] The evidence and its interpretations would then be vigorously debated until the early decades of the nineteenth century. Before the middle of the eighteenth century, however, the pertinent treatises of Celsius and Linnaeus—despite the great fame of the authors in other respects—aroused virtually no interest outside Sweden. Nor, in fact, did any of the theories of the earth discussed in the present chapter create a circle of adherents. Two in particular, by Hooke and Moro, seemed negligible if not actually perverse, as the authors had placed undue emphasis on phenomena that other naturalists considered to be local and uncommon. The more fundamental question broached by most writers began with their acknowledgment that the sea had, at various times, covered the present continents, as shown by the sequence of sedimentary strata that did not conform to John Woodward's theory. How to explain apparently changing sea levels was the rock on which all theorists either came to grief or abandoned their vessels by admitting their ignorance. When Antonio

[84] Discussion of Browallius is based on the text in Alléon-Dulac 1763, 4:94–184. The editor omitted the subject of mountain-building on the sea floor, pp. 181–82, because it had already been tackled by one of Buffon's critics, Targioni Tozzetti (see Chapter 8). There is more Christian outrage in Browallius than I have indicated; see Frängsmyr 1983, 139. Other Swedish critics are briefly discussed by Wegmann in Schneer 1969, 392–93.

[85] Some translations are indicated in Linnaeus 1972, 29 n. 1; others have been cited here, nn. 69, 84. A substantial discussion of the conflicting Baltic and Adriatic evidence was added to Frisi 1774, bk. 3, chap. 5, and does not occur in the 1762 edition. Desmarest 1794, 1:133–50, treats these problems in detail.

Conti in 1743 asked the Paris Academy for information on this subject, one may interpret the cautious reply to mean that the phenomenon itself required more investigation before any explanation could or should be hazarded.

CHAPTER 8

Buffon and the Rejection of History

B y the 1740s, when Buffon was at work on the first volumes
of his *Histoire naturelle,* naturalists had reached a consensus
about some topics concerning the earth's past. Most fundamental of these was the acknowledgment that the earth had had a past
and no longer resembled the globe that had emerged from the hand of
the Creator. Even here, however, one detects two schools of thought, as
diluvialists maintained that only negligible geological changes had
occurred since the Flood, while their opponents insisted that the strata
recorded a more complex sequence of historical events. What also united
naturalists was confidence that the Flood, described in Genesis and confirmed by pagan legends and myths, had really taken place. How to interpret the event geologically, however, produced disagreement: was the Flood
universal or local? was the event a literal matter of flooding, or had it been
essentially a great earthquake? did it have major effects, or was it only one
of several marine incursions? should one interpret the Flood naturalistically
or relegate it to the realm of miracle?

Widespread interest in marine sedimentation stemmed in part from
Woodward's *Essay* but had its earlier, more general roots in efforts to interpret fossil shells. On the whole, one may say that the fossil question was
resolved by about 1710 or 1715—although specific forms, such as belemnites, would continue for decades to be debated—and that attention thereafter focused increasingly on marine movements. Despite the accumulating

evidence that the earth possessed internal heat, even in nonvolcanic regions, naturalists found little use for this subject. In particular, as shown by their reactions to the theories of Hooke and Moro, they generally agreed that earthquakes and volcanoes were local phenomena, affecting only some parts of the earth's superficial crust. Elevation of land was thus considered to be a rare, non-universal occurrence, whereas the more mobile sea, and the ubiquity of marine deposits, required study. Here, diluvialists and their critics agreed on the central importance of marine sedimentation, but not on causal interpretation.

Recalling Hooke's remark that fossils could provide a "natural chronology" of the earth, it is pertinent here to ask if naturalists combined a sequential view of strata with a correlative sequence of organisms. (For diluvialists, no such question arose, as all marine forms in particular had existed simultaneously and were, in theory, sorted by floodwaters in order of their specific gravities.) Such writers as Ray, Plot, and Vallisneri had noted that fossil shells occur in curious groups, and not in random or confused assemblages. As they put it, shells seemed to be gathered in populations resembling communities to be found on the sea floor. Some had noted other peculiarities of the fossil evidence, such as the impressions of plants associated chiefly with the Coal Measures or found in slates that were interpreted as the remains of freshwater lakes or marshes. And some were aware that the great land animals had been buried relatively recently, the fossils still containing some organic material. Woodward himself had drawn up instructions for his correspondents and other collectors, urging that they record in detail the materials, orientation, lateral extent, and fossil content of the strata they examined. Advice comparable to Woodward's, but with the notable omission of fossils, may be found as early as 1684 in Martin Lister's proposal that "a Soil or Mineral Map" be constructed, charting the locations of chalks, clays, coal, and other deposits. Years later, in 1719, John Strachey would draw up just such a description of the strata in the coal mines of Mendip.[1]

What all these observations and proposals meant to the writers themselves is far from clear. To be sure, Woodward expected further observation to confirm his theory, and his disciples and critics at times engaged in detailed analyses of stratification with his theory in mind.[2] A decade before his *Essay*, however, Lister tentatively suggested that from his proposed maps

---

[1] Woodward 1696; Lister in *PT* 14 (1684b), 739–46. Strachey in *PT* 30 (1719), 968–73. See also Boud 1975.

[2] In addition to Chapter 5, above, see Rudwick 1972, 90, and Porter 1977, 58–61.

something more might be comprehended from the whole, and from every part, then I can possibly foresee . . . For I am of the opinion, such upper *Soiles,* if natural, infallibly produce such *under Minerals,* and for the most part in such order.[3]

Strachey had no declared aim, but his choice of coal mines implies the same economic interests shown by Lister. At the same time, both Strachey and Lister signaled the existence of regular patterns of stratification, and Strachey would eventually try to explain this phenomenon. His peculiar explanation—what Gordon Davies has called turning the earth into a "gigantic Catherine-wheel"—allowed him to conclude that every stratum must somewhere appear at the earth's surface, so that one does not need "specifick Gravitation to cause the lightest to be uppermost." This allusion to Woodwardian theory stands in contrast to Benjamin Holloway's, for Holloway, finding a "regular Disposition of the Earth into like *Strata* . . . commonly through vast Tracts," announced that his observations confirmed Woodward.[4]

Other writers, such as Ray and Vallisneri, were drawn to stratification by the peculiarities of fossil populations, and they speculated that each population had been buried by a local flood or marine incursion, to be replaced by another assemblage of organisms. (For Moro, successive faunas had been buried by volcanic eruptions.) Conclusions with such large implications were far from common, however, and it may be that only Leibniz— with Fontenelle following in his wake—glimpsed the possibility that the earth's history might be correlated with a history of life.

For most naturalists, no chronological pattern seemed evident: the same kinds of materials could be found at various places in the geological column, and no agreement could be reached on whether any forms of life were extinct (and hence peculiar to periods of the past). Focus on the sea thus meant essentially grappling with such problems as amounts of water, past and present, and changes in sea level. Less intractable were such topics as changing shorelines, the building of deltas, and the role of running water in eroding the landscape; these things could be glimpsed by the observer himself, and added information was readily available in ancient, medieval, and more recent written documents. Far more difficult than dealing with water, however, was any effort to examine those parts of the earth's crust that a later generation would label "primitive." In his singular

---

[3] Lister in *PT* 14 (1684b), 739–40. Not much clearer than Lister was Guettard whose much later explanations of mapping are discussed by Rappaport in Schneer 1969, 274–80.

[4] Strachey, above, n. 1, and in *PT* 33 (1725), 398. See also Davies 1969, 103–4 and pl. 3; Porter 1977, 119. Holloway in *PT* 32 (1723), 421.

fashion, Leibniz did broach the matter, and his quasi-Cartesian hypothesis of 1693 would be revived by Buffon. In the fifty years separating Leibniz from Buffon, naturalists often remained silent about the earth's "core" or simply stated that the oldest detectable nonsedimentary formations dated from the Creation.[5]

In crucial areas of methodology, the treatises and articles of this period display so many common traits that one may say their authors had arrived at a consensus. Most obvious was their commitment to laws of nature and the extension of that commitment to a philosophy that would later be dubbed "actualism." Laws and actualism did not always go hand in hand because, as any scientist knew, the ways in which nature works depend always on the existing physical conditions. For Thomas Burnet, therefore, Cartesian physics served as the analytical tool for describing the formation of the earth and its first crust; after the Flood, new physical conditions required a different kind of analysis. Cartesians and Newtonians alike wondered about the extent to which laws now in existence could apply to the most remote part of the earth's history. In the full text of the *Protogaea*, published in 1749, Leibniz summed up the dilemma inherent in actualism: those who judge "the early condition of the earth too much by its present state" cannot believe that the sea once covered high elevations—and yet the proof is there to be seen in the deposits of marine fossils.[6]

On the one hand, then, actualism had to be tempered by an awareness that past conditions might have been different. On the other hand, one had no alternative but to rely on known laws of nature. The grand metaphor of the cosmic "machine" became the "machine of the earth." As both "L.P." and Dr. Arbuthnot put it in their critiques of Woodward, the basic properties of bodies and fluids had not changed in the course of time; any other assumption would undermine all of natural philosophy. In studies of sedimentation, therefore, naturalists paid attention to hydraulics, commenting on topics ranging from the behavior of floodwaters to whether currents could exist on the sea floor. Volcanoes were simulated experimentally, on the assumption that one could approximate the proper ingredients and conditions. And probably the most persuasive argument against interpreting fossils as *lusus naturae* was that nature works in regular and predictable ways.

With such principles in hand, naturalists nonetheless felt impelled to seek human witnesses to past events. By 1700, erudite historians had

[5] The several "chemical cosmogonies" discussed by Laudan 1987, 44–46, 66–68, are mainly German.

[6] Leibniz [1749] 1993, chap. 26, p. 91; also chap. 3, p. 17, on the inadequacies of modern furnaces to duplicate the intensity and duration of the earth's igneous phase.

developed rules for dealing with the reliability of texts of all sorts, but naturalists faced a fundamentally different problem: they were trying to reconstruct an unobservable past, for which they constantly had to assume that *if* certain conditions had obtained, *then* certain results had followed. That they engaged in such inferential reasoning is abundantly clear; but clear, too, is that they disliked doing so, for they considered their conclusions to be merely conjectural. In this respect, Fontenelle tried to promote the view that nature was not only older than mankind but had also produced reliable "histories." His prose was echoed now and then, but even colleagues such as Réaumur and Jussieu displayed far more confidence in their observations than in the tentative conclusions based on those observations. The same attitude can be detected elsewhere, as naturalists like Hooke, Leibniz, and Swedenborg admitted both that human witnesses to some events did not exist (or could not be found) and that reconstructions of the past were thus probabilistic at best and conjectural at worst.

BUFFON'S THEORY OF THE EARTH

When Buffon composed his theory—much of it written, according to the author, in 1744—he had available four possible models: diluvialism, diminution theory, vulcanism, and what may be called the sedimentary alternative to diluvialism. Buffon came closest to the latter, but he had something to say about three of the four models, the exception being diminution theory.[7] On the whole, he cited his predecessors very selectively, paying more attention to travel literature. Probable reasons for these choices will be suggested in the next pages.

Buffon believed his own theory to be more naturalistic and more probable than any that had gone before. As he explained in his First Discourse, on "how to study natural history," the naturalist ought to proceed inductively, paying especial attention to repeated events or patterns rather than to isolated facts. Unlike the mathematical physicist, who dealt with a restricted part of nature, the naturalist had all of nature to contend with, and seeking regularities was the only route to the discovery of laws. If experiment was an important analytical aid, the investigator's confidence stemmed neither from this procedure nor from amassing facts; more fundamentally, the naturalist should recognize that he, unlike the

[7] Carozzi in Maillet 1968, 4, thinks it obvious that Buffon read and studied a manuscript of *Telliamed,* but gives no specific evidence. The foremost Buffon scholar, Jacques Roger, nowhere mentions the possibility. In general, one may contrast Maillet's gradual desiccation of the earth with the Buffonian steady-state cycles.

mathematician, made no "suppositions," but observed sequences and repetitions of facts and events. The resultant "physical truth" could then be described as possessing "a probability so great that it amounts to (*équivaut à*) a certainty."[8]

If the critique of the mathematical ideal echoes Gassendi and Bayle, more significant for Buffon is the elevation of physical truth so that one should have no hesitation in projecting observed patterns into the historical past. As he specified in the Second Discourse, "the history and theory of the earth," this would be his own method. His predecessors, he argued, had glimpsed the proper method, but had mistaken "local causes" and "accidents" for "general" ones: they had used unusual events (a comet passing near the earth) or local phenomena (volcanoes), thus producing debatable "systems." In his own search for "actual" processes and universal patterns, Buffon therefore relied heavily on the observations of an impressive array of travelers who had visited parts of the earth ranging from the interior of Russia to the islands of the Pacific. In addition, wherever possible, Buffon sought in branches of physics those laws applicable to the reconstruction of the geological past.

One can readily see Buffon's ideals at work in his criticisms of certain earlier writers, chief among them the diluvialists. Two or three decades earlier, he might well have tackled the matter of *lusus naturae,* but this issue, unlike diluvialism, no longer needed more than a rapid dismissal. In dealing with the Flood, Buffon paid sporadic attention to specific gravities and other topics examined by earlier critics, but he offered no detailed examination in the manner of a Vallisneri or a Moro. Rather, he focused on the use of miracles and single causes; both kinds of explanation invoked uniqueness, or lack of pattern, but miracles required some additional comment. Expressing diluted admiration of Woodward as an observer, Buffon scathingly pointed also to his failings:

> he adds to the miracle of the universal Flood other miracles, or at least physical impossibilities which do not agree either with the text of the Bible or with the mathematical principles of natural philosophy.[9]

---

[8] The quotation is from Buffon [1749] 1954, 24A, from the First Discourse. Also in the First Discourse is the famous attack on taxonomy in general, Linnaeus in particular. For analysis, see Roger 1989, chap. 6. Any French edition of Buffon will do, the author having made no changes in the many reprintings during his lifetime. In the footnotes that follow, I will therefore indicate both page numbers in the editions I have used and the section of the work from which these citations come. The only scholarly edition, as far as I know, is by Renzoni, referred to below as Buffon [1749] 1959.

[9] Buffon [1749] 1954, 83B (Article 4), and 88B–90B (Art. 5) for a discussion of miracles. Renzoni in Buffon [1749] 1959, 538, finds Buffon's arguments so similar to

The last five words are the title of Newton's *Principia,* which had no direct relevance to the particular impossibilities to be found in Woodward. These included God's opening of the abyss, the dissolution of all rocks, and the immunity of organisms that retained their "cohesion" in the face of the "universal solvent." Buffon by no means denied that a universal flood had occurred, but he became even more eloquent than usual in insisting that diluvialists had erred in two ways: their theory conflicted with geological evidence, and they distorted Genesis in their attempts to transform a miracle into a natural event. Unconsciously echoing Vallisneri and Moro, Buffon argued that miracles had no place in science. By the improper application of human reason to the ways of God, Buffon added, one undermined the simplicity and integrity of the biblical text.

Elements of William Whiston's theory seemed to Buffon more interesting and sophisticated than Woodward's and required a somewhat different critique. By giving his book a pseudo-geometrical form, Whiston had disguised a mere "hypothesis" and rendered it acceptable to readers who had a misplaced confidence in mathematics. Furthermore, Whiston's comets posed serious methodological problems.

> One cannot absolutely deny . . . that the earth, encountering the tail of a comet when the latter neared its perihelion, could have been flooded, especially if one grants to the author that a comet's tail can contain watery vapors. Nor can one deny as an absolute impossibility that the tail of a comet, returning from its perihelion, could burn the earth. . . . The same is true of the rest of this system; but, although there is no absolute impossibility, there is so little likelihood in each thing taken separately, that the result is an impossibility of the whole taken together.[10]

Whiston, in short, had built his system on a tissue of barely possible events and suppositions, single (and singular) causes, the whole decked out with a mathematical façade. If the true tests of a theory, rather than a mere system, were probability and verifiability, then Whiston's had failed both.

Among other writers, Leibniz deserves attention for the two kinds of criticism Buffon leveled at the sketch published half a century earlier in the *Acta eruditorum.* Here we read that Leibniz's

---

Vallisneri's that she concludes Buffon must have had the earlier work in mind, even if he never cited it. As Roger (1989, 192–93) indicates, Buffon was familiar with Vallisneri's views on biological generation.

[10] Buffon [1749] 1954, 86B (Art. 5).

ideas are interconnected, the hypotheses are not utterly impossible, and the consequences to be drawn from them are not contradictory: but the great defect of this theory is that it does not apply at all to the present state of the earth; it explains the past, and that past is so old and has left so few traces that one can say anything one likes about it, and the cleverer a man is, the more he will say things that seem very likely.[11]

The faint praise of the first lines gives no hint that Buffon would adapt the Leibnizian claim that the earth had once been a star, trying to make this more plausible than had Leibniz. As presented by Leibniz, the claim (according to Buffon) was not subject to proof or disproof. In the last lines, Leibniz had erred by not using the present as a key to the past.

Although he condemned cosmic speculation, Buffon presented a theory of the origin of the solar system, explicitly meant to replace Newton's suggestion that God had imparted inertial motion to the planets. More professedly naturalistic, Buffon proposed that a comet had detached pieces of the sun, sending these fragments into planetary orbits. He knew that he, like Whiston, was inventing a comet, but he offered as "confirmation" the curious fact that the planets all move from west to east and that their orbits all lie within the plane of the ecliptic. This pattern could not be coincidental and must be the product of a single cause: not God's design, but cometary impact. Further confirmation lay in the consequences, for, as Buffon noted in connection with Leibniz, the legitimacy of any theory depended on its being in accord with known laws and not in contradiction with subsequent observations.[12]

From the earth's origin, at least three consequences followed. For one, the earth's initial "fluidity"—it consisted of materials "liquefied" by fire—permitted it to assume the shape of an oblate spheroid, a shape Newton had predicted and later scientists had verified. For another, both its origin and the regularity of its rotation showed that the earth was approximately homogeneous in the density of its interior, that is to say, it possessed no central spaces, caverns, or abysses. Finally, there was Leibniz's claim that vitrified materials lay beneath the earth's sedimentary crust; this assertion had now been placed in a larger framework of astrophysical laws, lending greater plausibility to a mere Cartesian hypothesis.

[11] Ibid., 87B (Art. 5).

[12] Ibid., 66B, for God's role. This is in Article 1, the whole of which is devoted to the origin of the planets. Buffon warns the reader at the outset, and occasionally thereafter, that he is engaging in some conjecture, but that the present state of the earth is, by contrast, a fact. He concludes that he might have put this Article in mathematical form, but that to disguise hypotheses in such fashion smacked of charlatanism.

When he turned to the most visible features of the earth, Buffon abandoned reference to cosmology, his treatment of that subject having achieved his aim of showing how one might support seeming speculation within a context of laws and observations. Also achieved was a naturalistic system, lacking divine intervention and final causes. Incidentally eliminated were the earth's internal caverns, used by diluvialists and by some of their critics. From this point on, allusions to the earth's core occur only when Buffon indicates that vitrified debris can be found mingled with other materials in the earth's sedimentary crust.

In studying that crust, Buffon built his theory on present-day observations of three kinds: worldwide deposits of marine fossils, equally universal deposits of horizontal and parallel sedimentary strata, and innumerable examples of so-called salient and re-entering angles, called to his attention in the writings of Louis Bourguet. The first two of these were indeed so basic to his theory that he rejected the claims of French scientists, on their return from Peru, that no marine fossils could be found at the highest elevations of the Andes. For Buffon, the failure to detect should not be interpreted to mean that no evidence existed.[13] Equally vital to Buffon were Bourguet's observations. As Bourguet explained the matter, he had noted a curious correspondence of formations on either side of each Alpine valley he examined: "the salient [projecting] angles of each side correspond reciprocally to the re-entering angles which are alternatively opposed to them."[14] What Bourguet saw as dazzling structural symmetry, Buffon transferred (as we shall see) to landforms originating on the ocean floor.

These elements in hand, one may reconstruct Buffon's main lines of argument. He first considered it obvious that marine fossils and their enclosing sediments had been laid down in the sea. These deposits being universal, it seemed equally obvious that all land, now elevated, had formerly been under the sea. To explain the process of marine deposition and the submarine sculpting of landforms, Buffon made use of earlier observations that suggested to him a general east-to-west motion of the earth's seas; this claim could be substantiated by an analysis of tides. Well aware that tidal regularity depends on the earth-moon relationship, Buffon rejected the notion that cycles of ebb and flow merely resulted in an equally cyclical removal of material from shorelines during ebb and redeposition of *the same material on the same shores* during flow. Instead, the sea, following the path of the moon, removed material from European coasts and carried

[13] Ibid., 100A–B (Art. 8). Also Roger in Buffon 1962, xix–xx.
[14] Bourguet 1729, 182, 195–96, in Taylor 1974, 73.

it toward the Americas. As his most detailed example, he cited a report of cycles observed at the Straits of Magellan, where

> the tides near twenty feet, and they continue at this height six hours; but the reflux, or ebbing, lasts only two hours, and the waters run to the west. This incontestably proves, that the reflux is not equal to the flux, and that, from both there results a motion to the west, which is stronger during the flux than the reflux.[15]

Buffon in this argument reformulated the common maxim, familiar to contemporaries from the *Geography* of Varenius (what the sea gains in one place, it loses in another), into what he perceived as a pattern: the western edges of the continents were being eroded, and their eastern shores were being progressively built up. Barely acknowledged in one sentence was a phenomenon known to any reader of Jussieu's memoir of 1718: plants living in the West Indies had been found in the strata of France. Buffon could only reply to this embarrassing evidence by saying that there "must be . . . some irregular movements" that would explain transport in this contrary direction.[16]

The effects of tidal motion and other disturbances of the sea's surface were extended by Buffon to the sea floor. Like so many writers, he here had to struggle with Boyle's tract about the ocean depths, in which Boyle had assembled information from divers showing that storms at sea rarely if ever affected the deeper parts of the ocean. Buffon resolved the difficulty by ignoring Boyle's careful distinction between shallower and deeper seas, concluding that "the bottom of the sea" does suffer "agitation" that removes clay, shells, and other material from one location and deposits these elsewhere.[17] This reinterpretation of Boyle provided an essential ingredient of Buffon's theory.

Once over this hurdle, Buffon could say that sediments were being built up on the sea floor by the motions of the sea itself. Such accumulations in turn brought other forces into play. As little mounds and hills developed, these obstructions in effect channeled the waters and produced currents that further sculpted underwater topography. Here Bourguet's salient and re-entering angles became significant, as Buffon argued that bottom currents behave as do modern rivers: they carve their routes through

---

[15] Translation from Buffon [1749] 1812, 1:408–9 (Art. 12). On the east-west motion of the sea, sources in Varenius and Bourguet are cited by Renzoni in Buffon [1749] 1959, 561, 564; in addition, see the review of Dassié in *PT* 12 (1677), 879–83.

[16] Translation from Buffon [1749] 1812, 1:413 (Art. 12).

[17] Buffon [1749] 1954, 51B–52A (Second Discourse).

sediments, creating channels flanked on either side by the angles Bourguet had observed. As additional evidence, Buffon noted that the strata on either side of a valley—the English Channel being one such valley—correspond in the sequence of sedimentary deposits. Valleys had thus been formed under the sea by bottom currents, and they had been further deepened by running water after the land had somehow emerged above sea level.[18]

That land "somehow" emerged represents the key difficulty in Buffon's theory, as he admitted. To account for changing shorelines, using the sea's east-west motion, did not explain how those shores and continents had been elevated above sea level. Not available to Buffon were the central caverns into which water could subside; nor did the uniformities of nature allow a diminution of the earth's water supply. Buffon himself posed the question he could not answer.

> Why did the waters of the sea not remain on the land which they had covered for so long? What accident, what cause can have produced this change on the earth? is it even possible to conceive of a cause powerful enough to produce such a result?

In the manner of his predecessors, Buffon briefly considered the usual gamut of agents of uplift: earthquakes, volcanoes, and central heat. Nothing could adequately fulfill the role of the "general causes" that Buffon sought. He could only assert that, "the facts being certain, their cause can remain unknown without prejudicing the conclusions we should draw from them."[19]

Having developed a theory of sedimentation and the origin of landforms, Buffon also commented on the various materials of which sedimentary strata are composed. Early in the earth's history, the sea had detached fragments and particles of vitrified substances, like granites, to form the oldest stratified layers. The appearance of organisms then gave rise to new kinds of sediments: calcareous deposits, composed of whole and fragmented shells and of pulverized particles derived from shells. These younger strata, however, cannot always be found superimposed on older formations, because the older granites and clays, elevated above sea level, are being continuously eroded and the fragments transported into the sea. Thus, Buffon argued, any exposed series of sedimentary strata will show some alternation of calcareous materials with older substances. In support of this reasoning, Buffon produced a list of strata extending to a depth of

---

[18] Buffon 1749, 1:321–24 (Art. 9).

[19] Buffon [1749] 1954, 55A (Second Discourse). Analysis of volcanoes and earthquakes can be found in Article 16.

about 100 feet, observed at Marly-la-Ville, and he also cited the classic list of Varenius.[20]

When Buffon turned from sedimentation to such topics as volcanoes, earthquakes, and central heat, his discussion on the whole may be called traditional, with the addition of an occasional argument of his own. Nowhere did he debate the evidence of Boyle and Dortous de Mairan on behalf of the existence of heat within the earth; nor did he consider the earth's igneous origin to have had any long-term residual effects. Instead, he interpreted "heat" to mean "fire," and argued in customary fashion that subterranean fire could not be sustained without an air supply. Citing Lemery's simulation of a volcano, as well as Borelli's analysis of Etna, Buffon further insisted that eruptions can only occur where there is both a deposit of bitumens and essential air vents; these conditions obtained only at high elevations, and one might even conclude that most mountains had once been volcanic but had eventually exhausted their fuel supply. More generally, Buffon observed that eruptions and earthquakes are not universal phenomena, but "secondary" and "accidental" ones. They should not be used—and here he cites John Ray, not Hooke or Moro—for explanations of the principal features of the earth's crust. If shocks could indeed explain crustal collapses and faulting, no cause of this kind could account for the elevation of land masses. Nor did sedimentary strata display the signs of disorder that would be the necessary consequence of violent upheaval.[21]

Because Buffon would later attempt to calculate the age of the earth, it seems proper to ask here whether his earlier synthesis gives evidence of a sense of elapsed time. On the whole, the subject is only rarely broached and never discussed. In more than one passage, Buffon mentions Halley's suggestion that the salt content of oceans has increased in the course of time; agreeing with Halley, Buffon did not take this opportunity to follow Halley in speculating about the results of future calculations of the geological timescale. Elsewhere, Buffon cited instances of quite rapid change: the visible effects of denudation, the accumulation of more than one foot of vegetable mold on pavement dating from Roman times. His version of the earth's origin also meant a malleable early crust and the possibility of rapid geological change at that time. (As he carefully expressed it, *the same causes in operation today* had produced larger effects in a short time during the earth's earliest history.)

---

[20] Buffon 1749, vol. 1, esp. 235–38 (Art. 7).
[21] Ibid., 1:502–35 (Art. 16), esp. 502 (bitumens), 503 (central heat), 522–25 (Ray and critique of).

By way of contrast, he also used the common locutions of his predecessors, referring to the "remote" past and to the "long" periods when the sea had covered the continents. One might say, too, that his description of sedimentation and of the cyclical motion of the sea around the earth imply a vision of long and slow processes. In the very last sentences of his treatise, Buffon grandly concluded that our knowledge of the earth's past is imperfect:

> we lack experience and time, [but] we do not consider that the time we lack is not lacking in nature; we wish to find a relationship between the moment of our existence and the centuries past and future, without considering that this moment, human life, even extended as much as it can be by history, is only a point in duration, a single fact in the history of the facts of God.[22]

If the perspective here may remind us of Fontenelle's discussion of gardeners and roses, Buffonian prose cannot disguise the fact that his treatise lacks both a timescale and even a clearly developed sense of time.

The late Jacques Roger more than once argued that Buffon's theory of the earth was a physical system, not a historical one. As Roger put it, Buffon became a historian of the earth later, when he turned his attention to the succession of fossil forms in the strata. In 1749, however, Buffon offered readers an analysis of physical cycles, rather than a sequence of irreversible historical events.[23] Certainly, this describes well the main lines of argument of the earlier treatise. Even the two stages in the earth's history, the igneous and the aqueous, may be considered a negligible part of the treatise, for the igneous phase served almost no purpose in the theory taken as a whole. Buffon's insistence on the sufficiency of present causes led him to project into the past processes identical to those now observable, and the result was a cyclical pattern that he summarized as "the changes of land into sea and of sea into land."[24]

This pattern notwithstanding, Buffon also presented a few instances of historical sequences that may or may not have been irreversible. One such example is his suggested sequence of the emergence of continents from

---

[22] Buffon [1749] 1954, 105B (Conclusion). Also Buffon 1749, 1:360–61 (Art. 10), 425–26 (Art. 11), for discussion of Halley. On the early malleable crust, Buffon [1749] 1954, 49A (Second Discourse); also quoted and discussed by Roger 1989, 148, and Sloan in *Buffon 88*, p. 215.

[23] Perhaps best presented by Roger in *Buffon 88*, pp. 193–205. The brilliant lecture on this subject that I heard Roger deliver at the 26th International Geological Congress, held in Paris in 1980, seems not to have been published.

[24] The quotation is the title of Buffon's Article 19.

the sea, with Asia being the oldest, Europe and Africa newer, and the Americas the youngest. Whether these land masses should be considered merely the latest examples of "the changes . . . of sea into land" is nowhere clarified, but Buffon did remark on one occasion about the long era of marine sedimentation: "The waters thus have covered and *can again* successively cover all parts of the exposed continents."[25] Should this be read as a prediction of future cycles that have not occurred in the past? Elsewhere, Buffon would remark on minicycles of marine and freshwater deposits in a single locality, contrasting these recent strata with the older, harder, more consolidated ones formed at the bottom of the sea.[26] Again, the reader cannot be quite certain if cycles are here being combined with perhaps irreversible events.

In quite a different sense, however—and one also noted by Jacques Roger—Buffon may be considered unhistorical, that is to say, in his separation of geology from the human past. This perhaps stemmed from a vision of nature's past as longer than the human, but on this point, as we have seen, Buffon was allusive and grandiloquent.[27] More striking, in view of the habits of his predecessors, is Buffon's almost entire avoidance of the quest for human testimony about natural events dating from the early history of mankind. One detects here a disciple of Fontenelle—and Buffon often cited passages from the Academy's *Histoire*—but Buffon developed in his own fashion the view that nature is so reliable as to make human testimony unnecessary. To be sure, he made use of modern travel accounts, but he conducted no antiquarian searches for ancient documents to confirm his analysis of the earth's past.

Buffon applied his technique equally to the Bible and to pagan texts. The Flood narrative, of course, he found useless, for the Flood had been a miracle and was thus irrelevant to science. On one of the rare occasions when he did cite the Bible, it was for a conclusion he had already drawn from nature: "one cannot doubt, even independently of the sacred testimony of the Bible, that before the Flood the earth was composed of the same materials that it is today."[28] As for pagan texts, these he sometimes used alongside more modern ones to add to the available examples of certain phenomena, such as trade winds, the annual flooding by some

[25] The quotation, italics added, is from Buffon [1749] 1954, 58B (Second Discourse). The ages of the different continents are treated in Article 6.

[26] Buffon 1749, vol. 1, esp. 578–80 (Art. 18) for the observations of Modenese terrain by Bernardino Ramazzini.

[27] Cf. Roger 1989, 150–51.

[28] Buffon [1749] 1954, 99B–100A (Art. 8).

rivers, and the building of deltas. But when he briefly considered ancient reports of the appearance of new islands, he concluded: "On all this we have more reliable and newer information."[29]

Even more significant in revealing his method and his preferences are passages where Buffon treated the Black Sea, the Straits of Gibraltar, and the myth of Atlantis. For the eruption of the Black Sea into the Mediterranean, Buffon knew the discussion by Tournefort, who had cited ancient tradition to confirm his scientific "conjectures." Buffon reversed the process, emphasizing physical evidence by which one might judge the accuracy of tradition. For Gibraltar, earlier writers often used the Hercules myth to "confirm" the opening of the straits. Again reversing the order of proof, Buffon declared that comparable geological strata on both sides of the straits confirmed the ancient tradition that Europe and Africa had once been joined. Similarly, when Buffon asked whether one should accept the Atlantis myth to explain the formation of the Atlantic Ocean, he remarked that confirmation should be sought not in texts or traditions but by comparing the fossils of Ireland and North America.[30] Only once did Buffon accept ancient traditions, emanating from Southeast Asia, that various islands had formerly been attached to continents, but he immediately moved on to one instance, the opening of the English Channel, for which he found physical evidence firmer than mere human tradition.[31]

In short, Buffon consciously rejected the common method of naturalists who looked for verification of their science in Robert Plot's "records of time." To be sure, Buffon did employ an often quoted passage from Ovid as the epigraph for his Second Discourse, but that passage was nowhere discussed, much less used to bolster Buffon's science. In his effort to rely wholly on nature and present causes, Buffon sought to anchor his geology with liberal doses of mechanics and hydraulics. Insofar as his treatise relied on books, his literature of choice consisted mainly of the reports of modern travelers and a selection of some modern scientists. Ancient testimony was less reliable than modern observations and than the laws of nature.

[29] Buffon 1749, 1:536–37 (Art. 17). Other topics mentioned can be found in Articles 10 and 14.

[30] Buffon [1749] 1954, 55A–B (Atlantis), 56A–B (Gibraltar); both passages are in the Second Discourse. Buffon 1749, 1:410–14 (Black Sea), 605–6 (Atlantis); these passages respectively are in Articles 11 and 19.

[31] Buffon 1749, 1:586–88 (Article 19). Very little is known about Buffon's associations with the *érudits* of Dijon and Paris, so that one can only speculate about his exposure to studies of ancient history, tradition, and myth; see Hanks 1966, 15–16, 68–70, 263–66, for analysis of the available fragments.

EPILOGUE: THE AFTERMATH OF 1749

To assess the reception of Buffon's theory of the earth poses historical problems of some complexity. One must bear in mind that Buffon's cosmogony
and geology occupied only one of three volumes published simultaneously—the second volume and part of the third examined "the natural
history of man," and the third also contained Daubenton's description of
the "cabinet" at the Jardin du roi. Some readers understandably paid more
attention to one volume than another, although it can be said with certainty that the Buffon portions aroused far more comment than Daubenton's catalogue. Daniel Mornet long ago described what he called "la
querelle Buffon" that erupted in France, but no historian has detected—
at least, none has discussed—any comparable "querelle" elsewhere in
Europe.[32] The further one proceeds into the 1750s and 1760s, the more
difficult become the problems of assessment. In those decades, naturalists
did not remain idle, and some new observations, methods, and ideas challenged aspects of Buffon's synthesis. Above all, the dissemination in those
decades of German works focused attention on the relevance of chemistry
to geology and especially on ways to analyze the "primitive" formations of
the earth's crust. To at least some French and Italian naturalists, Buffon's
minimal use of chemistry made his work incomplete at best, obsolete at
worst.

For these several reasons, a thoroughgoing analysis of the reception of
Buffon's theory is not yet possible. The next pages will focus on immediate reactions of reviewers and naturalists, in the years before German chemistry became well known outside of Germany. Even after this chemical
invasion, however, Buffon's work was still being read. What naturalists continued to find of value in Buffon will be examined very briefly.

In his classic study, Daniel Mornet distinguished between the popular
and the "professional" reactions in France to Buffon's first volumes. On the
one hand, Mornet argued, Buffon was an unequivocal success; on the other,
his peers sometimes offered polite compliments but also doubted his scientific competence. For the popular side, Mornet's evidence included his
analysis of the content of five hundred private libraries where Buffon's work
had only one serious rival in the field of natural history, the abbé Pluche's

---

[32] Mornet 1911, pt. 2, chap. 2. Jacques Roger spent much of his scholarly life trying to
persuade historians that Buffon was a serious thinker; that he had a degree of success is
evident in the collection entitled *Buffon 88*. There nonetheless remains a strong tradition
among historians of geology that Buffon can be ignored or minimized because he did no
fieldwork; this point of view has no bearing on the historical fact that his writings were widely
read.

Spectacle de la nature: Buffon won that battle by a score of 220 copies to 206 for Pluche. (Both works outdid such best-sellers as Voltaire's *Henriade* and Rousseau's *Nouvelle Héloïse.*) Mornet also attributed primarily to Buffon the increase in numbers of natural history collections, the rising market for scientific dictionaries and handbooks, and the success of lecturers in attracting audiences to courses in natural history.[33] That natural history did experience a boom after 1750 is undeniable, and there is sufficient evidence to suggest that Buffon had much to do with this. The causal links, however, remain difficult to establish firmly, especially because Mornet ranged too widely, his examples culled from the forty years after 1749. Is it really possible that the initial enthusiasm and fashionable curiosity lasted so long? Remarkably, the answer may be *yes.* Both the *Imprimerie Royale* and profit-minded publishers continued for decades to reprint Buffon's work, and this phenomenon does much to support the general accuracy of Mornet's argument.[34]

Less scholarly research has so far been done to trace the dissemination of Buffon outside of France. As early as 1750, a German translation began to appear, with a preface by Albrecht von Haller, and Haller would also contribute reviews to the *Bibliothèque raisonnée* and probably to the then-new learned journal emanating from Göttingen.[35] If no Italian translation was published until 1782, naturalists such as Vitaliano Donati, Giovanni Targioni Tozzetti, Giovanni Arduino, and the younger Antonio Vallisneri quickly got hold of one of the French editions. In the small city of Vicenza, within the Venetian Republic, six of twelve libraries contained copies of Buffon.[36] Fragmentary information about Britain shows a considerable audience for Buffon, although here, too, translation was not undertaken until 1781. That there was a degree of dissent in Britain, but not a "querelle Buffon," is implied by the choice of successive translators, who produced "complete" Buffons without including the First Discourse.[37]

[33] Mornet 1911, esp. pt. 3, and Mornet 1910.

[34] See the bibliography of editions in Buffon 1954, 525–30.

[35] The Haller preface is in Buffon 1750, 1:ix–xxii. There was apparently another preface to a subsequent volume; see Lyon and Sloan 1981, 311. That Haller may have written the Göttingen review is indicated, ibid., 311; Haller was one of the editorial committee of the Göttingen journal (Reill 1975, 194). For Haller's admission of authorship of the anonymous review in *BR,* see Haller 1983, 822.

[36] The naturalists will be treated below. For Vicenza, see Piva 1980. Italian editions are listed in Buffon 1954, 527.

[37] On the four incomplete English translations, see Lyon 1976, 134; Lyon here produces for the first time a translation of the First Discourse. For samplings of British libraries, see Kaufman 1960, 1969; more revealing are the comments by G. Rousseau in Rivers 1982, 218–25, and Wood's analysis (1987) of Aberdeen.

On the whole, the first reviews, with some exceptions, were both friendly and cautious, the journalists faithfully reporting and distinguishing fact from speculation. In the latter category, they followed Buffon in admitting that the cosmogony attached to the geology was hypothetical. Singling out two reviews in particular can, in fact, put into perspective the Sorbonne's condemnation of Buffon (in 1751), an affair often misrepresented or misunderstood by modern writers. While the Sorbonne theologians pursued their deliberations, Buffon had already been complimented by the Jesuit periodical, the *Mémoires de Trévoux,* and excoriated by the clandestine Jansenist publication, the *Nouvelles ecclésiastiques*.[38] For the Jesuit reviewer, fresh from the horrors of *Telliamed,* Buffon had displayed a proper respect for the Bible, calling the Flood a miracle and eschewing any tortuous naturalistic explication of Genesis. Buffon's cosmogony, to be sure, was not in accord with Genesis, but the journalist noted that the cosmogony, unlike the geology, was merely hypothetical. In the Jansenist journal, the reviewer found nothing to admire, Buffon's respectful treatment of Scripture being considered a mere façade. By emphasizing that all knowledge was only probable—the reviewer did not quite understand Buffon on this point— Buffon had in effect destroyed any possibility that the study of nature could provide proofs of the existence of God. In these and other ways, Buffon had shown himself to be as subversive as the Academy of Sciences, which had for years permitted in its publications "systems which derogate from the truth and authority of the Sacred Books." In this conspiracy against religion, the *Journal des savants* might also be implicated (its review had been favorable) as might be even the royal government, which allowed such books "to be printed and sold with official knowledge."[39]

At the Sorbonne, the theologians found a middle ground between the accommodating Jesuits and the infuriated Jansenists. They singled out for censure three classes of assertions, about the human soul, certainty and probability, and cosmogony and geology. In the latter group, objections had to do with the origin of the solar system, the likelihood that the sun might

---

[38] The following comments are based on *Trévoux* (September 1749), 1853–72; (October 1749), 2226–45; (November 1749), 2362–78. (Subsequent numbers of the journal reviewed volumes 2 and 3 by Buffon.) For the Jansenist journal, see the introductory remarks and the translated text of the review in Lyon and Sloan 1981, 235–52. Jansenists were not a clerical order, but rather a conglomeration of clerics and laymen who tackled different issues at various times in their long history. A consistent Jansenist target was Jesuit "laxity" in dealing with Christian sinners and with potential converts (such as the Chinese).

[39] In Lyon and Sloan 1981, 250. See also Pappas 1983. Because Buffon's volumes were printed by the *Imprimerie Royale,* they did not have to go through the normal process of censorship, which, in any case, was quite erratic. And because he did not identify himself on his title-page as a member of the Academy of Sciences, he was not obliged to seek the Academy's "approbation" of his work.

one day exhaust its supply of combustibles, and the cyclical changes of land into sea and sea into land (as phrased by the theologians, the problem was the eventual disappearance of now-inhabited continents, designed for man, and their replacement by new continents emerging from the sea).[40] If, as has been suggested, Buffon's response was actually drafted by the theologians and presented to him for his signature, then the Sorbonne considered sufficient a rather bland statement that Buffon's cosmogony was intended as a hypothesis rather than a challenge to Moses.[41] Notable in Buffon's so-called "retraction" is the entire absence of reference to geology, the Flood, or the timescale; apart from a bow in the direction of Moses, the Buffonian text deals entirely with epistemology and the soul. Buffon did fulfill his promise to publish this curious apologia in volume 4 (1753) of the *Histoire naturelle*, but it should be noted that he never changed the language or ideas in any edition of the first three volumes. As Voltaire later remarked, the French public knew what such "retractions" by men of letters were worth.[42]

Among other early reviewers, some mentioned the religious implications of Buffon's theory, but without much urgency or emphasis. Not a whisper of criticism can be found in the *Journal des savants*, and the several writers of newsletters—Pierre Clément, Elie Fréron, F. M. Grimm—all allowed that Buffon had combined natural history with some metaphysics. To Fréron, it was easy to distinguish the hypothetical elements in Buffon from "the most satisfying proofs" of a geological kind.[43] Clément initially reported that Buffon was being accused of system-building, of being more metaphysician than observer. Two years later, in 1751, Clément would

[40] The Sorbonne's objections are printed in Buffon 1954, 106–7.

[41] For Buffon's response, ibid., 107–9. Detailed analysis of the whole episode is in Stengers 1974.

[42] The Voltaire remark is in Stengers 1974, 100. This version of the Sorbonne condemnation differs from the traditional notion that Buffon was charged with heresy, that he recanted, and that his example frightened later (French) scientists. This apparently logical sequence has no basis in fact and mistakes the way the Old Regime in France operated. In the pecking order of the Old Regime, Buffon was virtually untouchable, as he was a royal *fonctionnaire* (intendant of the Jardin du roi) and a person with connections in the Royal Household (the ministry called the Maison du roi). For non-traditional readings of the evidence, see, for example, Fellows and Milliken 1972, 82–83, and Solinas 1965, 287–89. For a realistic analysis of the French church and clergy, see the valuable work by Palmer 1939.

[43] Fréron 1749, 3:3–28, quotation on p. 28 where the writer contrasts the fantasies of *Telliamed* with the solidity of Buffon. See also *JS* (October 1749), 648–57, and O'Keefe 1974. The *JS* stopped mentioning volumes by Buffon until 1754, and the review then of volume 4 (June 1754, pp. 329–39) makes no reference to Buffon's apologia prefacing that volume. Sophisticated disdain can be found in Grimm 1877, 1:336–44, but contrast the assurance of LeCat in 1750 that his own theory of the earth could not be very far wrong if it agreed in some elements with both Buffon and *Telliamed;* see Artigas-Menant 1991, 354.

encounter detailed criticism of Buffon on both scientific and religious grounds. In Clément's view, the critic, the abbé Lignac, had been no more faithful to the Bible than had Buffon; as for the scientific objections, they seemed now and then to be correct, but Clément confessed that he nonetheless loved Buffon like a favorite mistress.[44] No such tone of frivolity can be found in the two reviews by Albrecht von Haller or in comments by some other contemporaries.

In turning to Haller, one enters the realm of the experienced naturalist. Whether one should place in this category Lignac's *Lettres à un amériquain* remains a moot point. The author did have some small competence as a naturalist, and he probably received some aid from both Réaumur and astronomer Pierre Bouguer. The tone of the *Lettres,* however, is not only fiercely polemical, but Clément was not alone in recognizing that Lignac at times twisted Buffon's words and ideas in doubtful ways.[45] For uncompromising violence, the *Lettres* bear a family resemblance to the attack in the clandestine Jansenist press.

In general, Lignac did all he could to present Buffon as anti-Christian. It was indeed simple enough to claim that Buffon's account of the origin of the solar system conflicted with Genesis. Citing in this connection Buffon's remark that chance could *not* have produced the regularities of planetary orbits, Lignac noted that Epicurus, too, had used chance, thus trying to tar Buffon with the brush of Epicurean atheism.[46] More pervasive in the *Lettres* is the theme of the timescale, Lignac spelling out the implications he perceived in Buffon. The initial phase of the earth's history, before any land emerged from the sea, must have lasted "thousands of centuries," and similarly long ages must have been necessary for the accumulation of marine shells in the strata. Here Lignac found one passage where Buffon, quoting Fontenelle, mentioned actual numbers, estimating the amount of time (12,000 years) it would take to destroy a coastal falaise; this passage confirmed Lignac's impression that Buffon had in mind a very long timescale.[47] Possibly most offensive to Lignac was Buffon's failure to use the Flood and final causes. If the Flood was miraculous in cause, Lignac insisted that the event had had natural effects. As for teleology, Lignac here indulged his taste for sarcasm. Buffon had proposed

[44] Clément 1755, 1:206–9; 2:114–18.
[45] Ibid., 2:116. See also the complaints of Condillac that he had been misread by Lignac, in *Mercure de France* (April 1756, vol. 2), 84–95, and the inept reply by Lignac (November 1756), 76–90. The Lignac affair is analyzed by Roger 1993, 691–704, and by Solinas 1973, 71–203.
[46] Lignac 1751, vol. 1, letter 2, pp. 28–29 (the several letters being separately paginated). Historians have tried to discover Buffon's views on religion in general, Christianity in particular; see Bremner 1982; Wattles 1989; and Roger in Buffon 1962, xciv–cxiv.
[47] Lignac 1751, vol. 1, letter 3, pp. 19–20.

that the earth was covered by water for thousands of centuries. And why? In order to produce mollusks and fish. For what purpose? To enclose them in stone, to produce chalk, etc. Ah, who would presume to think that God decided to create the earth so that mollusks and fish would for thousands of centuries be the only witnesses to the earth's wonders.[48]

Buffon's God had not acted very reasonably in waiting so long to create man.

Réaumur's role in this attack is difficult to identify with any precision. He had himself given little thought to the earth's pre-sedimentary history, remarking briefly on one occasion that seemingly unchanging features of the earth's structure (mountain ranges) probably dated from Creation.[49] At the same time, his memoir on the Touraine had provided nondiluvialists with crucial evidence for gradual and successive sedimentation, over a period of time implying the inadequacy of biblical chronology. Like his contemporaries, Réaumur did not question final causes, but he had also condemned the lack of discrimination shown by some naturalists who tried to make all the minutiae of nature testify to God's purposes. In other words, Réaumur could not have endorsed all of Lignac's arguments, but his antagonism to Buffon is patent in his enthusiastic promotion of Lignac's *Lettres.*[50]

Traces of Réaumur's dislike of Buffon date from at least as early as 1740, and he may well have been further offended by Buffon's disdain, expressed in 1749, for those small-minded naturalists whose love of detail for its own sake added little to the understanding of nature's laws. The most explicit comment by Réaumur about Buffon's first volumes occurs in a letter of 1750 to J.-B. Ludot, a provincial scholar with Parisian connections.

The three volumes could not but be harmful to the progress of natural history and of science in general, if the ideas there proposed were accepted. But I hear from every side that an outcry has arisen against this work, which leads me to believe that it will have no fearful consequences. Besides, one cannot have much confidence in the observations recorded by the author when he alone has seen them.[51]

---

[48] Ibid., pp. 15–17, the quotation from p. 15.

[49] Réaumur to Bourguet, 25 March 1741, Neuchâtel, MS 1278, fols. 48–49. This is a very offhand comment, meant to show that Réaumur did not seriously disagree with Bourguet on at least one point.

[50] Réaumur 1734, 1:23–24. Torlais 1958, 14. Roger 1993, 691–92.

[51] Réaumur 1886, 121. For Réaumur's earlier relations with Buffon, see Torlais 1936, 213–14.

The final sentence expresses well the mistrust of the professional for the amateur who had drawn large conclusions too hastily.

Comparable distrust can be found in Haller's preface to the first German translation of Buffon (1750). Haller by no means disdained hypotheses, much of his preface defending their use as a means of posing questions and opening new avenues to truth. Even the new and strange principles set forth by Buffon need not be viewed with repugnance, for history revealed cases of the fruitfulness of ideas once considered bizarre. In Buffon's volumes, "one finds much information about things, many experiments, and much insight. Yet the author always goes somewhat further than his information, experiments, and insight."[52] These cautionary words received considerable expansion in a long review Haller wrote for the *Bibliothèque raisonnée.* Here Haller argued that Buffon's actualism could be turned against the author himself. Could the sea build mountains, and had anyone ever observed such a phenomenon? Was it not obvious that the causes in operation today had produced only minor geological changes in the four thousand years of recorded history? Furthermore, Buffon's information was too often unreliable: Alpine valleys did not display as much regularity as Buffon asserted, and nor did North American coastlines display the effects of the supposed motion of the sea from east to west. Not only had Buffon ignored or misrepresented such facts, but he had engaged in the same groundless hypothesizing (use of a comet) for which he had criticized William Whiston. Buffon, in short, had erred in fundamental ways that a more careful observer could detect.[53] The pious Haller did not think that Buffon's theory would have evil consequences for religion and morals—genuinely evil books of this sort should be suppressed—but he did believe that Buffon had followed the fashion of "modern philosophers" in minimizing or excluding the role of the Flood. Nor did the passage of time soften in the least Haller's judgment. Years later, when his good friend Charles Bonnet engaged Haller in discussion of Buffon's theory, Haller insisted again that Buffon had erected a theory on poor foundations, ignoring evidence in conflict with the views he wished to promote.[54]

Quite different was the initial reaction of Italian naturalist and polymath Giovanni Targioni Tozzzetti. When Targioni was assembling for publication the detailed accounts of his own travels, he suddenly encountered Buffon's

[52] Translated by Lyon and Sloan 1981, 307, from Buffon 1750, xxii. As earlier fruitful results of hypothesizing, Haller mentions the voyages of Columbus and Magellan. Only the last pages of the preface deal explicitly with Buffon.

[53] This review is translated in Lyon and Sloan 1981, 253–82. For its authorship, see above, n. 35.

[54] Correspondence of 1764 and 1769, in Haller 1983, to be discussed below.

first three volumes, even while his own second volume was in press. He immediately announced Buffon's theory of the earth to be "the most beautiful, most judicious, and most probable of any proposed up to now," even if "some few details" did not quite correspond to facts.[55] Far from a mere "detail," Targioni immediately singled out—as did other critics—Buffon's theory of mountain-building: did there really exist marine currents capable of sculpting landforms, or were the ocean depths places of tranquillity?[56] As Targioni also noted, every theory of marine sedimentation had failed to account for what appeared to be an actual diminution of the amount of water as the seas retreated from present continents; in this respect, Buffon had also failed. Furthermore, Targioni could not reconcile his own observations with Buffon's claim that the early crust of the earth consisted of vitrified materials; such substances, he argued, were actually quite rare, and sandstones should not be considered the debris of materials produced by fire.[57]

As a critic of diluvialism, Targioni did not object to Buffon's omission of the Flood from geology. Nonetheless, his doubts about Buffon's theory are so substantial that any reader must wonder why Targioni also expressed so much enthusiasm. Indeed, Buffon's name continued to appear in subsequent volumes of Targioni's *Relazioni* and was inserted perhaps even more frequently in a later edition of the same work.[58] Commonly such citations begin with a compliment and then proceed either to use some information from the *Histoire naturelle* or to dispute matters large and small. What this pattern suggests is that, in Targioni's eyes, Buffon's work was as rich in information and as fertile in ideas as naturalists of the next decades would also acknowledge. One could not ignore Buffon.

Like Targioni, other naturalists found Buffon valuable, even while they criticized him in many ways. To an observer writing in 1779, it was hard to judge "whether his fame be owing to the solidity of his investigations and discoveries, or to those sublime flights, and that magic power, energy, and

---

[55] Targioni 1751, 2:236. See also the valuable study by Arrigoni 1987, esp. 34–41.

[56] Targioni 1751, 2:237–38. When he returned to the subject (4:24–25), he had to admit that he agreed on this point with Lignac, author of "una sanguinosa, e troppo sediziosa critica." See on the same point, Bertrand 1752, 64–66. Contrast the neutrality of *JS* (October 1749), 652.

[57] Targioni 1751, 2:237–41. Targioni had no difficulty with the idea of "long" time, although he, like Steno, wondered if elephant fossils should be dated from Hannibal. He would discuss such finds at length in a letter (addressed to Buffon?), Targioni 1755, later excerpted in Alléon-Dulac 1763, 2:337–44; he concluded that the animals probably antedated man.

[58] This is my impression from examining parts of the 1751 and 1768 editions, neither of which has an index.

grace of style, that have astonished and bewitched a considerable part of Europe."[59] Modern readers, too, can be seduced and bewitched by the Buffonian prose that aroused his contemporaries to admiration and condemnation. If the prose got people to read the books, it still behooves us to ask whether there was something more than style that continued to attract the attention of naturalists in particular.

Targioni offers a clue: this was the best theory so far proposed, and one could, so to speak, bounce one's own observations and ideas off the framework provided by Buffon. Similarly, the young Giovanni Arduino would find in Buffon the inspiration to move from mining technology into the contentious area of geological theory. Like Targioni and others, Arduino could not agree with aspects of Buffon's theory, but he nonetheless considered that Buffon offered a point of departure for his own thinking.[60] A comparable point of view would later be expressed in some detail in the correspondence of Haller and Charles Bonnet.

Not a follower of irreligious "modern philosophers," Bonnet had for some time been worried about issues long familiar to naturalists. As he complained to Haller in 1764,

> I cannot reconcile the parallelism of strata with the hypothesis of the upheaval (*bouleversement*) of the Flood. It seems to me that a universal upheaval excludes an almost universal parallelism. I well know that the latter is trustworthy evidence of deposition in water. The correspondence of salient and re-entering angles adds to this evidence.[61]

Even admitting the earth to be older than commonly thought, Bonnet continued, it seemed impossible to find a satisfactory explanation of how the sea had once deposited marine shells and then retreated from modern continents. To these queries and doubts, Haller replied that one should not exaggerate the parallelism of sedimentary strata because signs of disturbance also abounded. The earth had perhaps undergone more than one great upheaval (*révolution*) in addition to the Flood—"universal earthquakes" seemed a possibility. This proposal did not satisfy Bonnet, because scientists would have to assume, in the absence of written records, the

---

[59] *Monthly Review* 61 (1779), 531. Comments on Buffon's style were commonplace.

[60] Vaccari 1993, 84–106, 116–18. Some of Arduino's enthusiasm was expressed in letters to the younger Vallisneri. Less excited, but equally unable to ignore Buffon, was Vitaliano Donati who also wrote to Vallisneri's son on this subject; for examples, see his letters in Donati 1845, 123, 129, 134, 147.

[61] Haller 1983, 386–87. I have not reproduced all the capitalization and italics of the original.

several necessary upheavals. Any such assumption would require the gathering of much more information than anyone possessed.

In the surviving Bonnet-Haller correspondence, these questions then vanished for five years, until 1769 when Bonnet announced to his friend that he was re-reading Buffon's theory of the earth, which bore the marks of *un génie étendu.* "I do not," he quickly added, "refer to his Visions about the *Formation* of the earth." What Bonnet now had was a very clear conception of

> the accumulation and superposition of strata by the motion of water; the various deposits of shells and marine bodies; the assembling of heavy substances above light ones; the external configuration of mountains; the correspondence of their [salient and re-entering] angles . . . etc.[62]

Buffon had, to be sure, not answered all questions and had made some doubtful assumptions. Indeed, the main questions seemed now clearer than before: how had the sea covered the Alps, and how had it then retreated?

The ensuing lively and argumentative exchanges between Bonnet and Haller cannot be followed in detail here, but these letters do suggest why Buffon continued to be read by naturalists. When Haller remained immovably anti-Buffon, Bonnet delicately implied that his friend might be prejudiced. After all, it was not Buffon but Louis Bourguet who had announced the importance of salient and re-entering angles, and Haller "could not find suspect his [Bourguet's] principles in religious matters." Bonnet also wondered whether Haller had read Buffon recently enough to discern the value of the *Histoire naturelle,* for Haller had been raising objections to specific points that Buffon had treated in the same manner as Haller himself.[63]

Bonnet nowhere assembled his reasons for admiring Buffon, but these can be extracted or inferred from the letters to Haller. Setting aside matters clearly hypothetical—those "visions" about the earth's origin—Bonnet found in Buffon a massive presentation of data, even while he admitted that more was needed. More important than the data, Buffon had outlined with skill, clarity, and brilliance the fundamental problems that naturalists would have to address. He had *solved* only one of those problems: the ubiquity of marine fossils and the parallelism of sedimentary strata persuaded Bonnet that the sea had "for a long time" occupied the present con-

[62] Ibid., 819.
[63] Ibid., 821.

tinents. The vague, ill-formed questions that agitated Bonnet in his earlier letters (1764) had been clarified by his later (1769) re-reading of Buffon: he now had a better idea of what was known, what one might make of it, and what gaps existed.[64]

As intimated by Bonnet, Targioni, and Arduino, a chief reason for reading Buffon was that theory here was accompanied by massive information. The latter is easy to overlook, as modern historians, with good reason, have regarded Buffon more as a thinker than a compiler of data. To his first readers, however, Buffon was not only a theorist but also a wonderful source of detail on a huge range of subjects. In the Diderot *Encyclopédie*, for example, Buffon's text was one of the many pillaged for concrete details in such articles as "Caverne," "Continent," "Fleuve," "Inondation," "Noire mer," and "Rouge mer." This did not mean uncritical acceptance of Buffon's interpretations of his data, as is evident in a small group of articles by d'Alembert on the east-west motion of the sea ("Courant," "Détroit," "Flux et reflux"). D'Alembert seems to have been uneasy about the Buffonian scheme of interpretation, but the east-west motion of the sea had been put forth by respected geographers such as Varenius, so that d'Alembert could only express vague doubts on this subject (article "Courant").

Because the *Encyclopédie* is the work of many hands, one cannot expect a consistent point of view in matters geological. "Strata," for example, seems to have been lifted wholesale from the earlier encyclopedia of Ephraim Chambers, and here we find William Derham's baffling struggle with Woodwardian specific gravities. Completely Buffonian were contributions by Buffon's collaborator, Daubenton announcing in his article on ammonites ("Corne d'Ammon") that Buffon had "thoroughly" explained the transport of exotic organisms to European strata. Not deferential at all, articles by the Baron d'Holbach insisted that Buffon and other French scientists were behind the times: they had not taken into account the research of German chemists that had important implications for the history of the earth.[65] Even at his most unrelenting, d'Holbach nonetheless recognized that the main challenge to his proposed reforms of the study of geology was Buffon's synthesis.

In this sampling—and it is no more than a sampling—of reactions to Buffon, I have discussed chiefly the response of naturalists to difficulties that were not hard to detect. Some also raised questions of method, metaphysics, and religion. But what about *history*? As defined in earlier chap-

---

[64] Ibid., remarks scattered through 819–27.
[65] For d'Holbach's critique, see Rappaport 1994, 237–38.

ters, concern for "history" meant finding human witnesses to natural events. Buffon had rejected this approach, elaborating on Fontenelle's hint that "histories written by the hand of nature itself" were more reliable than human testimony. Physical evidence amounted to certainty, and human evidence should be confined to modern reports rather than ancient texts that are hard to interpret.

To ask if Buffon's readers recognized what is here called "the rejection of history" requires that one notice what is *absent* from the comments of early reviewers and naturalists. Rarely did these commentators raise the issue that seemed so clear to Father Castel in connection with Jussieu's memoir of 1718: if marine currents had brought exotic species to France, why was there no record of such events in historical writings? Why, as Hooke's critics asked, had chroniclers failed to record events as striking as earthquakes? Such questions are generally absent from even the unfavorable notices of Buffon's theory. What seems to have happened is that Buffon himself had defined the terms in which his theory should be discussed: this was a *physical* system, and it was so treated by the commentators. Journalistic "neutrality" here came into play, as this policy entailed a careful rehearsal of the content of a book under review; but reviewers also failed to ask whether Buffon's theory had any historical warrant.

This is not to suggest that the "historical" mode of proof instantly died, as may be illustrated in the writings of Nicolas Desmarest. In his essay on the former land-bridge connecting England with France (1753), Desmarest's performance was wholly traditional, as he examined in detail the kinds of human testimony assembled by William Camden and John Wallis. Very few years later, in 1757, Desmarest would announce in the pages of the *Encyclopédie* that one should avoid acceptance of information provided by ancient authors, for it was hard to distinguish "marvels" from facts in such texts. In this shift from "history" to the certainty of physical laws and processes, Desmarest has been described as a "Buffonian."[66]

Lest one attribute too much to Buffon, one must point out that ancient texts did not disappear from geological treatises. How such texts were used remains an important question for historians to investigate.[67] From the Buffon literature considered here, it seems possible that Buffon helped to divert attention from pagan and other texts, even while he inadvertently moved his readers to look again at the Bible. In other words, antiquarian scholarship and historical confirmation of nature had perhaps become

---

[66] I am indebted here to Kenneth Taylor (forthcoming). Desmarest's doubts about ancient texts are in his "Géographie physique," in the *Encyclopédie*, vol. 7 (1757), 616A.

[67] See, for example, the superb analysis of Boulanger by Cristani 1994.

irrelevant, even while it still was necessary to reconcile the Bible with geology. When Buffon himself turned in his *Epoques de la nature* (1778) from physical cycles to historical reconstruction, he announced that he was going to deal with "monuments" pertinent to the earth's history. That these could not include ancient texts was manifest in his scheme of "epochs," with human beings appearing only after most of the earth's history had taken place.

Abbreviations

AAS, P-V	Archives of the Academy of Sciences (Paris), Procès-verbaux des séances
Acta	*Acta eruditorum*
AOP	Archives of the Observatory of Paris
APS	American Philosophical Society
Bacchini	*Giornale de' letterati* (Parma, Modena)
Basnage	*Histoire des ouvrages des savans*
BJHS	*British Journal for the History of Science*
BN	Bibliothèque nationale, Paris
BR	*Bibliothèque raisonnée*
Commentarii	*De Bononiensi Scientiarum et Artium Istituto atque Academia Commentarii*
DBI	*Dizionario biografico degli Italiani* (1960–)
DHI	*Dictionary of the History of Ideas*
DSB	*Dictionary of Scientific Biography*
GLI	*Giornale de' letterati d'Italia*
HARS or *MARS*	*Histoire* or *Mémoires* of the Royal Academy of Sciences, Paris
HARS depuis 1666	Fontenelle, *Histoire de l'Académie royale des sciences. Depuis son établissement en 1666...*
HWL	*History of the Works of the Learned*
JB	Journal Books of the Royal Society of London
JHI	*Journal of the History of Ideas*
JS	*Journal des savants*
MARI	*Mémoires* of the Royal Academy of Inscriptions, Paris
MHN	Muséum national d'histoire naturelle (Bibliothèque centrale), Paris

Nazari	*Giornale de' letterati* (Rome)
Neuchâtel	Bibliothèquc publique et universitaire, Neuchâtel, Switzerland
Nîmes	Bibliothèque municipale, Nîmes, France
NRL	*Nouvelles de la République des lettres*
NRRS	*Notes and Records of the Royal Society*
PT	*Philosophical Transactions* of the Royal Society
RHS	*Revue d'histoire des sciences*
SVEC	*Studies on Voltaire and the Eighteenth Century*
Trévoux	*Mémoires de Trévoux*

The following conventions are also used in the notes and bibliography. *MARS* [1706] 1707, signifies the volume for the year 1706, actually published in 1707. Leibniz [1749] 1993, refers to the original date of publication (1749) and the date of the edition consulted (1993) in this book.

Bibliography

MANUSCRIPT COLLECTIONS

London, at the Royal Society
  Journal Books (consulted on microfilm)
  Early Letters and Classified Papers, 1660–1740 (microfilm)
Neuchâtel, Bibliothèque publique et universitaire
  Fonds Louis Bourguet: correspondence with Abauzit (MS 1259), Jallabert (MS 1273), Réaumur (MS 1278), Vallisneri (MS 1282)
Nîmes, Bibliothèque municipale
  Correspondence of Jean-François Séguier, MS 136
Paris, Archives de l'Académie des sciences
  Procès-verbaux des séances
  Dossiers of individual members
Paris, Archives de l'Observatoire
  Papers of Joseph-Nicolas Delisle, MSS B.1.1, B.1.4
Paris, Bibliothèque nationale
  MS fr 9774: "Nouveau Systeme du Monde ou Entretiens de Talliamed [sic] Philosophe Indien . . ."
  MS fr 22,229: Papers of Jean-Paul Bignon
  MS n.a.fr. 22,335: "Mélanges sur l'Histoire d'Afrique, d'Asie et d'Amérique"
Paris, Muséum national d'histoire naturelle (Bibliothèque centrale)
  MSS 175, 2187: Papers of Jean-Etienne Guettard
  MS 1391: Papers of Joseph Pitton de Tournefort
  MS 3501: Papers of Jussieu family

BOOK-REVIEW JOURNALS

Listed here are the main journals consulted, not all of them given equally detailed attention. In addition to city of origin and starting date for each, an editor's name is indicated only when a journal is especially associated with one person. The bibliography in the following pages will not list separately any reviews of books, but only a small number of original articles that these journals sometimes carried.

Acta eruditorum, Leipzig, 1682
Bibliothèque germanique, Berlin, 1720
Bibliothèque italique, Geneva, 1728
Bibliothèque raisonnée, Amsterdam, 1728
Bibliothèque universelle et historique, Amsterdam, 1686; founded by Jean LeClerc and edited mainly by him
Europe savante, The Hague, 1718
Galleria di Minerva, Venice, 1696
Giornale de' letterati, Parma, 1686; edited by Benedetto Bacchini
Giornale de' letterati, Rome, 1668; edited by Francesco Nazari
Giornale de' letterati d'Italia, Venice, 1710
Histoire des ouvrages des savans, Rotterdam, 1687; edited by Henri Basnage de Beauval
History of the Works of the Learned, London, 1699
Journal des savants, Paris, 1665
Mémoires de Trévoux, Trévoux, 1701 (sometimes called *Journal de Trévoux,* its complete title is *Mémoires pour l'histoire des sciences et des beaux-arts*)
Monthly Review, London, 1749
Nouvelles de la République des lettres, Amsterdam, 1684; founded and edited by Pierre Bayle until February, 1687, then continued by others
Novelle letterarie, Florence, 1740; edited by Giovanni Lami
Raccolta di opuscoli scientifici e filologici, Venice, 1728; edited by Angelo Calogerà
Weekly Memorials for the Ingenious, London, 1682

PUBLICATIONS OF SCIENTIFIC ACADEMIES AND SOCIETIES

Principal articles in these publications are listed by name of author in the next pages. No such procedure is possible for the Florentine *Saggi* (1667), listed below under Accademia del Cimento. For the Rouen Academy, listings by author seem inappropriate, the surviving materials being so fragmentary; the valuable publication by Gosseaume (1814) offers summaries and some quotations but rarely a glimpse of how any one writer developed his argument and evidence. The *Mémoires* of the Montpellier society contain in the first published volume (1766) a statement by the secretary that some articles will no doubt seem outdated. As this remark implies, the texts were not abridged or otherwise altered, as one may confirm by comparing the published version of Jean Astruc's memoir with the journalistic reports of 1708. I have therefore listed by author a few articles written by Montpelliérains early in the century and published only much later.

Other institutional publications present simpler or fewer problems. The Berlin or Prussian Academy, reasonably active in the days of Leibniz, needed revitalization in the 1740s. As a result, there were two series of publications, the earlier one in Latin, the later in French. Articles in both series are listed below by author. After Henry Oldenburg's death in 1677, his *Philosophical Transactions* suffered temporary death, then was replaced for a time by the *Philosophical Collections*; some items in the latter series are cited in the notes, but they are book reviews and are therefore not listed separately in the bibliography.

Potentially confusing are the publications of the Paris Academy, because their titles are so similar. The series called *Histoire et Mémoires* began with the year 1699 (the volume actually published in 1702), and this continued through the eighteenth century as the public record of academic research. There also exists a short-lived series, entitled *Mémoires* (without *Histoire*), published in 1692 and 1693. Although it is impossible in the latter volumes to know whether the academicians or the secretary wrote the texts, I have listed by author a small number of these articles. Yet another series labeled *Histoire* (but some volumes called *Mémoires*) is in part a compilation by Fontenelle of the early proceedings of the Academy, based both on J.-B. DuHamel's Latin history and on materials not used by DuHamel. One volume in this series is a collection of the early writings of the first of the Cassini dynasty, and a single article from that volume is listed below under the author's name.

Selected Original Sources and Secondary Works

Aarsleff, Hans. 1982. *From Locke to Saussure: Essays on the Study of Language and Intellectual History*. Minneapolis.

Abetti, Giorgio, ed. 1942. *Le Opere dei discepoli di Galileo Galilei, I: L'Accademia del Cimento*. Florence.

Académie française. 1695. *Le Grand Dictionnaire*. 2d ed. 2 vols. Paris.

Académie royale des sciences, Paris. 1729–34. *Histoire de l'Académie . . . depuis 1666. . . .* 11 vols. Paris. (Some volumes entitled *Mémoires*.)

Accademia del Cimento, Florence. 1684. *Essayes of Natural Experiments*. Translated by Richard Waller. London.

Accademie e cultura. Aspetti storici tra Sei e Settecento. 1979. Florence.

Accordi, Bruno. 1975. "Paolo Boccone (1633–1704)." *Geologica Romana* 14:353–59.

——. 1978. "Agostino Scilla, painter from Messina (1629–1700), and his experimental studies on the true nature of fossils." *Geologica Romana* 17:129–44.

Account of the late terrible earthquake in Sicily. 1693. London.

Acton, Alfred. 1929–1930. "Swedenborg and His Scientific Reviewers." *New Philosophy* 32:19–160; 33.

——. 1948. *The Letters and Memorials of Emanuel Swedenborg*. Bryn Athyn, Pa.

Adams, Frank Dawson. [1938] 1954. *The Birth and Development of the Geological Sciences*. New York.

Addison, Joseph. 1914. *Miscellaneous Works*. Edited by A. C. Guthkelch. 2 vols. London.

Adelmann, Howard B. 1966. *Marcello Malpighi and the Evolution of Embryology*. 5 vols. Ithaca.

Aiton, Eric J. 1985. *Leibniz: A Biography*. Bristol.

Akerman, Susanna. 1991. *Queen Christina of Sweden and Her Circle*. Leiden.

Albritton, Claude C., Jr., ed. 1963. *The Fabric of Geology*. Reading, Mass.

Aldridge, Alfred O. 1950. "Benjamin Franklin and Jonathan Edwards on Lightning and Earthquakes." *Isis* 41:162–64.

Alembert, Jean le Rond d'. 1753. "Chronologie." In Diderot, *Encyclopédie* 3:390–92.

Allen, David E. 1976. *The Naturalist in Britain: A Social History.* London.

Allen, Don Cameron. [1949] 1963. *The Legend of Noah.* Urbana.

———. 1964. *Doubt's Boundless Sea: Scepticism and Faith in the Renaissance.* Baltimore.

———. 1970. *Mysteriously Meant: The Rediscovery of Pagan Symbolism and Allegorical Interpretation in the Renaissance.* Baltimore.

Alléon-Dulac, Jean-Louis. 1763–65. *Mélanges d'histoire naturelle.* 6 vols. Lyons.

Almagor, Joseph. 1989. *Pierre Des Maizeaux (1673–1745), Journalist and English Correspondent for Franco-Dutch Periodicals, 1700–1720.* Amsterdam.

Altieri Biagi, Maria Luisa, and Bruno Basile, eds. 1980. *Scienziati del Seicento.* Milan.

———. 1983. *Scienziati del Settecento.* Milan.

Ambri Berselli, Paola. 1955. "Lettere di illustri francesi a F. M. Zanotti." *Strenna storica bolognese* 5:17–33.

Amico, Vito Maria. 1764. "Lettera . . . intorno a' testacei montani, che in Sicilia, ed altrove si trovano." In *Opuscoli di Autori Siciliani* (Palermo), 8:199–232.

Amontons, Guillaume. [1703] 1705. "Que les nouvelles expériences que nous avons du poids & du ressort de l'air, nous font connoître . . . [les causes] de tres-grands tremblements & bouleversements sur le Globe terrestre." *MARS* 101–8.

Andreoli, Aldo. 1922. "Il Morgagni, il Manfredi e il Muratori." *L'Archiginnasio* 17:216–21.

———. 1964. "L'Accademia degli Inquieti. Inediti di Eustachio Manfredi." *Convivium* 32:386–402.

Anelli, Vittorio, et al. 1986. *Leggere in provincia. Un censimento delle biblioteche private a Piacenza nel Settecento.* Bologna.

Anton Lazzaro Moro (1687–1987). Atti del Convegno di Studi. 1988. San Vito al Tagliamento.

Arbuthnot, John. 1697. *An Examination of Dr. Woodward's Account of the Deluge.* London.

Arderon, William. 1746. "Extract of a Letter . . . containing Observations on the Precipices or Cliffs on the North-East Sea-Coast of the County of Norfolk." *PT* 44:275–84.

Arnauld, Antoine, and Pierre Nicole. 1662. *La Logique, ou l'art de penser.* Paris.

———. 1964. *The Art of Thinking: Port-Royal Logic.* Translated with an introduction by J. Dickoff and P. James. Indianapolis.

Arrigoni, Tiziano. 1986. "Inventario del carteggio di Giovanni Targioni Tozzetti." *Nuncius* 1:59–139.

———. 1987. *Uno Scienziato nella Toscana del Settecento. Giovanni Targioni Tozzetti.* Florence.

Artigas-Menant, Geneviève. 1991. "La Vulgarisation scientifique dans *Le Nouveau Magasin français* de Mme Leprince de Beaumont." *RHS* 44:343–57.

Ascoli, Georges. 1930. *La Grande-Bretagne devant l'opinion française au XVIIe siècle.* 2 vols. Paris.

Astruc, Jean. [1707] 1766. "Mémoire sur les pétrifications de Boutonnet." In Montpellier, *Mémoires,* 1:48–74.

———. 1737. *Mémoires pour l'histoire naturelle de la province de Languedoc.* Paris.

Atkinson, Richard J. C. 1956. *Stonehenge.* London.

Badaloni, Nicola. 1961. *Introduzione a G. B. Vico.* Milan.

———. 1968. *Antonio Conti: un abate libero pensatore tra Newton e Voltaire.* Milan.

Baier, Johann Jacob. 1708. *Oryktographia norica.* Nuremberg.

Baillet, Adrien. 1691. *La Vie de Monsieur Des-cartes.* 2 vols. Paris.

Baker, Henry. 1745. "A Letter . . . concerning an extraordinary large fossil Tooth of an Elephant." *PT* 43:331–35.

Baker, Victor R. 1978. "The Spokane Flood Controversy and the Martian Outflow Channels." *Science* 202:1249–56.

Baldini, Massimo. 1975. *Teoria e storia della scienza.* Rome.

Baldini, Ugo. 1980. "La scuola galileiana" and "L'attività scientifica nel primo Settecento." In *Storia d'Italia, Annali 3.* Turin.

Baldini, Ugo, and Luigi Besana. 1980. "Organizzazione e funzione delle accademie." In *Storia d'Italia, Annali 3.* Turin.

Balfour, Andrew. 1700. *Letters Write [sic] to a Friend.* Edinburgh.

Baltrusaitis, Jurgis. 1967. *La Quete d'Isis: Introduction à l'Egyptomanie.* Paris.

Baltus, Jean-François. 1707. *Réponse à l'histoire des oracles, de Mr. de Fontenelle.* Strasbourg.

Banier, Antoine. 1739–40. *The Mythology and Fables of the Ancients, Explain'd from History.* Translated from the French. 4 vols. London.

Barber, William Henry. 1955. *Leibniz in France from Arnauld to Voltaire.* Oxford.

Barker, Peter, and Roger Ariew, eds. 1991. *Revolution and Continuity: Essays in the History and Philosophy of Early Modern Science.* Washington, D.C.

Barnes, Annie. 1938. *Jean LeClerc (1657–1736) et la République des Lettres.* Paris.

Barnes, Barry, and Steven Shapin, eds. 1979. *Natural Order: Historical Studies of Scientific Culture.* Beverly Hills.

Barnes, Sherman B. 1934. "The Scientific Journal, 1665–1730." *Scientific Monthly* 38:257–60.

——. 1936. "The Editing of Early Learned Journals." *Osiris* 1:155–72.

Barocchi, Paola, and Giovanna Ragionieri, eds. 1983. *Gli Uffizi. Quattro secoli di una galleria.* 2 vols. Florence.

Barr, James. 1985. "Why the World Was Created in 4004 B.C.: Archbishop Ussher and Biblical Chronology." *Bulletin of the John Rylands University Library* 67:575–608.

Barrère, Pierre. 1746. *Observations sur l'origine et la formation des pierres figurées.* Paris.

Barret-Kriegel, Blandine. 1988. *Les Académies de l'histoire.* Paris.

Barrière, Pierre. 1951. *L'Académie de Bordeaux.* Bordeaux.

Barrow, Isaac. 1734. *The Usefulness of Mathematical Learning.* Translated by John Kirby. London.

Basile, Bruno. 1987. *L'Invenzione del vero.* Rome.

Bastholm, Ejvind. 1950. *The History of Muscle Physiology.* Copenhagen.

Battail, Jean-François. 1982. "Essai sur le cartésianisme suédois." *NRL* (Naples), no. 2, pp. 25–71.

Bayle, Pierre. 1727–31. *Oeuvres diverses.* 4 vols. The Hague.

——. 1734–41. *A General Dictionary, Historical and Critical.* 10 vols. London.

Beall, Otho T., Jr. 1961. "Cotton Mather's Early 'Curiosa Americana' and the Boston Philosophical Society of 1683." *William and Mary Quarterly,* 3d ser., 18:360–72.

Beasley, William G., and E. G. Pulleyblank, eds. 1961. *Historians of China and Japan.* London.

Beattie, Lester M. 1935. *John Arbuthnot, Mathematician and Satirist.* Cambridge, Mass.

Beaufort, Louis de. 1738. *Dissertation sur l'incertitude des cinq premiers siècles de l'histoire romaine.* Utrecht.

Beaumont, John. 1676. "Two Letters . . . concerning Rock-Plants and their growth." *PT* 11:724–42.

——. 1683. "A further account of some Rock-plants growing in the Lead Mines of Mendip Hills." *PT* 13:276–80.

——. 1693a. "Considerations on a Book Entituled, *The Theory of the Earth.*" *PT* 17:888–92.

——. 1693b. *Considerations on a Book, Entituled The Theory of the Earth.* London.

Behrens, Georg Henning. [1703] 1730. *The Natural History of Hartz-Forest.* Translated from the German by John Andree. London.

Bellanger, Claude, et al. 1969–76. *Histoire générale de la presse française.* 5 vols. Paris.

Bellers, Fettiplace. 1712. "A Description of the several Strata . . . in a Coal-Pit . . . To which is added, a Table of the Specifick Gravity of each Stratum: By Mr. Fr. Hauksbee, F.R.S." *PT* 27:541–44.

Belozubov, Leonid. 1968. *L'Europe savante (1718–20).* Paris.

Benitez, Miguel. 1980. "Benoît de Maillet et la littérature clandestine: Étude de sa correspondance avec l'abbé Le Mascrier." *SVEC* 183:133–59.

——. 1984. "Benoît de Maillet et l'origine de la vie dans la mer: Conjecture amusante ou hypothèse scientifique?" *Revue de synthèse* 105:37–54.

Berengo, Marino, ed. 1962. *Giornali veneziani del Settecento.* Milan.

Beringer, J. B. A. 1963. *The Lying Stones of Dr. Johann Bartholomew Adam Beringer.* Translated and annotated by M. E. Jahn and D. J. Woolf. Berkeley.

Berkeley, Edmund, and Dorothy Berkeley, eds. 1965. *The Reverend John Clayton.* Charlottesville.

Berkeley, (Bishop) George. 1901. *Works.* Edited by Alexander Campbell Fraser. 4 vols. Oxford.

Berkvens-Stevelinck, Christiane, and Jeroom Vercruysse. 1993. "Le Métier de journaliste au dix-huitième siècle. Correspondance entre Prosper Marchand, Jean Rousset de Missy et Lambert Ignace Douxfils." *SVEC* 312.

Bernard, Jean-Frédéric, ed. 1740. *Dissertations mêlées, sur divers sujets importans et curieux.* 2 vols. Amsterdam.

Bertelli, Sergio. 1955. "La crisi dello scetticismo e il rapporto erudizione-scienza agl'inizi del secolo XVIII." *Società* 11:435–56.

——. 1960. *Erudizione e storia in Ludovico Antonio Muratori.* Naples.

——. 1973. *Ribelli, libertini e ortodossi nella storiografia barocca.* Florence.

—— , ed. 1980. *Il libertinismo in Europa.* Milan.

Bertrand, Elie. 1752. *Memoires sur la structure interieure de la terre.* Zurich.

——. 1757. *Mémoires historiques et physiques sur les tremblemens de terre.* The Hague.

——. 1763. *Dictionnaire universel des fossiles propres, et des fossiles accidentels.* Avignon.

——. 1766. *Recueil de divers traités sur l'histoire naturelle de la terre et des fossiles.* Avignon.

Betts, C. J. 1984. *Early Deism in France.* The Hague.

Bevis, John, ed. 1757. *The History and philosophy of Earthquakes, from the remotest to the present times.* London.

Bianchini, Francesco. 1697. *La Istoria universale provata con monumenti, e figurata con simboli de gli antichi.* Rome.

Birch, Thomas. 1756–57. *The History of the Royal Society of London.* 4 vols. London.

Birembaut, Artur. 1957. "Fontenelle et la géologie." *RHS* 10:360–74.

Birembaut, Artur, Pierre Costabel, and Suzanne Delorme. 1966. "La Correspondance Leibniz-Fontenelle et les relations de Leibniz avec l'Académie Royale des Sciences en 1700–1701." *RHS* 19:115–32.

Biswas, Asit K. "Edmond Halley, F.R.S., hydrologist extraordinary." *NRRS* 25:47–57.

Blair, Patrick. 1710. "Osteographia Elephantina." *PT* 27:51–168.

Bléchet, Françoise. 1979. "L'Abbé Bignon, Bibliothécaire du Roy, et les milieux savants en France au début du XVIIIe siècle." In *Buch und Sammler. Private und öffentliche Bibliotheken im 18. Jahrhundert.* Heidelberg.

Bligny, Michel. 1973. "Il mito del diluvio universale nella coscienza europea del Seicento." *Rivista storica italiana* 85:47–63.

Bloom, Edward A. 1957. "'Labors of the Learned': Neoclassic Book Reviewing Aims and Techniques." *Studies in Philology* 54:537–63.

Blunt, Wilfrid. 1986. *Linné, 1707–1778.* Translated by Françoise Robert. Paris.

Boccone, Paolo. 1671. *Recherches et observations naturelles.* Paris.

Bodemann, Eduard. 1895. *Die Leibniz-Handschriften der Königlichen Öffentlichen Bibliothek ш Hannover.* Hannover.

Boehm, Laetitia, and Ezio Raimondi, eds. 1981. *Università, Accademie e Società scientifiche in Italia e in Germania dal Cinquecento al Settecento.* Bologna.

Boerhaave, Hermann. 1735. *Elements of Chemistry.* Translated by Timothy Dallowe. 2 vols. in 1. London.

Bolingbroke, Henry St. John, viscount. 1809. *Works.* 8 vols. London.

Bollème, Geneviève, et al. 1965. *Livre et société dans la France du XVIIIe siècle.* Paris.

Bonno, Gabriel. 1943. "Hans Sloane et les relations intellectuelles franco-anglaises au dix-huitième siècle." *Romanic Review* 34:40–49.

———. 1948. "La Culture et la civilisation britanniques devant l'opinion française . . . , 1713–1734." APS, *Transactions,* n.s., 38, pt. 1.

Borel, Pierre. 1649. *Les Antiquitez, Raretez, Plantes, Mineraux, & autres choses considerables de la Ville, & Comté de Castres d'Albigeois, & des lieux qui sont à ses enuirons.* Castres.

Borghero, Carlo. 1983. *La certezza e la storia. Cartesianesimo, pirronismo e conoscenza storica.* Milan.

Bork, Kennard B. 1974. "The geological insights of Louis Bourguet (1678–1742)." *Journal of the Scientific Laboratories, Denison University* 55:49–77.

———. 1991. "Elie Bertrand (1713–1797) sees God's order in nature's record." *Earth Sciences History* 10:73–88.

Bos, H. J. M., et al. 1980. *Studies on Christiaan Huygens.* Lisse, The Netherlands.

Boscherini Giancotti, Emilia. 1963. "Nota sulla diffusione della filosofia di Spinoza in Italia." *Giornale critico della filosofia italiana,* 3d ser., 17:339–62.

Bosdari, Filippo. 1928. "Francesco Maria Zanotti nella vita bolognese del Settecento." *Atti e Memorie della R. Deputazione di storia patria per le provincie di Romagna,* 4th ser., 18:157–222.

Bots, Hans, ed. 1988. *La Diffusion et la lecture des journaux de langue française sous l'ancien régime.* Amsterdam.

Bots, Hans, and Lenie van Lieshout. 1984. *Contribution à la connaissance des réseaux d'information au début du XVIIIe siècle. Henri Basnage de Beauval et sa correspondance à propos de l' "Histoire des Ouvrages des Savans" (1687–1709).* Amsterdam.

Boud, R. C. 1975. "The Early Development of British Geological Maps." *Imago Mundi* 27:73–96.

Bouillier, Francisque. 1854. *Histoire de la philosophie cartésienne.* 2 vols. Paris.

Bourguet, Louis. 1729. *Lettres philosophiques.* Amsterdam.

Bourguet, Louis, and Pierre Cartier. 1742. *Traité des pétrifications.* 2 vols. in 1. Paris.

Bourguignon [possibly Bourgnon]. 1708. "Relation de l'isle nouvelle de Santorin." *Trévoux* (July): 1261–76.

Bovet, Richard. [1684] 1951. *Pandaemonium.* Introduction and notes by Montague Summers. Aldington, Kent.

Boyle, Robert. 1665. "Of a place in England, where, without petrifying Water, Wood is turned into Stone." *PT* 1:101–2.

———. 1666. "General Heads for a Natural History of a Countrey, Great or small." *PT* 1:186–89, 315–16, 330–43.

———. 1671. *Tracts written by the Honourable Robert Boyle.* Oxford.

———. 1772. *Works.* Edited by Thomas Birch. 6 vols. London.

Bredvold, Louis I. [1934] 1956. *The Intellectual Milieu of John Dryden.* Ann Arbor.

Bremner, Geoffrey. 1982. "Buffon and the casting out of fear." *SVEC* 205:75–88.

Breyne, Johann Philipp. 1737. "A Letter . . . with Observations, and a Description of some Mammoth's Bones dug up in Siberia, proving them to have belonged to Elephants." *PT* 40:124–38.

Brink, C. O. 1986. *English Classical Scholarship.* Cambridge.

Broc, Numa. 1969. *Les Montagnes vues par les géographes et les naturalistes de langue française au XVIIIe siècle.* Paris.

———. 1974. *La Géographie des philosophes: Géographes et voyageurs français au XVIIIe siècle.* Paris.

Brockliss, Lawrence W. B. 1987. *French Higher Education in the Seventeenth and Eighteenth Centuries.* Oxford.

Bromley, John S., and E. H. Kossmann, eds. 1960. *Britain and the Netherlands.* London.

Brooke, John Hedley. 1989. "Science and the Secularisation of Knowledge: Perspectives on Some Eighteenth Century Transformations." *Nuncius* 4:43–65.

Brown, Harcourt. 1933. "Un Cosmopolite du grand siècle: Henri Justel." *Bulletin de la Société de l'histoire du protestantisme français* 82:187–201.

———. 1934. *Scientific Organizations in Seventeenth Century France (1620–1680).* Baltimore.

———. 1948. "Jean Denis and Transfusion of Blood, Paris, 1667–1668." *Isis* 39:15–29.

———. 1972. "History and the Learned Journal." *JHI* 33:365–78.

———. 1976. *Science and the Human Comedy.* Toronto.

Brugmans, Henri L. 1935. *Le Séjour de Christian Huygens à Paris.* Paris.

Brunelli Bonetti, Bruno. 1938. *Figurine e costumi nella corrispondenza di un medico del Settecento.* Milan.

Brunet, Pierre. 1926. *Les Physiciens hollandais et la méthode expérimentale en France au XVIIIe siècle.* Paris.

———. 1947. "La Méthodologie de Mariotte." *Archives internationales d'histoire des sciences* 1:26–59.

Buffon, Georges-Louis Leclerc, comte de. 1749. *Histoire naturelle, générale et particulière.* Vol. 1. Paris.

———. [1749] 1750. *Allgemeine Historie der Natur.* Vol. 1. Foreword by Albrecht von Haller. Notes by Abraham Kästner. Hamburg.

———. [1749] 1812. *Natural History, general and particular.* New ed. Vols. 1–2. Translated by William Smellie. London.

———. [1749] 1959. *Storia naturale.* Translated with notes by Marcella Renzoni. Turin.

———. 1954. *Oeuvres philosophiques de Buffon.* Edited by Jean Piveteau. Paris.

———. 1962. *Les Epoques de la nature.* Edited with notes and introduction by Jacques Roger. Paris.

Buffon 88. 1992. Actes du Colloque international pour le bicentenaire de la mort de Buffon. Paris.

Buonanni, Filippo. 1681. *Ricreatione dell'occhio e della mente.* Rome.

Burnet, Thomas. 1684. *The Theory of the Earth.* London. (The original Latin edition and subsequent English ones have the word "sacred" in the title.)

——. 1726. *The Sacred Theory of the Earth.* 6th ed. 2 vols. London.

Burns, Robert M. 1981. *The Great Debate on Miracles: From Joseph Glanvill to David Hume.* London.

Bynum, William F., et al., eds. 1981. *Dictionary of the History of Science.* Princeton.

Byrd, William. 1921. "Letters of William Byrd II, and Sir Hans Sloane relative to plants and minerals of Virginia." *William and Mary Quarterly,* 2d ser., 1:186–200.

Calinger, Ronald S. 1968. "Frederick the Great and the Berlin Academy of Sciences (1740–1766)." *Annals of Science* 24:239–49.

Calmet, Augustin. 1707–16. *Commentaire littéral sur tous les livres de l'Ancien et du Nouveau Testament.* 23 vols. Paris.

——, 1715. *Discours et dissertations sur tous les livres de l'Ancien et du Nouveau Testament* 5 vols. Paris.

——. 1720–21. *Dictionnaire historique, critique, chronologique, géographique et littéral de la Bible.* 4 vols. Paris.

Camden, William. 1695. *Camden's Britannia, Newly Translated into English: With large additions and improvements.* Edited by Edmund Gibson. London.

Campbell, John L., and Derick Thomson. 1963. *Edward Lhuyd in the Scottish Highlands, 1699–1700.* Oxford.

Camusat, Denis-François. 1734. *Histoire critique des journaux.* 2 vols. in 1. Amsterdam.

Cantelli, Gianfranco. 1972. "Mito e storia in J. LeClerc, Tournemine e Fontenelle." *Rivista critica di storia della filosofia* 27:269–86, 385–400.

Canziani, Guido, and G. Paganini, eds. 1986. *Le edizioni dei testi filosofici e scientifici del '500 e del '600.* Milan.

Capra, Carlo, et al. 1986. *La Stampa italiana dal cinquecento all' ottocento.* Bari.

Carozzi, Albert V. 1971. "Une nouvelle interprétation du soi-disant catastrophisme de Cuvier." *Archives des sciences* 24:367–77.

Carozzi, Albert V., and Gerda Bouvier. 1994. *The Scientific Library of Horace-Bénédict de Saussure (1797).* Geneva.

Carozzi, Marguerite. 1983. "Voltaire's Attitude toward Geology." *Archives des sciences* 36:3–145.

——. 1986. "From the concept of salient and reentrant angles by Louis Bourguet to Nicolas Desmarest's description of meandering rivers." *Archives des sciences* 39:25–51.

Carozzi, Marguerite, and A. V. Carozzi. 1984. "Elie Bertrand's changing theory of the earth." *Archives des sciences* 37:265–300.

Carpanetto, Dino, and Giuseppe Ricuperati. 1987. *Italy in the Age of Reason, 1685–1789.* Translated by Caroline Higgitt. London.

Carré, J. R. 1932. *La Philosophie de Fontenelle ou le sourire de la raison.* Paris.

Carroll, Robert T. 1975. *The Common-Sense Philosophy of Religion of Bishop Edward Stillingfleet, 1635–1699.* The Hague.

Cary, Robert. 1677. *Palaeologia Chronica: A Chronological Account of Ancient Time.* London.

Casini, Paolo. 1978. "Les Débuts du Newtonianisme en Italie, 1700–1740." *Dix-huitième siècle* 10:85–100.

Cassini, Jean-Dominique (or Giandomenico), called Cassini I. 1730. "Réflexions sur la chronologie chinoise." *MARS/HARS depuis 1666* 8:300–11.

Castel, Louis-Bertrand. 1722. "Conjecture sur les pierres figurées qu'on trouve à Saint Chaumont dans le Lyonnois." *Trévoux* (June): 1089–1102.

Catesby, Mark. [1731–43] 1771. *The Natural History of Carolina, Florida and the Bahama Islands.* Rev. ed. 2 vols. London.

Cavadini-Canonica, Tiziana. 1970. *Le Lettere di Scipione Maffei e la Bibliothèque italique.* Lugano.

Cavazza, Marta. 1980. "Bologna and the Royal Society in the Seventeenth Century." *NRRS* 35:105–23.

———. 1981. "Accademie scientifiche a Bologna dal 'Coro anatomico' agli 'Inquieti' (1650–1714)." *Quaderni storici* 16:884–921.

———. 1985. "Impact du concept baconien d'histoire naturelle dans les milieux savants de Bologne." *Les Etudes philosophiques,* 405–14.

———. 1990. *Settecento Inquieto: Alle origini dell'Istituto delle Scienze di Bologna.* Bologna.

Caylus, Anne-Claude-Philippe, comte de. 1914. *Voyage d'Italie, 1714–1715.* Edited by A.-A. Pons. Paris.

Cestoni, Giacinto. 1940–41. *Epistolario ad Antonio Vallisnieri.* Edited by S. Baglioni. 2 vols. Rome.

Chabbert, Pierre. 1968. "Pierre Borel (1620?–1671)." *RHS* 21:303–43.

Chambers, Ephraim. 1751–52. *Cyclopaedia.* 7th ed. 2 vols. London.

———. 1753. *Supplement.* 2 vols. London.

Charas, Moïse. 1673. *New Experiments upon Vipers . . . Also a Letter of Francesco Redi.* London.

———. 1692. "Réflexions sur les causes de la chaleur des sources chaudes." *MARS,* 155–58.

Cheyne, George. 1715. *Philosophical Principles of Religion: Natural and Reveal'd.* London.

Childrey, Joshua. 1660. *Britannia Baconica.* London.

———. 1667. *Histoire des singularitez naturelles d'Angleterre, d'Escosse, et du pays de Galles.* Translated by P. Briot. Paris.

Churchill, John, ed. 1732. *A Collection of Voyages and Travels.* 2d ed. 6 vols. London.

Clair, Pierre. 1978. *Jacques Rohault (1618–1672). Bio-bibliographie, avec l'édition critique des entretiens sur la philosophie.* Paris.

Clarke, Desmond M. 1989. *Occult Powers and Hypotheses: Cartesian Natural Philosophy under Louis XIV.* Oxford.

Clément, Pierre. 1755. *Les Cinq années littéraires.* 2 vols. Berlin.

Cobban, Alfred. 1964. *The Social Interpretation of the French Revolution.* Cambridge.

Cochrane, Eric. 1958. "The Settecento Medievalists." *JHI* 19:35–61.

———. 1965. "Muratori: The Vocation of a Historian." *Catholic Historical Review* 51:153–72.

———. 1973. *Florence in the Forgotten Centuries, 1527–1800.* Chicago.

Cohen, Claudine. 1991. "Benoît de Maillet et la diffusion de l' histoire naturelle à l'aube des lumières." *RHS* 44:325–42.

———. 1994. *Le Destin du mammouth.* Paris.

Cohen, I. Bernard, ed. 1958. *Isaac Newton's Papers and Letters on Natural Philosophy.* Cambridge, Mass.

Cohen, Murray. 1977. *Sensible Words: Linguistic Practice in England, 1640–1785.* Baltimore.

Cole, Francis Joseph. 1937. "Leeuwenhoek's Zoological Researches—Part II. Bibliography and Analytical Index." *Annals of Science* 2:185–235.

———. 1944. *A History of Comparative Anatomy from Aristotle to the Eighteenth Century.* London.

Colie, Rosalie L. 1957. *Light and Enlightenment: A Study of the Cambridge Platonists and the Dutch Arminians.* Cambridge.

———. 1963. "Spinoza in England, 1665–1730." *APS, Proceedings,* 107:183–219.

Collection académique, Partie étrangère. 1755–79. 13 vols. Dijon.

Collier, Katharine Brownell. 1934. *Cosmogonies of Our Fathers.* New York.

Collina, Maria D. 1957. *Il carteggio letterario di uno scienziato del Settecento, Janus Plancus* [Giovanni Bianchi]. Florence.

Collis, Patrick A. 1920–21. "The Preface of the 'Acta Sanctorum'." *Catholic Historical Review* 6:294–307.

Comparato, Vittor Ivo. 1970. *Giuseppe Valletta.* Naples.

Conti, Antonio. 1972. *Scritti filosofici.* Edited by N. Badaloni. Naples.

Cope, Jackson I. 1956. *Joseph Glanvill, Anglican Apologist.* St. Louis.

Copenhaver, Brian P. 1991. "A Tale of Two Fishes: Magical Objects in Natural History from Antiquity through the Scientific Revolution." *JHI* 52:373–98.

Costantini, Giuseppe Antonio. 1747. *La verità del diluvio universale vindicata dai dubbj, e dimostrata nelle sue testimonianze.* Venice.

Costard, G. 1747. "A Letter . . . concerning the Chinese Chronology and Astronomy." *PT* 44:476–93.

Cotgrave, Randle. [1611] 1950. *A Dictionarie of the French and English Tongues.* Introduction by W. S. Woods. Columbia, S.C.

Cotoni, Marie-Hélène. 1984. "L'Exégèse du Nouveau Testament dans la philosophie française du dix-huitième siècle." *SVEC* 220.

Couperus, Marianne. 1972. *L'Etude des périodiques anciens.* Paris.

Cousin, Victor. 1865–66. *Fragments philosophiques.* 5th ed. 5 vols. Paris.

Couturat, Louis. 1901. *La Logique de Leibniz.* Paris.

Craig, John. 1964. "Craig's Rules of Historical Evidence." *History and Theory* 3, Beiheft 4, pp. 1–31.

Crane, Ronald S. 1933–34. "Anglican Apologetics and the Idea of Progress, 1699–1745." *Modern Philology* 31:273–306, 349–82.

Crane, Ronald S., and F. B. Kaye. 1927. "A Census of British Newspapers and Periodicals, 1620–1800." *Studies in Philology* 24:1–205.

Cremante, Renzo, and Walter Tega, eds. 1984. *Scienza e letteratura nella cultura italiana del Settecento.* Bologna.

Cristani, Giovanni. 1994. "Tradizione biblica, miti e rivoluzioni geologiche negli 'Anecdotes de la nature' di Nicolas-Antoine Boulanger." *Giornale critico della filosofia italiana* 14:92–123.

Croft, Herbert. 1685. *Some Animadversions Upon a Book Intituled The Theory of the Earth.* London.

Crombie, Alistair C. 1975. "Marin Mersenne (1588–1648) and the Seventeenth-Century Problem of Scientific Acceptability." *Physis* 17:186–204.

Crosland, Maurice P., ed. 1976. *The Emergence of Science in Western Europe.* New York.

Crucitti Ullrich, Francesca Bianca. 1974. *La 'Bibliothèque italique': Cultura 'italianisante' e giornalismo letterario.* Milan.

Dagen, Jean. 1966. "Pour une histoire de la pensée de Fontenelle." *Revue d'histoire littéraire de la France* 66:619–41.

Dahm, John J. 1970. "Science and Apologetics in the Early Boyle Lectures." *Church History* 39:172–86.

Dale, Samuel. 1704. "A Letter . . . concerning Harwich Cliff, and the Fossil Shells there." *PT* 24:1568–78.

Dance, S. Peter. 1966. *Shell Collecting: An Illustrated History.* London.

Daniel, Gabriel. 1729. *Histoire de France.* New ed. 10 vols. Paris.

Daniel, Glyn. 1967. *The Origins and Growth of Archaeology.* Harmondsworth.

Darnton, Robert. 1985. *The Great Cat Massacre*. New York.

Daston, Lorraine. 1988. *Classical Probability in the Enlightenment*. Princeton.

——. 1991. "The Ideal and Reality of the Republic of Letters in the Enlightenment." *Science in Context* 4:367–86.

Daumas, Maurice, ed. 1957. *Histoire de la science*. Encyclopédie de la Pléiade. Paris.

David, Madeleine V. 1965. *Le Débat sur les écritures et l'hiéroglyphe aux XVIIe et XVIIIe siècles*. Paris.

Davies, Gordon L. 1966. "The Eighteenth-Century Denudation Dilemma and the Huttonian Theory of the Earth." *Annals of Science* 22:129–38.

——. 1969. *The Earth in Decay: A History of British Geomorphology, 1578–1878*. New York.

Davillé, Louis. 1909. *Leibniz historien*. Paris.

Deacon, Margaret. 1965. "Founders of Marine Science in Britain: The Work of the Early Fellows of the Royal Society." *NRRS* 20:28–50.

——. 1971. *Scientists and the Sea, 1650–1900*. London.

Dean, Dennis R. 1981. "The Age of the Earth Controversy: Beginnings to Hutton." *Annals of Science* 38:435–56.

——. 1985. "The Rise and Fall of the Deluge." *Journal of Geological Education* 33:84–93.

Dear, Peter. 1985. "*Totius in verba*: Rhetoric and Authority in the Early Royal Society." *Isis* 76:145–61.

——. 1988. *Mersenne and the Learning of the Schools*. Ithaca.

——. 1990. "Miracles, Experiments, and the Ordinary Course of Nature." *Isis* 81:663–83.

DeBeer, Gavin R. 1948. "Johann Gaspar Scheuchzer, F.R.S., 1720–1729." *NRRS* 6:56–66.

——. 1950. "Johann Heinrich Hottinger's Description of the Ice-Mountains of Switzerland, 1703." *Annals of Science* 6:327–60.

Debus, Allen G. 1970. *Science and Education in the Seventeenth Century: The Webster-Ward Debate*. London.

——. 1977. *The Chemical Philosophy: Paracelsian Science and Medicine in the Sixteenth and Seventeenth Centuries*. 2 vols. New York.

Delair, J. B., and W. A. S. Sarjeant. 1975. "The Earliest Discoveries of Dinosaurs." *Isis* 66:5–25.

Delehaye, Hippolyte. 1959. *L'Oeuvre des Bollandistes à travers trois siècles, 1615–1915*. 2d ed. Brussels.

Delorme, Suzanne. 1947–48. "Pierre Perrault, auteur d'un traité *De l'Origine des Fontaines* et d'une théorie de l'expérimentation." *Archives internationales d'histoire des sciences* 1:388–94.

——. 1951. "L'Académie Royale des Sciences: Ces Correspondants en Suisse." *RHS* 4:159–70.

Delumeau, Jean. 1992. *Une Histoire du paradis: Le jardin des délices*. Paris.

DeMichelis, Cesare. 1968. "L'epistolario di Angelo Calogerà." *Studi Veneziani* 10:621–704.

——. 1979. *Letterati e lettori nel Settecento veneziano*. Florence.

Denis, Jean-Baptiste. 1672 (–1683). *Recueil des memoires et conferences*. Paris.

Derham, William. 1712. "Observations concerning the Subterraneous Trees in Dagenham." *PT* 27:478–84.

——. [1713] 1727. *Physico-Theology*. 7th ed. London.

Desautels, Alfred R. 1956. *Les Mémoires de Trévoux et le mouvement des idées au XVIIIe siècle, 1701–1734*. Rome.

Descartes, René. 1984–85. *The Philosophical Writings.* Translated by J. Cottingham, R. Stoothoff, and D. Murdoch. 2 vols. Cambridge.

Descartes et le cartésianisme hollandais. 1950. Paris.

Deshayes, J. 1963. "De l'abbé Pluche au citoyen Dupuis: À la recherche de la clef des fables." *SVEC* 24:457–86.

Desmarest, Nicolas. 1753. *Dissertation sur l'ancienne jonction de l'Angleterre à la France.* Amiens.

——. 1757. "Fontaine" and "Géographie physique." In Diderot, *Encyclopédie* 7:81–101, 613–26.

——. 1794. *Géographie physique.* Vol. 1. Paris.

Deyron, Jacques. 1663. *Des Antiquités de la ville de Nismes.* Nîmes.

Dezallier d'Argenville, Antoine-Joseph. 1727. "Lettre sur le choix & l'arrangement d'un Cabinet curieux." *Mercure de France* (June): 1294–1330.

——. 1742. *L'Histoire naturelle éclaircie dans deux de ses parties principales.* Paris.

——. 1755. *L'Histoire naturelle éclaircie dans une de ses parties principales, l'oryctologie.* Paris.

——. 1757. *L'Histoire naturelle éclaircie dans une de ses parties principales, la conchyliologie.* Paris.

Dibon, Paul, ed. 1959. *Pierre Bayle le philosophe de Rotterdam.* Paris.

——. 1978. "Communication in the Respublica Literaria of the Seventeenth Century." *Res Publica Litterarum* 1:43–55.

——. 1984. "Naples et l'Europe savante dans la seconde moitié du XVIIe siècle." *NRL* (Naples): 27–45.

Dictionnaire de théologie catholique. 1923–46. Edited by A. Vacant et al. 15 vols. Paris.

Dictionnaire de Trévoux. 1721. New ed. 5 vols. Trévoux.

Diderot, Denis. 1751–80. *Encyclopédie.* 35 vols. of text, supplement, plates, and index. Paris.

Dini, Alessandro. 1985. *Filosofia della natura, medicina, religione. Lucantonio Porzio (1639–1724).* Milan.

Donati, Vitaliano. 1845. *Lettere inedite scientifico-letterarie.* Edited by Antonio Roncetti. Milan.

Doni Garfagnini, Manuela. 1977. "Antonio Magliabechi fra erudizione e cultura." *Critica storica* 14:371–403.

——, ed. 1981. *Lettere e carte Magliabechi: Regesto.* 2 vols. Rome.

Dooley, Brendan. 1982. "The *Giornale de' letterati d'Italia* (1710–40): Journalism and 'Modern' Culture in the Early Eighteenth Century Veneto." *Studi Veneziani,* n.s., 6:229–70.

——. 1984. "Science Teaching as a Career at Padua in the Early Eighteenth Century: The Case of Giovanni Poleni." *History of Universities* 4:115–51.

——. 1988. "La scienza in aula nella rivoluzione scientifica: Dallo Sbaraglia al Vallisneri." *Quaderni per la storia dell'università di Padova* 21:23–41.

——. 1990. "Revisiting the Forgotten Centuries: Recent Work on Early Modern Tuscany." *European History Quarterly* 20:519–50.

——. 1991. *Science, Politics, and Society in Eighteenth-Century Italy: The Giornale de' Letterati d'Italia and Its World.* New York.

Douglas, David C. 1939. *English Scholars.* London.

Drake, Ellen T. 1981. "The Hooke Imprint on the Huttonian Theory." *American Journal of Science* 281:963–73.

Dulieu, Louis. 1958a. "Le mouvement scientifique montpelliérain au XVIIIe siècle." *RHS* 11:227–49.

——. 1958b. "La contribution montpelliéraine aux recueils de l'Académie Royale des Sciences." *RHS* 11:250–62.

——. 1973. "Jean Astruc." *RHS* 26:113–35.

Dumas, Gustave. 1936. *Histoire du Journal de Trévoux depuis 1701 jusqu'en 1762.* Paris.

Dupront, Alphonse. 1930. *Pierre-Daniel Huet et l'exégèse comparatiste au XVIIe siècle.* Paris.

DuVeil, Charles Marie. 1683. *A Letter To the Honourable Robert Boyle, Esq. . . . In Answer to Father Simon's Critical History of the Old Testament.* London.

Egerton, Frank N. 1966. "The Longevity of the Patriarchs: A Topic in the History of Demography." *JHI* 27:575–84.

Ehrard, Jean. 1963. *L'Idée de nature en France dans la première moitié du XVIIIe siècle.* 2 vols. Paris.

Eisenstein, Elizabeth L. 1992. *Grub Street Abroad: Aspects of the French Cosmopolitan Press from the Age of Louis XIV to the French Revolution.* Oxford.

Ellenberger, François. 1975–77. "A l'aube de la géologie moderne: Henri Gautier (1660–1737)." *Histoire et nature,* no. 7, pp. 3–58; nos. 9–10:.

——. 1980. "De l'influence de l'environnement sur les concepts: L'exemple des théories géodynamiques au XVIIIe siècle en France." *RHS* 33:33–68.

——. 1988, 1994. *Histoire de la géologie.* 2 vols. Paris.

——. 1989. "Etude du terme *révolution.*" *Documents pour l' histoire du vocabulaire scientifique* 9:69–90.

Emery, Clark. 1948. "John Wilkins and Noah's Ark." *Modern Language Quarterly* 9:286–91.

Emery, Frank V. 1958. "English Regional Studies from Aubrey to Defoe." *The Geographical Journal* 124:315–25.

——. 1971. *Edward Lhuyd, F.R.S., 1660–1709.* Caerdydd, Wales.

Encyclopaedia Britannica. 1771. 3 vols. Edinburgh.

Engfer, Hans-Jürgen. 1982. *Philosophie als Analysis: Studien zur Entwicklung philosophischer Analysiskonzeptionen unter dem Einfluss mathematischer Methodenmodelle im 17. und frühen 18. Jahrhundert.* Stuttgart.

Erasmus, Hendrick J. 1962. *The Origins of Rome in Historiography from Petrarch to Perizonius.* Assen, The Netherlands.

Esper, Johann Friedrich. 1774. *Description des zoolithes nouvellement decouvertes d'animaux quadrupedes inconnus.* Translated by J. F. Isenflamm. Nuremberg.

Evans, Arthur William. 1932. *Warburton and the Warburtonians.* London.

Evans, R. J. W. 1977. "Learned Societies in Germany in the Seventeenth Century." *European Studies Review* 7:129–51.

Evelyn, John. 1883–87. *Diary and Correspondence.* Edited by William Bray. 4 vols. London.

Ewan, Joseph, and Nesta Ewan. 1970. *John Banister and His Natural History of Virginia, 1678–1692.* Urbana.

Eyles, Victor A. 1958. "The Influence of Nicolaus Steno on the Development of Geological Science in Britain." *Acta Historica Scientiarum naturalium medicinalium* 15:167–88.

——. 1971. "John Woodward, F.R.S., F.R.C.P., M.D. (1665–1728): A Bio-Bibliographical Account of His Life and Work." *Journal of the Society for the Bibliography of Natural History* 5:399–427.

Ezell, Margaret J. M. 1984. "Richard Waller, S.R.S.: 'In the pursuit of nature'." *NRRS* 38:215–33.

Fantuzzi, Giovanni. 1781–94. *Notizie degli scrittori bolognesi.* 9 vols. Bologna.

Farber, Paul L. 1975. "Buffon and Daubenton: Divergent Traditions within the *Histoire naturelle.*" *Isis* 66:63–74.

Fardella, Michelangelo. 1986. *Pensieri scientifici e lettera antiscolastica.* Edited by S. Femiano. Naples.

Favre, Robert, et al. 1976. "Bilan et perspectives de recherches sur les *Mémoires de Trévoux.*" *Dix-huitième siècle* 8:237–55.

Febvre, Lucien. [1942] 1982. *The Problem of Unbelief in the Sixteenth Century: The Religion of Rabelais.* Translated by Beatrice Gottlieb. Cambridge, Mass.

Feldhay, Rivka. 1987. "Knowledge and Salvation in Jesuit Culture." *Science in Context* 1:195–213.

Feldhay, Rivka, and Yehuda Elkana, eds. 1989. "'After Merton'; Protestant and Catholic Science in Seventeenth-Century Europe." *Science in Context* 3:3–302.

Fellows, Otis, and Stephen Milliken. 1972. *Buffon.* New York.

Ferrone, Vincenzo. 1981. "Galileo, Newton e la libertas philosophandi nella prima metà del XVIII secolo in Italia." *Rivista storica italiana* 93:143–85.

———. 1982. *Scienza, Natura, Religione: Mondo newtoniano e cultura italiana nel primo Settecento.* Naples.

Feuillée, Louis. 1714. *Journal des observations physiques, mathématiques et botaniques.* 2 vols. Paris.

Fiering, Norman S. 1976. "The Transatlantic Republic of Letters: A Note on the Circulation of Learned Periodicals to Early Eighteenth-Century America." *William and Mary Quarterly* 3d ser., 33:642–60.

Filleau de la Chaise, Jean. 1672. *Discours sur les pensées de M. Pascal.* Paris.

Findlen, Paula. 1994. *Possessing Nature: Museums, Collecting, and Scientific Culture in Early Modern Italy.* Berkeley.

Fischer, Hans. 1973. *Johann Jakob Scheuchzer (2. August 1672–23. Juni 1733): Naturforscher und Arzt.* Zurich.

Fischer, Heinz-Dietrich, ed. 1973. *Deutsche Zeitschriften des 17. bis 20. Jahrhunderts.* Pullach.

Fletcher, Dennis J. 1966. "The Fortunes of Bolingbroke in France in the Eighteenth Century." *SVEC* 47:207–32.

———. 1967. "Bolingbroke and the Diffusion of Newtonianism in France." *SVEC* 53:29–46.

Fontenelle, Bernard le Bovier de. [1709, 1720] 1969. *Histoire du renouvellement de l'Académie royale des sciences . . . et les éloges historiques.* 2 vols. Brussels.

———. 1825. *Oeuvres.* 5 vols. Paris.

———. 1966. *Entretiens sur la pluralité des mondes.* Critical edition by Alexandre Calame. Paris.

———. 1971. *Nouveaux dialogues des morts.* Critical edition by Jean Dagen. Paris.

———. 1973. *Entretiens sur la pluralité des mondes.* Collection "Marabout université." Verviers, Belgium.

Fontenelle. 1989. Actes du Colloque tenu à Rouen du 6 au 10 octobre 1987. Edited by Alain Niderst. Paris.

Force, James E. 1985. *William Whiston: Honest Newtonian.* Cambridge.

Forster, Antonia. 1990. *Index to Book Reviews in England, 1749–1774.* Carbondale.

Foucquet, J. F. 1730. "An Explanation of the new Chronological Table of the Chinese History." *PT* 36:397–424.

Fox, Levi, ed. 1956. *English Historical Scholarship in the Sixteenth and Seventeenth Centuries.* London.

Frängsmyr, Tore, ed. 1983. *Linnaeus: The Man and His Work*. Berkeley.

Frank, Robert G., Jr. 1980. *Harvey and the Oxford Physiologists*. Berkeley.

Franklin, Benjamin. 1959–. *Papers*. Edited by L. W. Labaree et al. New Haven.

Frantz, Ray William. 1934. *The English Traveller and the Movement of Ideas, 1660–1732*. Lincoln, Nebr.

Fréret, Nicolas. [1724] 1729. "Réflexions sur l'étude des anciennes histoires, & sur le degré de certitude de leurs preuves." *MARI* 6:146–89.

Fréron, Elie. 1749–54. *Lettres sur quelques écrits de ce temps*. 13 vols. Paris.

Friedländer, Paul. 1937. "Athanasius Kircher und Leibniz." *Atti della Pontificia Accademia Romana di Archeologia*, 3d ser., *Rendiconti* 13:229–47.

Friedmann, Georges. 1962. *Leibniz et Spinoza*. Rev. ed. Paris.

Frisi, Paolo. 1762. *Del modo di regolare i fiumi e i torrenti*. Lucca.

———. 1774. *Traité des rivières et des torrens*. Translated from the Italian. Paris.

Funkenstein, Amos. 1986. *Theology and the Scientific Imagination from the Middle Ages to the Seventeenth Century*. Princeton.

Furetière, Antoine. 1690. *Dictionnaire universel*. 2 vols. The Hague.

Galilei, Galileo. 1957. *Discoveries and Opinions*. Translated and edited by Stillman Drake. Garden City,

———. 1962. *Dialogue concerning the Two Chief World Systems*. Translated by Stillman Drake. Berkeley.

Galluzzi, Paolo. 1981. "L'Accademia del Cimento: 'Gusti' del principe, filosofia e ideologia dell'esperimento." *Quaderni storici* 16:788–844.

Garagnon, Jean. 1976. "Les *Mémoires de Trévoux* et l'événement, ou Jean-Jacques Rousseau vu par les Jésuites." *Dix-huitième siècle* 8:215–35.

Gardair, Jean-Michel. 1984. *Le "Giornale de' letterati" de Rome (1668–1681)*. Florence.

Garin, Eugenio. 1966. *Storia della filosofia italiana*. 3 vols. Turin.

Gaubil, Antoine. 1970. *Correspondance de Pékin, 1722–1759*. Edited by Renée Simon. Geneva.

Gaukroger, Stephen, ed. 1991. *The Uses of Antiquity: The Scientific Revolution and the Classical Tradition*. Dordrecht.

Generali, Dario. 1984. "Il 'Giornale de' letterati d'Italia' e la cultura veneta del primo Settecento." *Rivista di storia della filosofia* 39:243–81.

Geoffroy, Etienne-François. 1731. *Catalogus librorum cl. D. Stephani-Francisci Geoffroy, Doctoris Medici*. Paris.

George, Philip. 1952. "The Scientific Movement and the Development of Chemistry in England, as seen in the Papers Published in the *Philosophical Transactions* from 1664/5 until 1750." *Annals of Science* 8:302–22.

Gersaint, Edme-François. 1736. *Catalogue raisonné de coquilles, et autres curiosités naturelles*. Paris.

Gimma, Giacinto. 1730. *Della storia naturale delle gemme, delle pietre, e di tutti minerali, ovvero della fisica sotterranea*. 2 vols. Naples.

Glanvill, Joseph. [1689] 1966. *Saducismus Triumphatus; or, Full and Plain Evidence Concerning Witches and Apparitions*. 3d ed., 1689. Facsimile reprint, with an introduction by C. O. Parsons. Gainesville, Fla.

Gleditsch, Johann Gottlieb. [1748] 1750. "Observations sur la veritable osteocolle de la Marche de Brandenbourg." *Mémoires de l'Académie royale des sciences et belles lettres* (Berlin), 32–51.

Gliozzi, Giuliano. 1977. *Adamo e il Nuovo Mondo*. Florence.

Godwin, Joscelyn. 1979. *Athanasius Kircher: A Renaissance Man and the Quest for Lost Knowledge.* London.

Goguet, Antoine-Yves. [1758] 1775. *The Origin of laws, arts and sciences, and their progress among the most ancient nations.* Translated from the French. 3 vols. Edinburgh.

Gohau, Gabriel. 1987. *Histoire de la géologie.* Paris.

———. 1990. *Les Sciences de la terre aux XVIIe et XVIIIe siècles.* Paris.

Goldgar, Anne. 1995. *Impolite Learning: Conduct and Community in the Republic of Letters, 1680–1750.* New Haven.

Gosseaume, P.-L.-G. 1814–21. *Précis analytique des travaux de l'Académie des sciences, belles-lettres et arts de Rouen, depuis sa fondation en 1744 jusqu'à l'époque de sa restauration, le 29 juin 1803.* 5 vols. in 3. Rouen.

Gouhier, Henri. 1926. *La Vocation de Malebranche.* Paris.

Gould, Stephen Jay. 1985. *The Flamingo's Smile.* New York.

———. 1987. *Time's Arrow, Time's Cycle.* Cambridge, Mass.

———. 1989. *Wonderful Life: The Burgess Shale and the Nature of History.* New York.

———. 1993. *Eight Little Piggies.* New York.

Graf, Arturo. 1911. *L'Anglomania e l'influsso inglese in Italia nel secolo XVIII.* Turin.

Grafton, Anthony. 1975. "Joseph Scaliger and Historical Chronology: The Rise and Fall of a Discipline." *History and Theory* 14:156–85.

———. 1991. *Defenders of the Text: The Traditions of Scholarship in an Age of Science, 1450–1800.* Cambridge, Mass.

Graham, Walter. 1926. *The Beginnings of English Literary Periodicals: A Study of Periodical Literature, 1665–1715.* New York.

Greene, John C. [1959] 1961. *The Death of Adam.* New York.

Greene, Mott T. 1985. "History of Geology." *Osiris,* 2d ser., 1:97–116.

Grew, Nehemiah. 1681. *Musaeum Regalis Societatis. Or a Catalogue & Description of the Natural and Artificial Rarities Belonging to the Royal Society and preserved at Gresham Colledge.* London.

Grimm, Friedrich Melchior. 1877–82. *Correspondance littéraire, philosophique et critique.* Edited by Maurice Tourneux. 16 vols. Paris.

Grondona, Felice. 1969. "Basilischi artificiali all'esame radiografico. Contributo agli aspetti storico-culturali della teratologia." *Physis* 11:249–66.

Guerlac, Henry. 1954. "The Poets' Nitre: Studies in the Chemistry of John Mayow—II." *Isis* 45:243–55.

———. 1981. *Newton on the Continent.* Ithaca.

Guettard, Jean-Etienne. [1753] 1757. "Mémoire sur les poudingues." *MARS,* 63–96, 139–92.

———. 1768–86. *Mémoires sur différentes parties de la physique, de l'histoire naturelle, des sciences et des arts, &c.* 5 vols. in 6. Paris.

Guglielmini, Domenico. 1697. *Della natura de' fiumi.* See below, *Raccolta.*

Guntau, Martin. 1989. "Concepts of Natural Law and Time in the History of Geology." *Earth Sciences History* 8:106–10.

Gunther, Robert T. *Early Science in Oxford* series, as follows:

———. 1925. IV: *The Philosophical Society.* Oxford.

———. 1930. VI–VII: *The Life and Work of Robert Hooke.* Oxford.

———. 1939. XII: *Dr. Plot and the Correspondence of the Philosophical Society of Oxford.* Oxford.

———. 1945. XIV: *The Life and Letters of Edward Lhwyd.* Oxford.

Guy, Basil. 1963. "The French Image of China before and after Voltaire." *SVEC* 21.

Haber, Francis C. 1959. *The Age of the World: Moses to Darwin*. Baltimore.

Hacking, Ian. 1975. *The Emergence of Probability*. London.

Hahn, Roger. 1971. *The Anatomy of a Scientific Institution: The Paris Academy of Sciences, 1666–1803*. Berkeley.

Hale, Matthew. 1677. *The Primitive Origination of Mankind, considered and examined according to the light of nature*. London.

Hall, A. Rupert. 1952. *Ballistics in the Seventeenth Century*. Cambridge.

——. 1979. "Galileo nel XVIII secolo." *Rivista di filosofia* 70:367–90.

——. 1990. *Henry More: Magic, Religion, and Experiment*. Oxford.

Hall, A. Rupert, and Marie Boas Hall. 1976. "Les Liens publics et privés dans les relations franco-anglaises (1660–1720)." *Revue de synthèse* 97:51–75.

Hall, Marie Boas. 1966. *Robert Boyle on Natural Philosophy*. Bloomington.

——. 1975. "The Royal Society's Role in the Diffusion of Information in the Seventeenth Century." *NRRS* 29:173–92.

——. 1991. *Promoting Experimental Learning: Experiment and the Royal Society, 1660–1727*. Cambridge.

Haller, Albrecht von. 1983. *The Correspondence between Albrecht von Haller and Charles Bonnet*. Edited by Otto Sonntag. Bern.

Halley, Edmond. 1691. "An Account of the Circulation of the watry Vapours of the Sea, and of the Cause of Springs." *PT* 16:468–73.

——. 1695. "Some Account of the Ancient State of the City of Palmyra." *PT* 19:160–75.

——. 1715. "A short Account of the Cause of the Saltness of the Ocean." *PT* 29:296–300.

——. 1724. "Some Considerations about the Cause of the universal Deluge." *PT* 33:118–25.

Hankins, Thomas L. 1967. "The Influence of Malebranche on the Science of Mechanics during the Eighteenth Century." *JHI* 28:193–210.

Hanks, Lesley. 1966. *Buffon avant l'"Histoire naturelle."* Paris.

Harnack, Adolf von. 1900. *Geschichte der königlich preussischen Akademie der Wissenschaften zu Berlin*. 3 vols. in 4. Berlin.

Harris, John. 1697. *Remarks on some Late Papers, Relating to the Universal Deluge*. London.

——. 1708–10, 1736. *Lexicon Technicum*. 2d ed. 2 vols. London, 1708–10. Also, 5th ed. London, 1736.

Harth, Phillip. 1968. *Contexts of Dryden's Thought*. Chicago.

Hartmann, Fritz, and Rudolf Vierhaus, eds. 1977. *Der Akademiegedanke im 17. und 18. Jahrhundert*. Bremen.

Hartsoeker, Nicolaas. 1706. *Conjectures physiques*. Amsterdam.

——. 1710. *Eclaircissemens sur les conjectures physiques*. Amsterdam.

Hatin, Eugène. 1866. *Bibliographie historique et critique de la presse périodique française*. Paris.

Hatley, G. 1683/84. "A Letter concerning some form'd Stones." *PT* 14:463–65.

Hauksbee, Francis. See Bellers.

Hay, Denys. 1977. *Annalists and Historians*. London.

Hazard, Paul. 1963. *The European Mind (1680–1715)*. Translated by J. Lewis May. Cleveland.

Hearne, Thomas. 1698. *Ductor Historicus: or, A Short System of Universal History*. London.

Heizer, Robert F., ed. 1969. *Man's Discovery of His Past: A Sourcebook of Original Articles*. Palo Alto.

Henckel, Johann Friedrich. 1744. *Kleine Mineralogische und Chymische Schrifften.* Edited by C. F. Zimmermann. Dresden.

———. 1760. *Pyritologie.* Translated by d'Holbach. 2 vols. Paris.

Henry, John. 1986. "Occult Qualities and the Experimental Philosophy: Active Principles in Pre-Newtonian Matter Theory." *History of Science* 24:335–81.

Herries Davies. See Davies.

Hesse, Mary. 1966. "Hooke's Philosophical Algebra." *Isis* 57:67–83.

Hetherington, Noriss S. 1972. "The Hevelius-Auzout Controversy." *NRRS* 27:103–6.

Heyd, Michael. 1982. *Between Orthodoxy and the Enlightenment: Jean-Robert Chouet and the Introduction of Cartesian Science in the Academy of Geneva.* The Hague.

———. 1987. "The New Experimental Philosophy: A Manifestation of 'Enthusiasm' or an Antidote to It?" *Minerva* 25:423–40,

Hill, John, 1748–50. A *General Natural History.* 3 vols. London.

Hobbs, William. 1981. *The Earth Generated and Anatomized.* Edited with an introduction and notes by Roy Porter. *Bulletin of the British Museum (Natural History).* Historical Series, no. 8.

Hochstetter, Erich, ed. 1951. *Leibniz zu seinem 300. Geburtstag.* Berlin.

Holloway, Benjamin. 1723. "An Account of the Pits for Fullers-Earth in Bedfordshire." *PT* 32:419–23.

Honoré de Sainte-Marie, B.-V. 1713. *Reflexions sur les regles et sur l'usage de la critique.* 2 vols. in l. Paris.

Hooke, Robert. [1665] 1961. *Micrographia.* Facsimile reprint, with a preface by R. T. Gunther. New York.

———. 1686. "Some Observations, and Conjectures Concerning the Chinese Characters." *PT* 16: 63–78. (Author identified only as "R.H.")

———. [1705] 1971. *Posthumous Works.* Edited by Richard Waller. Facsimile reprint, with an introduction by T. M. Brown. London.

Hooykaas, R. 1981–82. "Pitfalls in the Historiography of Geological Science." *Histoire et nature* 19–20:21–34.

Hoppen, K. Theodore. 1970. *The Common Scientist in the Seventeenth Century: A Study of the Dublin Philosophical Society, 1683–1708.* Charlottesville.

———. 1976. "The Nature of the Early Royal Society." *BJHS* 9:1–24, 243–73.

Horn, Caspar. 1629. *Elephas, Das ist: Historischer unnd [sic] Philosophischer Discurs von dem grossen Wunderthier dem Elephanten.* Nuremberg.

Houghton, Walter E., Jr. 1942. "The English Virtuoso in the Seventeenth Century." *JHI* 3:51–73, 190–219.

Houtteville, C.-F. 1722. *La Religion chrétienne prouvée par les faits.* Paris.

Huddleston, Lee E. 1967. *Origins of the American Indians: European Concepts, 1492–1729.* Austin.

Huet, Pierre-Daniel. 1810. *Memoirs.* Translated with notes by John Aikin. 2 vols. London.

Hughes, Arthur. 1951–53, 1955. "Science in English Encyclopaedias, 1704–1875." *Annals of Science* 7:340–70; 8:323–67; 9:233–64; 11:74–92.

Hunter, Michael. 1971. "The Royal Society and the Origins of British Archaeology." *Antiquity* 45:113–21, 187–92.

———. 1975. *John Aubrey and the Realm of Learning.* New York.

———. 1981. *Science and Society in Restoration England.* Cambridge.

———. 1988. "Promoting the New Science: Henry Oldenburg and the Early Royal Society." *History of Science* 26:165–81.

———. 1989. *Establishing the New Science: The Experience of the Early Royal Society.* Woodbridge, Suffolk.

———, ed. 1994. *Robert Boyle Reconsidered.* Cambridge.

Hunter, Michael, and Simon Schaffer, eds. 1989. *Robert Hooke: New Studies.* Woodbridge, Suffolk.

Hunter, Michael, and Paul B. Wood. 1986. "Towards Solomon's House: Rival Strategies for Reforming the Early Royal Society." *History of Science* 24:49–108.

Hunter, William B., Jr. 1950. "The Seventeenth Century Doctrine of Plastic Nature." *Harvard Theological Review* 43:197–213.

Hutchison, Keith. 1982. "What Happened to Occult Qualities in the Scientific Revolution?" *Isis* 73:233–53.

Hutton, Sarah, ed. 1990. *Henry More (1614–1687): Tercentenary Studies.* Dordrecht.

Huygens, Christiaan. [1690] 1945. *Treatise on Light.* Translated by S. P. Thompson. Chicago.

———. 1888–1950. *Oeuvres complètes.* 22 vols. The Hague.

Impey, Oliver, and Arthur MacGregor, eds. 1985. *The Origins of Museums.* Oxford.

Index biographique des membres et correspondants de l'Académie des sciences. 1954. Paris.

Iofrida, Manlio. 1983. *La Filosofia di John Toland.* Milan.

Ito, Yushi. 1988. "Hooke's Cyclic Theory of the Earth in the Context of Seventeenth Century England." *BJHS* 21:295–314.

Iversen, Erik. 1961. *The Myth of Egypt and Its Hieroglyphs in European Tradition.* Copenhagen.

Jackson, Benjamin D. 1874. "A Sketch of the Life of William Sherard." *Journal of Botany* 12:129–38.

Jacob, Margaret C., and W. A. Lockwood. 1972. "Political Millenarianism and Burnet's *Sacred Theory.*" *Science Studies* 2:265–79.

Jacquet, Augustin-Joseph. 1886. *La Vie littéraire dans une ville de province sous Louis XIV.* Paris.

Jacquot, Jean. 1953. "Sir Hans Sloane and French Men of Science." *NRRS* 10:85–98.

———. 1954. *Le Naturaliste Sir Hans Sloane (1660–1753) et les échanges scientifiques entre la France et l'Angleterre.* Paris.

Jahn, Melvin E. 1972. "A Bibliographical History of John Woodward's *An Essay toward a natural history of the Earth.*" *Journal of the Society for the Bibliography of Natural History* 6:181–213.

Janssens-Knorsch, U. E. M. 1975. *Matthieu Maty and the Journal britannique, 1750–1755.* Amsterdam.

Jobe, Thomas H. 1981. "The Devil in Restoration Science: The Glanvill-Webster Witchcraft Debate." *Isis* 72:343–56.

Joy, Lynn S. 1987. *Gassendi the Atomist: Advocate of History in an Age of Science.* Cambridge.

Jussieu, Antoine de. [1718] 1719. "Examen des causes des impressions des plantes marquées sur certaines pierres des environs de Saint-Chaumont dans le Lionnois." *MARS,* 287–97.

———. [1721] 1723. "Recherches physiques sur les petrifications qui se trouvent en France de diverses parties de plantes & d'animaux étrangers." *MARS,* 69–75.

———. [1722] 1724. "De l'origine et de la formation d'une sorte de pierre figurée que l'on nomme Corne d'Ammon." *MARS,* 235–43.

———. [1723] 1725. "De l'origine et des usages de la pierre de foudre." *MARS,* 6–9.

——. [1723] 1725. "De l'origine des pierres appellées yeux de serpents et cra-paudines." *MARS*, 205–10.

——. [1724] 1726. "Observations sur quelques ossements d'une teste d'hip-popotame." *MARS*, 209–15.

Kafker, Frank A., ed. 1981. "Notable Encyclopedias of the Seventeenth and Eighteenth Centuries: Nine Predecessors of the *Encyclopédie*." *SVEC* 194.

——. 1994. "William Smellie's edition of the *Encyclopaedia Britannica*." *SVEC* 315:145–82.

Kargon, Robert H. 1966. *Atomism in England from Hariot to Newton*. Oxford.

Kaufman, Paul. 1960. *Borrowings from the Bristol Library, 1773–84*. Charlottesville.

——. 1969. *Libraries and Their Users: Collected Papers in Library History*. London.

Keill, John. 1699. *An Examination of the Reflections on "The Theory of the Earth." Together with A Defence of the Remarks on Mr. Whiston's "New Theory."* Oxford.

Kennedy, J. E., and W. A. S. Sarjeant. 1982. "'Earthquakes in the Air': The Seismo-logical Theory of John Flamsteed (1693)." *Journal of the Royal Astronomical Society of Canada* 76:213–23.

King, Charles [pseud. for Henry Rowlands]. 1705. *An Account of the Origin and Forma-tion of Fossil-Shells, &c*. London.

King, James, and Bernadette Lynn. 1980. "The *Metamorphoses* in English Eighteenth-Century Mythological Handbooks and Translations." *SVEC* 185:131–79.

King, Lester S. 1963. "Rationalism in Early Eighteenth Century Medicine." *Journal of the History of Medicine* 18:257–71.

Kircher, Athanasius. 1670. *La Chine d'Athanase Kirchere*. Amsterdam.

Klibansky, Raymond, and H. J. Paton, eds. 1936. *Philosophy and History*. Oxford.

Knowles, David. 1963a. *Great Historical Enterprises*. London.

——. 1963b. *The Historian and Character*. Cambridge.

Knowlson, James. 1975. *Universal Language Schemes in England and France, 1600–1800*. Toronto.

Kors, Alan C., and Paul J. Korshin, eds. 1987. *Anticipations of the Enlightenment in England, France, and Germany*. Philadelphia.

Koyré, Alexandre. 1965. *Newtonian Studies*. Chicago.

Kroll, Richard, et al., eds. 1992. *Philosophy, Science, and Religion in England, 1640–1700*. Cambridge.

Kronick, David A. 1976. *A History of Scientific & Technical Periodicals: The Origins and Development of the Scientific and Technical Press, 1665–1790*. 2d ed. Metuchen, N.J.

Kubrin, David. 1967. "Newton and the Cyclical Cosmos: Providence and the Mechan-ical Philosophy." *JHI* 28:325–46.

Kurmann, Walter. 1976. *Presenze italiane nei giornale elvetici del primo Settecento*. Bern.

L. P. 1695. See P., L. 1695.

Labrousse, Elisabeth. 1963–64. *Pierre Bayle*. 2 vols. The Hague.

Lach, Donald F. 1945. "Leibniz and China." *JHI* 6:436–55.

LaCondamine, Charles-Marie de. [1757] 1762. "Extrait d'un journal de voyage en Italie." *MARS*, 336–410.

Laeven, Augustinus Hubertus. 1986. *De "Acta eruditorum" onder Redactie van Otto Mencke (1644–1707)*. Amsterdam.

Lafitau, Joseph-François. 1974–77. *Customs of the American Indians Compared with the Customs of Primitive Times*. Translated and edited by W. N. Fenton and E. L. Moore. 2 vols. Toronto.

Lagarrigue, B. P. L. 1990. "Les coulisses de la presse de langue française dans les Provinces-Unies pendant la première moitié du XVIIIe siècle d'après la correspondance inédite de Charles de la Motte (1667?–1751), correcteur à Amsterdam." *Documentatieblad werkgroep achttiende eeuw* 22:77–110.

LaHarpe, Jacqueline de. 1941. *Le Journal des savants et l'Angleterre, 1702–1789.* Berkeley.

———. 1955. *Jean-Pierre Crousaz (1663–1750) et le conflit des idées au siècle des lumières.* Berkeley.

LaHire, Philippe de. 1692. "Description d'un tronc de palmier pétrifié, & quelques réflexions sur cette pétrification." *MARS,* 122–25.

———. [1703] 1705. "Remarques sur l'eau de la pluie, & sur l'origine des fontaines." *MARS,* 56–69.

Lamprecht, Sterling P. 1918–35. "The Role of Descartes in Seventeenth-Century England." In *Studies in the History of Ideas,* ed. Columbia University, Department of Philosophy. 3 vols. 3:181–240. New York.

Lamy, Bernard. 1966. *Entretiens sur les sciences.* Critical edition by F. Girbal and P. Clair. Paris.

Lamy, Edouard. 1929. "Deux Conchyliologistes français du XVIIIe siècle: Les Geoffroy oncle et neveu." *Journal de Conchyliologie* 73:129–32.

Lange, John. 1966. "The Argument from Silence." *History and Theory* 5:288–301.

Laudan, Rachel. 1982. "Tensions in the Concept of Geology: Natural History or Natural Philosophy?" *Earth Sciences History* 1:7–13.

———. 1987. *From Mineralogy to Geology: The Foundations of a Science, 1650–1830.* Chicago.

LeCat, Claude-Nicolas. 1750. "An Account of several Systems, particularly that of the ingenious Mr. Le Cat, with regard to the Formation of Mountains, and the Origin of fossile Shells and Animals." *Monthly Review* 3:375–93, 444–59.

LeClerc, Jean. 1696. *De l'Incredulité.* Amsterdam.

———. 1991–. *Epistolario.* Edited by Maria Grazia and Mario Sina. Florence.

Leclercq, Henri. 1953–57. *Dom Mabillon.* 2 vols. Paris.

LeComte, Louis. 1697. *Memoirs and Observations . . . Made in a late Journey Through the Empire of China.* Translated from the French with an introduction by [Tancred Robinson, F.R.S.]. London.

Leeuwenhoek, Antoni van. 1704. "A Letter . . . Concerning some Fossils of Swisserland." *PT* 24:1774–84.

———. 1800–1807. *Select Works.* Translated by Samuel Hoole. 2 vols. London.

LeGallois, Pierre. 1674. *Conversations academiques, tirées de l'academie de Monsieur l'abbé Bourdelot.* 2 vols. Paris.

Leibniz, Gottfried Wilhelm. 1693. "Protogaea." *Acta eruditorum* (January): 40–42.

———. 1710. "Epistola . . . ad Autorem Dissertationis de figuris animalium quae in lapidibus observantur, & Lithozoorum nomine venire possent." *Miscellanea Berolinensia* 1:118–20.

———. [1710] 1952. *Theodicy.* Translated by E. M. Huggard. Edited, with an introduction by A. Farrer. New Haven.

———. [1749] 1993. *Protogaea.* Translated by B. de Saint-Germain. Edited, with an introduction and notes by J.-M. Barrande. Toulouse.

———. 1859–75. *Oeuvres.* Edited by Foucher de Careil. 7 vols. Paris.

———. 1875–99. *Die philosophischen Schriften.* Edited by C. J. Gerhardt. 7 vols. Berlin.

———. 1903. *Opuscules et fragments inédits.* Edited by Louis Couturat. Paris.

——. 1926–. *Sämtliche Schriften und Briefe.* Herausgegeben von der Preussischen Akademie der Wissenschaften. Darmstadt.

——. 1969. *Philosophical Papers and Letters.* Translated and edited by Leroy Loemker. 2d ed. Dordrecht.

——. 1972. *Political Writings.* Translated and edited by Patrick Riley. Cambridge.

——. 1977. *Discourse on the Natural Theology of the Chinese.* Translated and edited by H. Rosemont and D. J. Cook. Honolulu.

——. 1981. *New Essays on Human Understanding.* Translated and edited by P. Remnant and J. Bennett. Cambridge.

Leibniz: Tradition und Aktualität. 1988. Hannover.

Leibniz, 1646–1716: Aspects de l'homme et de l'oeuvre. 1968. Paris.

Leigh, Charles. 1700. *Natural History of Lancashire Cheshire, and the Peak, in Derbyshire.* Oxford.

Lemery, Nicolas. [1700] 1703. "Explication physique & chymique des Feux souterrains, des Tremblemens de terre, des Ouragans, des Eclairs & du Tonnerre." *MARS,* 101–10.

Lennon, Thomas L., et al., eds. 1982. *Problems of Cartesianism.* Kingston/Montreal.

Leopold, Johann Friedrich. 1720. *Relatio Epistolica de Itinere suo Suecico Anno MDCCVII factô.* London.

Levine, Joseph M. 1977. *Dr. Woodward's Shield: History, Science, and Satire in Augustan England.* Berkeley.

——. 1987. *Humanism and History: Origins of Modern English Historiography.* Ithaca.

Lhwyd, Edward. 1693. "Ad Clariss. V. D. Christophorum Heimmer, Epistola." *PT* 17:746–54.

——. 1698. "Part of a Letter . . . concerning several regularly Figured Stones lately found by him." *PT* 20:279–80.

——. 1707. *Archaeologia Britannica.* Oxford.

——. 1708. "A Letter . . . giving an Account of . . . [Scheuchzer's] Itinera Alpina Tria." *PT* 26:143–67.

Lignac, J.-A. Lelarge de. 1751. *Lettres à un Amériquain.* 5 vols. Hamburg.

Limiers, H.-P. de. 1723. *Histoire de l'Academie appelée l'Institut des sciences et des arts, Etabli à Boulogne en 1712.* Amsterdam.

Lindberg, David C., and R. S. Westman, eds. 1990. *Reappraisals of the Scientific Revolution.* Cambridge.

Lindeboom, G. A. 1959. *Bibliographia Boerhaaviana.* Leiden.

Lindroth, Sten, ed. 1952. *Swedish Men of Science, 1650–1950.* Translated by Burnett Anderson. Stockholm.

Linnaeus, Carolus. 1781. "On the Increase of the Habitable Earth." In *Select Dissertations from the Amoenitates Academicae* 1:71–127. Translated by F. J. Brand. 2 vols. London.

——. 1972. *L'Equilibre de la nature.* Translated by B. Jasmin, with an introduction by C. Limoges. Paris.

——. 1973. *Linnaeus's Öland and Gotland Journey, 1741.* Translated by M. Asberg and W. T. Stearn, with an introduction by W. T. Stearn. London.

Lister, Martin. 1671. "A Letter." *PT* 6:2281–84.

——. 1684a. "Three Papers . . . the first of the Nature of Earth-quakes." *PT* 14:512–19.

——. 1684b. "An Ingenious proposal for a new sort of Maps of Countrys." *PT* 14:739–46.

——. [1699] 1967. *A Journey to Paris In the Year 1698*. Facsimile reprint of the 3d ed. Edited by R. P. Stearns. Urbana.

Locke, John. 1693. *Some Thoughts concerning Education*. London.

——. 1953. *Locke's Travels in France, 1675–1679, as related in his Journals, Correspondence and other papers*. Edited by John Lough. Cambridge.

——. 1958. *The Reasonableness of Christianity, with A Discourse of Miracles*. Edited by I. T. Ramsey. Stanford.

——. 1975. *An Essay concerning Human Understanding*. Edited by Peter H. Nidditch. Oxford.

——. 1976–89. *Correspondence*. Edited by E. S. de Beer. 8 vols. Oxford.

Lods, Adolphe. 1924. *Jean Astruc et la critique biblique au XVIIIe siècle*. Strasbourg.

Loemker, Leroy. 1955. "Boyle and Leibniz." *JHI* 16:22–43.

Loewe, Victor. 1924. *Ein Diplomat und Gelehrter: Ezechiel Spanheim (1629–1710)*. Berlin.

Lombard, Alfred. 1913. *L'Abbé Du Bos: Un initiateur de la pensée moderne*. Paris.

Lough, John. 1953. "Locke's Reading during his Stay in France (1675–79)." *The Library*, 5th ser., 8:229–58.

Lovell, Archibald. 1696. *A Summary of Material Heads Which may be Enlarged and Improved into a Compleat Answer to Dr. Burnet's Theory of the Earth*. London.

Luffkin, John. 1701. "Part of a Letter . . . concerning some large Bones, lately found in a Gravel-pit near Colchester." *PT* 22:924–26.

Lux, David S. 1989. *Patronage and Royal Science in Seventeenth-Century France: The Académie de Physique in Caen*. Ithaca.

Lyell, Charles. 1830–33. *Principles of Geology*. 3 vols. London.

Lyon, John. 1976. "The 'Initial Discourse' to Buffon's *Histoire naturelle*: The First Complete English Translation." *Journal of the History of Biology* 9:133–81.

Lyon, John, and Phillip R. Sloan, eds. 1981. *From Natural History to the History of Nature: Readings from Buffon and His Critics*. Notre Dame.

Mabillon, Jean. 1691. *Traité des études monastiques*. Paris.

McClaughlin, Trevor. 1975. "Sur les rapports entre la Compagnie de Thévenot et l'Académie royale des Sciences." *RHS* 28:235–42.

——. 1977. "Le Concept de science chez Jacques Rohault." *RHS* 30:225–40.

McClellan, James E., III. 1985. *Science Reorganized: Scientific Societies in the Eighteenth Century*. New York.

McConnell, Anita. 1986. "L. F. Marsigli's Voyage to London and Holland, 1721–1722." *NRRS* 41:39–76.

McCutcheon, Roger P. 1922. "The Beginnings of Book-Reviewing in English Periodicals." *PMLA* 37:691–706.

McKee, David R. 1941. *Simon Tyssot de Patot and the Seventeenth-Century Background of Critical Deism*. Baltimore.

McKeon, Richard. 1928. *The Philosophy of Spinoza*. New York.

McKeon, Robert M. 1965. "Une lettre de Melchisédech Thévenot sur les débuts de l'Académie royale des Sciences." *RHS* 18:1–6.

Maffei, Scipione. 1747. *Della formazione de' fulmini*. Verona.

Maffioli, Cesare S., and L. C. Palm, eds. 1989. *Italian Scientists in the Low Countries in the XVIIth and XVIIIth Centuries*. Amsterdam.

Magalotti, Lorenzo. 1741. *Lettere familiari*. Venice.

——. 1968. *Relazioni di viaggio in Inghilterra, Francia, e Svezia*. Edited by Walter Moretti. Bari.

Maillet, Benoît de. 1735. *Description de l'Egypte*. Edited by Le Mascrier. 2 vols. Paris.

———. 1750. *Telliamed*. Translated from the French. London.

———. 1968. *Telliamed*. Translated and edited by A. V. Carozzi. Urbana.

Mairan, J.-J. Dortous de. [1719] 1721. "Mémoire sur la cause generale du froid en hiver, & de la chaleur en été." *MARS*, 104–35.

———. 1749. *Dissertation sur la glace*. Paris.

Malebranche, Nicolas. 1958–70. *Oeuvres complètes*. Edited by André Robinet. 21 vols. Paris.

Mallatt, Jon M. 1982. "Dr. Beringer's Fossils: A Study in the Evolution of Scientific World View." *Annals of Science* 39:371–80.

Malpighi, Marcello. 1975. *Correspondence*. Edited by H. B. Adelmann. 5 vols. Ithaca.

Mandrou, Robert. 1968. *Magistrats et sorciers en France au XVIIe siecle*. Paris.

Manfredi, Eustachio. 1731. *Relazione per la diversione dei fiumi Ronco, e Montone della Città di Ravenna*. Ravenna.

———. [1732] 1746. "De aucta maris altitudine." *Commentarii*, vol. 2, pt. 2, pp. 1–19.

Mangenot, Eugène. 1890. "L'Universalité restreinte du déluge à la fin du XVIIe siècle." *Science catholique* 4:148–58, 227–39.

Manuel, Frank E. 1967. *The Eighteenth Century Confronts the Gods*. New York.

Mariotte, Edme. 1686. *Traité du mouvement des eaux*. Edited by Philippe de LaHire. Paris.

———. 1717. *Oeuvres de Mr. Mariotte*. 2 vols. Leiden.

Mariotte, savant et philosophe. 1986. Paris.

Markley, Robert. 1985. "Robert Boyle on Language: *Some Considerations Touching the Style of the Holy Scriptures*." *Studies in Eighteenth-Century Culture* 14:159–71.

Marsak, Leonard M. 1959. "Bernard de Fontenelle: The Idea of Science in the French Enlightenment." APS, *Transactions*, n.s., 49, pt. 7.

———, ed. 1961. *French Philosophers from Descartes to Sartre*. Cleveland.

Marsili [Marsigli], Luigi Ferdinando. 1711. *Brieve ristretto del saggio fisico intorno alla storia del mare*. Venice.

———. 1725. *Histoire physique de la mer*. Amsterdam.

[———]. 1728. *Atti legali per la fondazione dell'istituto delle scienze, ed arti liberali*. Bologna.

———. 1744. *Description du Danube*. Translated from the Latin. 6 vols. The Hague.

Martin, Ernest. 1880. *Histoire des monstres depuis l'antiquité jusqu'à nos jours*. Paris.

Martin, Henri-Jean. 1969. *Livre, pouvoirs et société à Paris au XVIIe siècle (1598–1701)*. 2 vols. Geneva.

Marx, Jacques. 1968. "Une revue oubliée du XVIIIe siècle: *La Bibliothèque impartiale*." *Romanische Forschungen* 80:281–91.

Mather, Cotton. 1714. "An Extract of several Letters . . . to John Woodward." *PT* 29:62–71.

Mather, Increase. [1684] 1856. *Remarkable Providences*. Introduction by George Offor. London.

Mattauch, Hans. 1968. *Die literarische Kritik der frühen französischen Zeitschriften (1665–1748)*. Munich.

Maugain, Gabriel. 1909. *Etude sur l'évolution intellectuelle de l'Italie de 1657 à 1750 environ*. Paris.

———. 1923. "Fontenelle et l'Italie." *Revue de littérature comparée* 3:541–603.

Maury, L.-F. Alfred. 1864. *L'Ancienne Académie des inscriptions et belles-lettres*. Paris.

Maylender, Michele. 1926–30. *Storia delle accademie d'Italia*. 5 vols. Bologna.

Mayow, John. 1926. *Medico-physical Works.* Translated by A. Crum Brown and Leonard Dobbin. Oxford.

Mélanges Alexandre Koyré. 1964. Edited by I. B. Cohen and R. Taton. 2 vols. Paris.

Melle, Jacob von [Jacob a Melle]. 1718. *De Echinitis wagricis.* Lubeck.

——. 1720. *De Lapidibus figuratis agri litorisque lubecensis.* Lubeck.

Mencken, Johann Burkard. [1715] 1937. *The Charlatanry of the Learned.* Translated by F. E. Litz, with an introduction and notes by H. L. Mencken. New York.

Mendelsohn, Everett I. 1971. "Philosophical Biology vs. Experimental Biology: Spontaneous Generation in the Seventeenth Century." *Actes du XIIe Congrès international d'histoire des sciences* (Paris), 1, B, 201–26.

Mendyk, Stan A. E. 1985. "Robert Plot: Britain's 'Genial Father of County Natural Histories'." *NRRS* 39:159–77.

——. 1989. *"Speculum Britanniae": Regional Study, Antiquarianism, and Science in Britain to 1700.* Toronto.

Mersenne, Marin. 1945–86. *Correspondance.* Edited by Cornelis de Waard et al. 16 vols. Paris.

Metzger, Hélène. 1987. *La Méthode philosophique en histoire des sciences.* Edited by Gad Freudenthal. Paris.

Middlekauff, Robert. 1971. *The Mathers: Three Generations of Puritan Intellectuals, 1596–1728.* New York.

Middleton, W. E. Knowles. 1971. *The Experimenters: A Study of the Accademia del Cimento.* Baltimore.

——. 1975. "Science in Rome, 1675–1700, and the Accademia Fisicomatematica of Giovanni Giustino Ciampini." *BJHS* 8:138–54.

Mirabaud, Jean-Baptiste. [1751] 1778. *Le Monde, son origine, et son antiquité.* 2d ed. 2 vols. London.

Mirto, Alfonso. 1984. *Stampatori, editori, librai nella seconda metà del Seicento.* Florence.

Misson, François-Maximilien. 1717. *Nouveau voyage d'Italie.* 5th ed. 3 vols. The Hague.

Molyneux, Thomas. 1685. "Part of 2 Letters . . . concerning a Prodigious Os Frontis in the Medicine School at Leyden." *PT* 15:880–81.

——. 1697. "A Dissertation concerning the Large Horns frequently found under Ground in Ireland." *PT* 19:489–512.

——. 1700. "An Essay concerning Giants." *PT* 22:487–508.

Momigliano, Arnaldo. 1950. "Ancient History and the Antiquarian." *Journal of the Warburg and Courtauld Institutes* 13:285–315.

——. 1960. *Secondo contributo alla storia degli studi classici.* Rome.

——. 1966. *Terzo contributo alla storia degli studi classici.* Rome.

Monconys, Balthasar de. 1695. *Voyages de M. de Monconys.* 4 vols. Paris.

Montesquieu, C.-L. de Secondat, baron de. 1949. *Oeuvres complètes.* Edited by Roger Caillois. Vol. 1. Paris.

——. 1950–55. *Oeuvres complètes.* Edited by André Masson. 3 vols. Paris.

Monti, Giuseppe. 1719. See Sarti 1988.

——. [1729] 1746. "De balanis fossilibus." *Commentarii,* vol. 2, pt. 2, pp. 52–56.

Moralec. 1713. "Lettre . . . contenant ses conjectures sur la nature des coquillages qui se trouvent dans les terres." *Trévoux* (January): 50–63.

Morello, Nicoletta. 1979a. *La Macchina della Terra: Teorie geologiche dal Seicento all'Ottocento.* Turin.

——. 1979b. *La Nascita della paleontologia nel Seicento: Colonna, Stenone e Scilla.* Milan.

——. 1981. "De Glossopetris Dissertatio: The Demonstration by Fabio Colonna of the True Nature of Fossils." *Archives internationales d'histoire des sciences* 31:63–71.

Morgagni, Giovanni Battista. 1875. *Carteggio tra Giambattista Morgagni e Francesco M. Zanotti*. Bologna.

——. 1964–69. *Opera postuma*. Rome.

Morgan, Betty Trebelle. 1928. *Histoire du Journal des Sçavants depuis 1665 jusqu'en 1701*. Paris.

Mornet, Daniel. 1910. "Les Enseignements des bibliothèques privées (1750–1780)." *Revue d'histoire littéraire de la France* 17:449–96.

——. 1911. *Les Sciences de la nature en France au XVIIIe siècle*. Paris.

Moro, Anton Lazzaro. 1740. *De' Crostacei e degli altri marini corpi che si truovano ne' monti*. Venice.

——. 1987. *Epistolario*. Edited by P. G. Sclippa. Pordenone, Italy.

Moro, 1988. See *Anton Lazzaro Moro*.

Morton, John. 1706. "A Letter . . . Containing a Relation of River and other Shells digg'd up, together with various Vegetable Bodies, in a Bituminous Marshy Earth, near Mears-Ashby, in Northamptonshire." *PT* 25:2210–14.

——. 1712. *Natural History of Northampton-shire; with Some Account of the Antiquities*. London.

Moureau, F., ed. 1988. *Les Presses grises. La contrefaçon du livre (XVIe–XIXe siècles)*. Paris.

Mouy, Paul. 1934. *Le Développement de la physique cartésienne, 1646–1712*. Paris.

Mullen, Allan. 1682. *An Anatomical Account of the Elephant Accidentally Burnt in Dublin*. London.

Mungello, David E. 1977. *Leibniz and Confucianism: The Search for Accord*. Honolulu.

——. 1985. *Curious Land: Jesuit Accommodation and the Origins of Sinology*. Studia Leibnitiana Supplementa, vol. 25 (Stuttgart).

Muratori, Ludovico Antonio. 1978. *Carteggi con Ubaldini . . . Vannoni*. Edited by M. L. Nichetti Spanio, as vol. 44 of the national edition of Muratori's letters. Florence.

Nangle, Benjamin C. 1934. *The Monthly Review, first series, 1749–1789: Indexes of contributors and articles*. Oxford.

Neubert, Fritz. 1920. "Einleitung in eine kritische Ausgabe von B. de Maillets Telliamed." *Romanische Studien* 19.

Neumeister, Sebastian, and Conrad Wiedemann, eds. 1987. *Res Publica Litteraria: Die Institutionem der Gelehrsamkeit in der frühen Neuzeit*. 2 vols. Wiesbaden.

Newton, Isaac. 1952. *Opticks*. Edited by I. B. Cohen et al. New York.

——. 1959–77. *Correspondence*. Edited by H. W. Turnbull et al. 7 vols. Cambridge.

Niccolò Stenone nella Firenze e nell'Europa del suo tempo. 1986. Florence.

Niccolò Stenone 1638–1686. 1988. Florence.

Niceron, Jean-Pierre. 1727–45. *Mémoires pour servir à l'histoire des hommes illustres dans la république des lettres*. 44 vols. Paris.

Nicholls, William. 1696. *A Conference with a Theist*. London.

Nicolson, Marjorie Hope. [1959] 1963. *Mountain Gloom and Mountain Glory*. New York.

Niderst, Alain. 1972. *Fontenelle à la recherche de lui-même (1657–1702)*. Paris.

——. 1984. "Fontenelle et la science de son temps." *SVEC* 228:171–78.

——. 1991. *Fontenelle*. Paris.

North, John David, and J. J. Roche, eds. 1985. *The Light of Nature: Essays in the History and Philosophy of Science presented to A. C. Crombie*. Dordrecht.

Northeast, Catherine M. 1991. "The Parisian Jesuits and the Enlightenment, 1700–1762." *SVEC* 288.

Oakley, Francis. 1961. "Christian Theology and the Newtonian Science: The Rise of the Concept of the Laws of Nature." *Church History* 30:433–57.

O'Keefe, Cyril B. 1974. *Contemporary Reactions to the Enlightenment (1728–1762)*. Geneva.

Oldenburg, Henry. 1965–86. *Correspondence*. Edited by A. R. Hall and M. B. Hall. 13 vols. Madison.

Oldroyd, David R. 1972. "Robert Hooke's Methodology of Science as Exemplified in his 'Discourse of Earthquakes'." *BJHS* 6:109–30.

Oldroyd, David R., and J. B. Howes. 1978. "The First published version of Leibniz's *Protogaea*." *Journal of the Society for the Bibliography of Natural History* 9:56–60.

Osbat, Luciano. 1974. *L'Inquisizione a Napoli: Il processo agli ateisti, 1688–1697*. Rome.

Osler, Margaret J. 1994. *Divine Will and the Mechanical Philosophy: Gassendi and Descartes on Contingency and Necessity in the Created World*. Cambridge.

Owen, Nicholas. 1777. *British Remains; or, A Collection of Antiquities Relating to the Britons*. London.

P., L. 1695. *Two Essays sent in a Letter from Oxford*. London.

Palissy, Bernard. 1844. *Oeuvres complètes*. Edited by P.-A. Cap. Paris.

Palmer, Robert R. 1939. *Catholics & Unbelievers in Eighteenth Century France*. Princeton.

Pappas, John. 1957. "Berthier's Journal de Trévoux and the philosophes." *SVEC* 3.

———. 1983. "Buffon vu par Berthier, Feller et les *Nouvelles ecclésiastiques*." *SVEC* 216:26–28.

Paragallo, Gaspare. 1705. *Istoria naturale del Monte Vesuvio*. Naples.

Parks, Stephen. 1968. "John Dunton and *The works of the learned*." *Library*, 5th ser., 23:13–24.

Parret, Herman, ed. 1976. *History of Linguistic Thought and Contemporary Linguistics*. Berlin.

Pascal, Blaise. 1941. *Pensées and the Provincial Letters*. Translated by W. F. Trotter and Thomas M'Crie. New York.

Pasini, Mirella. 1981. *Thomas Burnet. Una storia del mondo tra ragione, mito e rivelazione*. Florence.

Patey, Douglas Lane. 1984. *Probability and Literary Form: Philosophic Theory and Literary Practice in the Augustan Age*. Cambridge.

Patin, Charles. 1665. *Introduction a l'histoire, par la connoissance des medailles*. Paris.

———. 1695. *Histoire des medailles ou introduction a la connoissance de cette science*. Paris.

Paul, Charles B. 1980. *Science and Immortality: The 'Eloges' of the Paris Academy of Sciences (1699–1791)*. Berkeley.

Pellegrini, Carlo. 1940. "Giovanni Lami, le 'Novelle Letterarie' e la cultura francese." *Giornale storico della letteratura italiana* 116:1–17.

Penney, Norman, ed. 1926. "The Correspondence of James Logan and Thomas Story, 1724–1741." *Bulletin of Friends' Historical Association* 15:1–100.

Pepys, Samuel. 1926. *Private Correspondence and Miscellaneous Papers*. Edited by J. R. Tanner. 2 vols. London.

Perkins, Merle L. 1953. "The Abbé de Saint-Pierre and the Seventeenth-Century Intellectual Background." *APS, Proceedings*, 97:69–76.

———. 1958. "Late Seventeenth-Century Scientific Circles and the Abbé de Saint-Pierre." *APS, Proceedings*, 102:404–12.

Perrault, Claude. 1671. *Mémoires pour servir à l'histoire naturelle des animaux*. Paris.

———. [1671] 1688. *Memoir's [sic] for a Natural History of Animals*. Translated by Alexander Pitfeild. London.

———. 1680–84. *Essais de physique.* 4 vols. Paris.

Perrault, Claude, and Pierre Perrault. 1721. *Oeuvres diverses.* 2 vols. Leiden.

Perrault, Pierre. See preceding item.

Petty, William. 1928. *The Petty-Southwell Correspondence, 1676–1687.* Edited by the Marquis of Lansdowne. London.

Pezron, Paul-Yves. 1687. *L'Antiquité des tems rétablie et défenduë. Contre Les Juifs & les Nouveaux Chronologistes.* Paris.

Piggott, Stuart. 1989. *Ancient Britons and the Antiquarian Imagination.* London.

Pighetti, Clelia. 1988. *L'influsso scientifico di Robert Boyle nel tardo '600 italiano.* Milan.

Pigot, Thomas. 1683. "An Account of the Earthquake that happened at Oxford." *PT* 13:311–21.

Pinot, Virgile. 1932. *La Chine et la formation de l'esprit philosophique en France (1640–1740).* 2 vols. Paris.

Pintard, René. 1983. *Le Libertinage érudit dans la première moitié du XVIIe siècle.* New ed. Geneva.

Pitassi, Maria Cristina. 1987. *Entre croire et savoir. Le problème de la méthode critique chez Jean Le Clerc.* Leiden.

Piva, Franco. 1980. "La cultura francese nelle biblioteche venete del Settecento: Vicenza." *Archivio Veneto* 115:33–83.

Placet, François. 1668. *La Corruption du grand et petit monde.* 3d ed. Paris.

Platelle, Henri, ed. 1968. *Les Chrétiens face au miracle: Lille au XVIIe siècle.* Paris.

Plot, Robert. 1677. *Natural History of Oxford-shire.* Oxford.

———. 1686. *Natural History of Stafford-shire.* Oxford.

Pluche, Noël-Antoine. 1749–56. *Le Spectacle de la nature.* First published in eight volumes, 1732–50, but individual volumes were already being reprinted before 1750. Many sets are composites of varied editions of the individual volumes. The set I consulted (Paris, 1749–56) was just such a mélange.

———. [1739] 1748. *Histoire du ciel.* 2 vols. Paris.

Pocock, J. G. A. 1957. *The Ancient Constitution and the Feudal Law.* Cambridge.

———. 1973. *Politics, Language, and Time.* New York.

Pomian, Krzysztof. 1976. "Médailles/coquilles = érudition/philosophie." *SVEC* 154:1677–1703.

———. 1987. *Collectionneurs, amateurs et curieux. Paris, Venise: XVIe–XVIIIe siècle.* Paris.

Popkin, Richard H. 1968. *The History of Scepticism from Erasmus to Descartes.* Rev. ed. New York.

———. 1987. *Isaac La Peyrère (1596–1676), his life, work and influence.* Leiden.

Porcia, Giovanni Artico, conte di. 1986. *Notizie della vita, e degli studi del kavalier Antonio Vallisneri.* Critical edition by Dario Generali. Bologna.

Porter, Roy. 1977. *The Making of British Geology: Earth Science in Britain, 1660–1815.* Cambridge.

———. 1979. "John Woodward: 'A Droll Sort of a Philosopher'." *Geological Magazine* 116:335–43.

Postigliola, Alberto, ed. 1988. *Libro Editoria Cultura nel Settecento italiano.* Rome.

Pouilly, L.-J. Lévesque de. [1722] 1729. "Dissertation sur l'incertitude de l'histoire des quatre premiers siècles de Rome." *MARI* 6:14–29.

———. [1724] 1729. "Nouveaux essais de critique sur la fidélité de l'histoire." *MARI* 6:71–114.

Poulsen, Jacob E., and Egill Snorrason, eds. 1986. *Nicolaus Steno, 1638–1686: A Reconsideration by Danish Scientists.* Gentofte, Denmark.

Power, Henry. [1664] 1966. *Experimental Philosophy.* Introduction by M. B. Hall. New York.

Predaval Magrini, M. V., ed. 1990. *Scienza, filosofia e religione tra '600 e '700 in Italia.* Milan.

Price, David. 1989. "John Woodward and a Surviving British Geological Collection from the Early Eighteenth Century." *Journal of the History of Collections* 1:79–95.

Prior, Moody E. 1932. "Joseph Glanvill, Witchcraft, and Seventeenth-Century Science." *Modern Philology* 30:167–93.

Prutz, Robert Edward. 1845. *Geschichte des deutschen Journalismus.* Hannover.

Pryme, Abraham de la. 1700. "A Letter . . . concerning Broughton in Lincolnshire, with his observations on the Shell-fish observed in the Quarries about that place." *PT* 22:677–87.

——. 1870. *The Diary of Abraham de la Pryme, the Yorkshire Antiquary.* Publications of the Surtees Society. Vol. 54.

Purcell, Rosamond W., and Stephen Jay Gould. 1992. *Finders, Keepers: Eight Collectors.* New York.

Quondam, Amedeo, and Michele Rak, eds. 1978. *Lettere dal regno ad Antonio Magliabechi.* 2 vols. Naples.

Raccolta d'autori che trattano del moto dell'acque. 1723. 3 vols. Florence.

Raistrick, Arthur. 1950. *Quakers in Science and Industry.* London.

Ramazzini, Bernardino. 1964. *Epistolario.* Edited by Pericle Di Pietro. Modena.

Rappaport, Rhoda. 1982. "Borrowed Words: Problems of Vocabulary in Eighteenth-Century Geology." *BJHS* 15:27–44.

——. 1986. "Hooke on Earthquakes: Lectures, Strategy, and Audience." *BJHS* 19:129–46.

——. 1991a. "Fontenelle Interprets the Earth's History." *RHS* 44:281–300.

——. 1991b. "Italy and Europe: The Case of Antonio Vallisneri (1661–1730)." *History of Science* 29:73–98.

——. 1994. "Baron d'Holbach's Campaign for German (and Swedish) Science." *SVEC* 323:225–46.

——. 1997. "Questions of Evidence: An Anonymous Tract Attributed to John Toland." *JHI* 58:339–48.

Raspe, Rudolf Erich. [1763] 1970. *An Introduction to the Natural History of the Terrestrial Sphere.* Translated and edited by A. N. Iversen and A. V. Carozzi. New York.

Raven, Charles E. [1950] 1986. *John Ray Naturalist: His Life and Works.* 2d ed. Cambridge.

Ray, John. 1670. *A Collection of English Proverbs.* Cambridge.

——. 1673. *Observations Topographical, Moral, & Physiological; Made in a Journey Through part of the Low-Countries, Germany, Italy, and France.* London.

——. 1674. *A Collection of English Words.* London.

——. 1693. *Three Physico-Theological Discourses.* 2d ed. London. (First ed., 1692, entitled *Miscellaneous Discourses.* . . .)

——. 1713. *Three Physico-Theological Discourses.* 3d ed. Preface by William Derham. London.

——. 1718. *Philosophical Letters.* Edited by William Derham. London.

——. 1848. *Correspondence of John Ray.* Edited by Edwin Lankester. London.

——. 1928. *Further Correspondence of John Ray.* Edited by Robert W. T. Gunther. London.

Réaumur, René-Antoine Ferchault de. [1715] 1717. "Observations sur les Mines de Turquoises du Royaume." *MARS,* 174–202.

———. [1720] 1722. "Remarques sur les Coquilles fossilles de quelques cantons de la Touraine, & sur les utilités qu'on en tire." *MARS,* 400–16.

———. [1721] 1723. "Sur la nature et la formation des cailloux." *MARS,* 255–76. Continued in [1723] 1725: 273–84.

———. 1734–42. *Mémoires pour servir à l'histoire des insectes.* 6 vols. Paris.

———. 1886. *Lettres inédites.* Edited by G. Musset. LaRochelle.

Record of the Royal Society. 1912. 3d ed. London.

Recueil des Mémoires . . . dans les Actes de l'Académie d'Upsal, et dans les Mémoires de l'Académie royale des sciences de Stockolm. 1764. Translated by d'Holbach. 2 vols. Paris.

Reedy, Gerard. 1985. *The Bible and Reason: Anglicans and Scripture in Late Seventeenth-Century England.* Philadelphia.

Reesink, H. J. 1931. *L'Angleterre et la littérature anglaise dans les trois plus anciens périodiques français de Hollande de 1684 à 1709.* Zutphen and Paris.

(La) Régence. 1970. Paris.

Reill, Peter Hanns. 1975. *The German Enlightenment and the Rise of Historicism.* Berkeley.

Reilly, Conor. 1958. "A Catalogue of Jesuitica in the 'Philosophical Transactions of the Royal Society of London' (1665–1715)." *Archivum historicum Societatis Iesu* 27:339–62.

Reiske, Johann. 1687. *Commentatio physica aeque ac historica de glossopetris luneburgensibus.* Nuremberg.

Religion, érudition et critique à la fin du XVIIe siècle et au début du XVIIIe siècle. 1968. Paris.

Renaudot, Théophraste. 1666. *Recueil general des questions traitées és Conferences du Bureau d'Adresse.* 6 vols. Lyons.

Rétat, Pierre. 1976. *"Mémoires pour l'histoire des sciences et des beaux-arts:* Signification d'un titre et d'une entreprise journalistique." *Dix-huitième siècle* 8:167–87.

Rhodes, Dennis E. 1964. "Libri inglesi recensiti a Roma, 1668–1681." *Studi secenteschi* 5:151–60.

Riccati, Jacopo. 1761–65. *Opere.* 4 vols. Lucca.

———. 1985. *Carteggio (1719–1729). Jacopo Riccati, Antonio Vallisneri.* Edited by M. L. Soppelsa. Florence.

Ricupcrati, Giuseppe. 1970. *L'Esperienza civile e religiosa di Pietro Giannone.* Milan.

———. 1972. "A proposito dell'Accademia Medina Coeli." *Rivista storica italiana* 84:57–79.

———. 1984. "I giornali italiani del XVIII secolo: studi e ipotesi di ricerca." *Studi storici* 25:279–303.

Righi, Gaetano. 1933. "L'idea enciclopedica del sapere in L. A. Muratori." *Miscellanea di Studi Muratoriana. Atti e Memorie della R. Deputazione di Storia Patria per le Provincie Modenesi,* 7th ser., 8:61–102.

Righini Bonelli, Maria Luisa, and William R. Shea, eds. 1975. *Reason, Experiment, and Mysticism in the Scientific Revolution.* New York.

Rivers, Isabel, ed. 1982. *Books and their Readers in Eighteenth-Century England.* Leicester.

Rivière, Guillaume. [1708] 1766. "Mémoire sur les dents pétrifiées de divers poissons." In Montpellier, *Mémoires,* 1:75–84.

Rivosecchi, Valerio. 1982. *Esotismo in Roma barocca. Studi sul Padre Kircher.* Rome.

Robinet, André. 1988. *G. W. Leibniz, Iter Italicum (Mars 1689–Mars 1690).* Florence.

Robinson, Thomas. 1694. *The Anatomy of the Earth.* London.

———. 1696. *New Observations on the Natural History of this World of Matter, and this World of Life.* London.

——. 1709. *An Essay towards a Natural History of Westmorland and Cumberland.* London.

Roche, Daniel. 1988. *Les Républicains des lettres: Gens de culture et lumières au XVIIIe siècle.* Paris.

Rochot, Bernard. 1953. "Roberval, Mariotte et la Logique." *Archives internationales d'histoire des sciences* 6:38–43.

——. 1957. "Gassendi et les mathématiques." *RHS* 10:69–78.

Rodolico, Francesco. 1963. *L'Esplorazione naturalistica dell' Appennino.* Florence.

Roger, Jacques. 1973. "La Théorie de la Terre au XVIIe siècle." *RHS* 26:23–48.

——. 1984. "Per una storia storica delle scienze." *Giornale critico della filosofia italiana,* 6th ser., 4:285–314.

——. 1989. *Buffon.* Paris.

——. [1963] 1993. *Les Sciences de la vie dans la pensée française du XVIIIe siècle.* New ed. Paris.

Roger, Jacques, et al. 1988. *Transfert de vocabulaire dans les sciences.* Paris.

Rohault, Jacques. 1671. *Traité de physique.* Paris.

——. [1671] 1728. *A System of Natural Philosophy.* Translated with notes by John and Samuel Clarke. 2d ed. 2 vols. Facsimile reprint, New York, 1987.

Rome, Remacle. 1956. "Nicolas Sténon et la 'Royal Society of London'." *Osiris* 12:244–68.

Rooke, Lawrence. 1666. "Directions for Sea-men, bound for far Voyages." *PT* 1:140–43.

Rosa, Mario. 1956. "Atteggiamenti culturali e religiosi di Giovanni Lami nelle 'Novelle letterarie'." *Annali della Scuola Normale Superiore di Pisa,* 2d ser., 25:260–333.

Rosenfield, Leonora Cohen. 1957. "Peripatetic Adversaries of Cartesianism in Seventeenth-Century France." *Review of Religion* 22:14–40.

Rossetti, Lucia, ed. 1988. *Rapporti tra le università di Padova e Bologna: Ricerche di filosofia, medicina e scienza.* Trieste.

Rossi, Paolo. 1969. *Le Sterminate antichità. Studi vichiani.* Pisa.

——. 1970. *Philosophy, Technology, and the Arts in the Early Modern Era.* Translated by A. Attanasio. Edited by B. Nelson. New York.

——. 1984. *The Dark Abyss of Time.* Translated by Lydia G. Cochrane. Chicago.

Rothschild, Harriet D. 1964–65, 1968, 1977. "Benoît de Maillet's . . . letters." *SVEC* 30:351–75; 37:109–45; 60:311–38; 169:115–85.

Rouse, Hunter, and Simon Ince. 1957. *History of Hydraulics.* Iowa City.

Rowbotham, Arnold H. 1956. "The Jesuit Figurists and Eighteenth-Century Religious Thought." *JHI* 17:471–85.

Rowlands, Henry. 1766. *Mona Antiqua Restaurata.* 2d ed. London.

Rudwick, Martin J. S. 1972. *The Meaning of Fossils.* London.

——. 1990. "The Emergence of a New Science." *Minerva* 28:386–97.

Sack, A. F. W. [1745] 1746. Summary entitled "Sur de nouvelles pétrifications marines." In *Histoire de l'Académie royale . . . de Berlin,* 67–70.

St. Clair, Robert. 1697. *The Abyssinian Philosophy confuted: or, "Telluris Theoria" Neither Sacred, nor agreeable to Reason.* London.

Sallier, Claude. [1723] 1729. "Discours sur les premiers monuments historiques des Romains." *MARI* 6:30–51. Other articles on related themes, delivered by Sallier in 1724 and 1725, are in the same volume, pp. 52–70, 115–35.

Salomon-Bayet, Claire. 1978. *L'Institution de la science et l'expérience du vivant: Méthode et expérience à l'Académie royale des sciences, 1666–1793.* Paris.

Sarti, Carlo. 1988. *I fossili e il Diluvio Universale.* Bologna.

——. 1993. "Giuseppe Monti and Palaeontology in Eighteenth-Century Bologna." *Nuncius* 8:443–55.

Sauvages de la Croix, Pierre-Augustin de. [1743] 1746. "Mémoire sur différentes pétrifications tirées des animaux et des végétaux." *MARS*, 407–18.

——. [1746] 1751. "Mémoire contenant des observations de lithologie, pour servir à l'histoire naturelle du Languedoc, & à la théorie de la Terre." *MARS*, 713–58.

Schaffer, Simon. 1977. "Halley's Atheism and the End of the World." *NRRS* 32:17–40.

Scherz, Gustav, ed. 1958. *Nicolaus Steno and His Indice*. Copenhagen.

Scheuchzer, Johann Jakob. 1732–37. *Physique sacrée, ou histoire naturelle de la Bible*. Translated from the Latin. 8 vols. Amsterdam.

Schier, Donald S. 1941. *Louis Bertrand Castel, Anti-Newtonian Scientist*. Cedar Rapids.

Schnapper, Antoine 1900a. "The King of France as Collector in the Seventeenth Century." *Journal of Interdisciplinary History* 17:185–202.

——. 1986b. "Persistance des géants." *Annales. Economies. Sociétés* 41:177–200.

——. 1988. *Le géant, la licorne et la tulipe*. Paris.

Schneer, Cecil J. 1954. "The Rise of Historical Geology in the Seventeenth Century." *Isis* 45:256–68.

——, ed. 1969. *Toward a History of Geology*. Cambridge, Mass.

Scienze, credenze occulte, livelli di cultura. 1982. Florence.

Scilla, Agostino. 1670. *La Vana speculazione disingannata dal senso*. Naples.

Sédileau. 1693. "De l'origine des rivières." *MARS*, 81–93.

Seneuze, Laurent, ed. 1710. *Bibliotheca D. Joannis Galloys*. Paris.

Sgard, Jean, ed. 1976. *Dictionnaire des journalistes (1600–1789)*. Grenoble.

——. 1991. *Dictionnaire des journaux, 1600–1789*. 2 vols. Paris.

Shapin, Steven. 1987. "O Henry." *Isis* 78:417–24.

——. 1988. "Robert Boyle and Mathematics: Reality, Representation, and Experimental Practice." *Science in Context* 2:23–58.

Shapin, Steven, and Simon Schaffer. 1985. *Leviathan and the Air-Pump: Hobbes, Boyle, and the Experimental Life*. Princeton.

Shapiro, Barbara J. 1969. *John Wilkins, 1614–1672*. Berkeley.

——. 1983. *Probability and Certainty in Seventeenth-Century England*. Princeton.

Sherlock, Thomas. 1729. *The Tryal of the Witnesses of the Resurrection of Jesus*. 5th ed. London.

Shou-yi, Ch'ên. 1935. "John Webb: A Forgotten Page in the Early History of Sinology in Europe." *Chinese Social and Political Science Review* 19:295–330.

Simon, Renée. 1961. "Nicolas Fréret, académicien." *SVEC* 17.

Simon, Richard. 1685. *Histoire critique du Vieux Testament*. New ed. Rotterdam.

——. 1689. *A Critical History of the Text of the New Testament*. Translated from the French. London.

Simpson, George Gaylord. 1963. *This View of Life*. New York.

Skinner, Quentin. 1969. "Thomas Hobbes and the Nature of the Early Royal Society." *Historical Journal* 12:217–39.

Slaughter, M. M. 1982. *Universal Languages and Scientific Taxonomy in the Seventeenth Century*. Cambridge.

Sloane, Hans. [1727] 1729. "Mémoire sur les dents et autres ossemens de l'éléphant, trouvés dans terre." *MARS*, 305–34.

——. 1728. "An Account of Elephants Teeth and Bones found under Ground." *PT* 35:457–71, 497–514.

Smith, Adam. [1759] 1976. *The Theory of the Moral Sentiments*. Edited by D. D. Raphael and A. L. Macfie. In vol. 1 of the Glasgow Edition of the works and correspondence of Adam Smith. Oxford.

Solinas, Giovanni. 1965. "Illuminismo e storia naturale in Buffon." *Rivista critica di storia della filosofia* 20:267–312.

———, ed. 1973. *Saggi sull'Illuminismo*. Cagliari.

Solomon, Howard M. 1972. *Public Welfare, Science, and Propaganda in Seventeenth-Century France*. Princeton.

Sorbière, Samuel. 1660. *Lettres et discours*. Paris.

Souciet, Etienne-Augustin. 1729. "Dissertation sur les Coquillages que l'on trouve dans la Terre." *Trévoux* (February): 308–34; (March): 459–84; (April): 654–78.

Spanheim, Ezechiel. 1679. *Lettre à un amy*. Amsterdam.

Spener, C. M. 1710. "Disquisitio de Crocodilio in Lapide Scissili Expresso aliisque Lithozois." *Miscellanea Berolinensia* 1:99–118.

Spinoza, Benedict de (or Baruch). 1951–55. *Chief Works*. Translated and edited by R. H. M. Elwes. 2 vols. New York.

Sprat, Thomas. [1667] 1958. *History of the Royal Society*. Facsimile reprint, with critical apparatus by J. I. Cope and H. W. Jones. St. Louis.

Stark, Carl Bernhard. 1880. *Systematik und Geschichte der Archäologie der Kunst*. Leipzig.

Steinmann, Jean. 1960. *Richard Simon et les origines de l'exégèse biblique*. Paris.

Stengers, Jean. 1974. "Buffon et la Sorbonne." *Etudes sur le XVIIIe siècle* 1:97–127.

Steno, Nicolaus. [1669] 1916. *The Prodromus of Nicolaus Steno's Dissertation concerning a solid body enclosed by process of nature within a solid*. Translated and edited by J. G. Winter. New York.

———. 1671. *The Prodromus to a Dissertation Concerning Solids Naturally Contained within Solids*. Translated by Henry Oldenburg. London.

———. 1958. *The Earliest Geological Treatise (1667) by Nicolaus Steno*. Translated and edited by Axel Garboe. London.

———. 1969. *Geological Papers*. Translated by Alex J. Pollock. Edited by Gustav Scherz. Odense.

———. 1994. "Steno on Muscles." APS, *Transactions*, 84.

Sticker, Bernhard. 1967. "Leibniz' Beitrag zur Theorie der Erde." *Sudhoffs Archiv* 51:244–59.

———. 1971. "Leibniz et Bourguet: Quelques lettres inconnues sur la théorie de la terre." *Actes du XIIe Congrès international d'histoire des sciences* (Paris), 3, B, 143–47.

Stillingfleet, Edward. 1662. *Origines sacrae, or a Rational Account of the Grounds of Christian Faith*. London.

Stokes, Evelyn. 1969. "The Six Days and the Deluge: Some Ideas on Earth History in the Royal Society of London, 1660–1775." *Earth Science Journal* 3:13–39.

Storia della cultura veneta. Il Settecento. 1985–86. 5/I–II. Vicenza.

Stoye, John. 1994. *Marsigli's Europe, 1680–1730: The Life and Times of Luigi Ferdinando Marsigli, Soldier and Virtuoso*. New Haven.

Strachey, John. 1719. "A curious Description of the Strata observ'd in the Coal-Mines of Mendip in Somersetshire." *PT* 30:968–73.

———. 1725. "An Account of the Strata in Coal-Mines." *PT* 33:395–98.

Stroup, Alice. 1987. "Royal Funding of the Parisian Académie Royale des Sciences during the 1690s." APS, *Transactions*, 77.

———. 1990. *A Company of Scientists: Botany, Patronage, and Community at the Seventeenth-Century Parisian Royal Academy of Sciences*. Berkeley.

Stukeley, William. 1719. "An Account of the Impression of the almost Entire Sceleton of a large Animal." *PT* 30:963–98.

———. 1724. *Itinerarium curiosum. Or, an Account of the Antiquitys and remarkable Curiositys In Nature or Art, Observ'd in Travels thro' Great Brittan.* London.

Suppa, Silvio. 1971. *L'Accademia di Medinacoeli.* Naples.

Swedenborg, Emanuel. 1722. *Miscellanea observata.* Leipzig.

———. 1847. *Miscellaneous Observations.* Translated by C. E. Strutt. London.

Table générale des matiéres contenues dans le Journal des savans de l'édition de Paris, depuis l'année 1665 . . . jusqu'en 1750. 1753–54. 10 vols. Paris.

Targioni Tozzetti, Giovanni. 1751–52. *Relazioni d'alcuni viaggi.* 5 vols. Florence.

———. 1755. "Lettre . . . à M. de B***." *Journal étranger* (December): 228–35.

Taton, René, ed. 1964. *Enseignement et diffusion des sciences en France au XVIIIe siècle.* Paris.

Taylor, Kenneth L. 1974. "Natural Law in Eighteenth-Century Geology: The Case of Louis Bourguet." XIIIth International Congress of the History of Science, *Proceedings* (Moscow) 8:72–80.

———. "La Genèse d'un naturaliste: Desmarest, la lecture et la nature." In *De la géologie à son histoire,* edited by Gabriel Gohau et al. Forthcoming.

Taylor, Silas. 1730. *History and Antiquities of Harwich and Dovercourt.* London.

Tega, Walter, ed. 1986–87. *Anatomie accademiche.* 2 vols. Bologna.

Tentzel, Wilhelm Ernst. [1697?]. *Epistola de sceleto elephantino Tonnae nuper effosso.* 2d ed. Jena.

Thomas, Keith. 1971. *Religion and the Decline of Magic.* New York.

———. 1983. *Man and the Natural World.* New York.

Thorndike, Lynn. 1941–58. *A History of Magic and Experimental Science.* 8 vols. New York.

Tisserand, Roger. 1936. *Au temps de l'Encyclopédie: L'Académie de Dijon de 1740 à 1793.* Paris.

Tocanne, Bernard. 1978. *L'Idée de nature en France dans la seconde moitié du XVIIe siècle.* Paris.

Tolmer, Léon. 1949. *Pierre-Daniel Huet (1630–1721) humaniste-physicien.* Bayeux.

Torlais, Jean. 1936. *Réaumur.* Paris.

———. 1958. "Réaumur philosophe." *RHS* 11:13–33.

Torrini, Maurizio. 1977. *Tommaso Cornelio e la ricostruzione della scienza.* Naples.

———. 1981. "L'Accademia degli Investiganti, Napoli, 1633–70." *Quaderni storici* 16:845–83.

Toulmin, Stephen, and June Goodfield. 1965. *The Discovery of Time.* London.

Tournefort, Joseph Pitton de. [1702] 1704. "Description du Labirinthe de Candie." *MARS,* 217–34.

———. 1717. *Relation d'un voyage du Levant.* 3 vols. Lyons.

Tuan, Yi-Fu. 1968. *The Hydrologic Cycle and the Wisdom of God.* Toronto.

Tucci, Francesco Saverio. 1983 "'Il parlare della S. Scrittura e l' operare della natura': Gli interrogativi della geologia storica nella riflessione di Antonio Vallisnieri." *Contributi* 7:5–37.

Tuveson, Ernest Lee. 1950. "Swift and the World-Makers." *JHI* 11:54–74.

Tyssot de Patot, Simon. 1722. "Discours . . . au sujet de la Chronologie." *Journal littéraire,* 158–89.

Underwood, Edgar Ashworth, ed. 1953. *Science, Medicine, and History.* 2 vols. London.

(An) Universal History, from the Earliest Account of Time to the Present. 1740. Vol. 1. 2d ed. London.

Vaccari, Ezio. 1993. *Giovanni Arduino (1714–1795).* Florence.

Vallisneri (or Vallisnieri), Antonio. 1715. *Opere diverse.* 3 vols. in 1. Venice.

——. 1728. *De' corpi marini, che su' Monti si trovano.* 2d ed. Venice.

——. 1733. *Opere fisico-mediche.* Edited by Antonio Vallisneri, Jr. 3 vols. Venice.

——. 1965. *Carteggio inedito di Antonio Vallisneri con Giovanni Bianchi (Jano Planco).* Edited by Alessandro Simili. Turin.

——. 1991. *Epistolario. Vol. I: 1679–1710.* Edited by Dario Generali. Milan.

Van Kley, Edwin J. 1971. "Europe's Discovery of China and the Writing of World History." *American Historical Review* 76:358–85.

Van Leeuwen, Henry G. 1963. *The Problem of Certainty in English Thought, 1630–1690.* The Hague.

Varenius, Bernhard. 1681. *Geographia generalis.* Edited by Isaac Newton. 2d ed. Cambridge.

——. 1736. *A Compleat System of General Geography.* 3d ed. 2 vols. London.

Vaussard, Maurice. 1954. "Les Lettres inédites de Giovanni Lami à sa famille sur la France du XVIIIe siècle." *Revue des études italiennes,* n.s., 1:72–94.

Vernière, Paul. 1954. *Spinoza et la pensée française avant la Révolution.* 2 vols. Paris.

Vitaliano, Dorothy B. 1973. *Legends of the Earth: Their Geologic Origins.* Bloomington.

Voltaire. 1877–85. *Oeuvres complètes.* Edited by Louis Moland. 52 vols. Paris.

Voss, Jürgen. 1980. "Die Akademien als Organisationsträger der Wissenschaften im 18. Jahrhundert." *Historische Zeitschrift* 231:43–74.

Wade, Ira O. 1938. *The Clandestine Organization and Diffusion of Philosophic Ideas in France from 1700 to 1750.* Princeton.

Walker, Daniel P. 1972. *The Ancient Theology.* Ithaca.

Waller, R. D. 1937. "Lorenzo Magalotti in England, 1668–9." *Italian Studies* 1:49–66.

Wallis, John (d. 1703). 1701. "A Letter . . . Relating to that Isthmus, or Neck of Land, which is supposed to have joyned England and France in former Times." *PT* 22:967–79, 1030–38.

Wallis, John (d. 1793). 1769. *Natural History and Antiquities of Northumberland.* 2 vols. London.

Waquet, Françoise. 1982. "Antonio Magliabechi: Nouvelles interprétations, nouveaux problèmes." *NRL* (Naples), 173–88.

——. 1983. "De la lettre érudite au périodique savant: Les faux semblants d'une mutation intellectuelle." *XVIIe siècle* 35:347–59.

——. 1989a. *Le Modèle français et l'Italie savante. Conscience de soi et perception de l'autre dans la république des lettres (1660–1750).* Rome.

——. 1989b. "Qu'est-ce que la république des lettres? Essai de sémantique historique." *Bibliothèque de l'Ecole des Chartes* 147:473–502.

Waquet, Françoise, and Jean-Claude Waquet. 1979. "Presse et société: Le public des 'Novelle Letterarie' de Florence (1749–1769)." *Revue française d'histoire du livre,* n.s., 22:39–60.

Warburton, William. 1744. *Essai sur les hiéroglyphes des Egyptiens.* Translated by M.-A. Léonard de Malpeines. 2 vols. Paris.

Ward, John. 1740. *The Lives of the Professors of Gresham College.* London.

Warren, Erasmus. 1690. *Geologia.* London.

Wasserman, Earl R., ed. 1965. *Aspects of the Eighteenth Century.* Baltimore.

Wattles, Gurdon. 1989. "Buffon, d'Alembert and materialist atheism." *SVEC* 266: 285–341.

Webster, Charles. 1967. "Henry Power's Experimental Philosophy." *Ambix* 14:150–78.

———. 1982. *From Paracelsus to Newton: Magic and the Making of Modern Science.* Cambridge.

Whiston, William. 1696. *A New Theory of the Earth.* London.

———. 1698. *A Vindication of the New Theory of the Earth.* London.

———. 1702. *A Short view of the chronology of the Old Testament and of the harmony of the Four Evangelists.* Cambridge.

Wilkins, John. 1668. *An Essay towards a real character, and a philosophical language.* London.

———. 1969. *Of the Principles and Duties of Natural Religion.* First ed. 1675. Facsimile reprint of 1693 ed., with an introduction by H. G. Van Leeuwen. New York.

Williams, J. 1969. "An Edition of the Correspondence of John Aubrey with Anthony à Wood and Edward Lhuyd, 1667–1696." Ph.D. diss., University of London.

Winter, Eduard, ed. 1957. *Die Registres der Berliner Akademie der Wissenschaften 1746–1766.* Berlin.

———. 1962. *Lomonosov, Schlözer, Pallas.* Berlin.

Wodrow, Robert. 1937. *Early Letters of Robert Wodrow, 1698–1709.* Edited by L. W. Sharp. Edinburgh.

Wolfart, Peter. 1719. *Historiae naturalis Hassiae inferioris pars prima.* Cassel.

Wood, Paul B. 1980. "Methodology and Apologetics: Thomas Sprat's *History of the Royal Society.*" *BJHS* 13:1–26.

———. 1987. "Buffon's Reception in Scotland: The Aberdeen Connection." *Annals of Science* 44:169–90.

Woodbridge, John D. 1976. "Censure royale et censure épiscopale: Le conflit de 1702." *Dix-huitième siècle* 8:333–55.

Woodfin, Maude H. 1932. "William Byrd and the Royal Society." *Virginia Magazine of History and Biography* 40:23–34, 111–23.

Woodward, John. 1695. *An Essay toward a Natural History of the Earth.* London.

———. 1696. *Brief Instructions For Making Observations in all Parts of the World.* London.

———. 1726. *The Natural History of the Earth, Illustrated, Inlarged, and Defended.* Translated by Benjamin Holloway. 2 vols. in 1. London.

———. 1728. *Fossils of all kinds Digested into a Method.* London.

———. 1735. *Géographie physique, ou essay sur l'histoire naturelle de la terre.* Translated by Noguez. Paris.

Wurtzelbaur, P. 1687. "An Account of some Observations lately made at Nurenburg." *PT* 16:403–6.

Yardeni, Myriam. 1973. "Journalisme et histoire contemporaine à l'époque de Bayle." *History and Theory* 12:208–29.

Yourgrau, Wolfgang, and Allen D. Breck, eds. 1977. *Cosmology, History, and Theology.* New York.

Index

Footnotes are included in the index only when they contain substantial material not discussed in the text.

Academies, 19–39; Berlin, 14, 266–67; Bologna, 20, 30–31, 34–36, 39, 91, 167–68; Bordeaux, 31, 93, 130–31; Caen, 27–28; Dublin, 25; Florence, 24, 28, 33–34, 43–44, 60–61; Milan, 37; Montpellier, 15, 31, 93, 130–31, 266; Naples, 36–37, 63; Oxford, 25, 98, 205–6; Padua, 188; Rome, 24, 37; Rouen, 20n, 31, 190, 266. *See also* Academy of Inscriptions, Paris; Academy of Sciences, Paris; Royal Society of London

Academy of Inscriptions, Paris, 18–19, 51, 80–81, 92n

Academy of Sciences, Paris, 28–32; astronomy, 207–8; China, 77, 91–92; geology, 115–16, 129–30, 168, 201, 211–18, 226, 233–34; and journalism, 14–15; practices and philosophy, 34, 52, 60–62, 92–93; publications, 267; Sloane and, 24

Acta eruditorum, 9–13, 23, 25; geology, 144, 152, 169

Actualism, 95, 238, 240, 242, 246–47. *See also* Uniformity

Addison, Joseph, 69n

Alembert, Jean le Rond d', 260

Amontons, Guillaume, 185

Antiquarianism, antiquities. *See* Collections; History (human)

Arbuthnot, John, 164, 238

Arderon, William, 208

Arduino, Giovanni, 251, 258

Aristotle, Aristotelianism, 33–34, 41–43, 50–51, 125–27; landlocked seas, 186n; spontaneous generation, 106, 128; timescale, 198. *See also* Scholasticism

Arnauld, Antoine, 45, 56–57, 67

Astruc, Jean, 93–94, 131; shorelines, 98, 159, 219–20; timescale, 195

Aubrey, John, 206

Augustine (St.), 138, 143, 149

Aulisio, Domenico d', 148

Auzout, Adrien, 56–57

Bacchini, Benedetto: his journal, 13, 15, 16–17; review of Tentzel, 112; of Ramazzini, 146

Bacon, Francis, 28, 41, 53, 70; and France, 39; and Italy, 33, 35, 37

Baier, Johann Jacob, 133–34, 159–60, 165

Baillet, Adrien, 65

Baker, Henry, 118, 208
Balfour, Andrew, 107–8
Banister, John, 85, 125
Barrère, Pierre, 168, 217
Basnage de Beauval, Henri, 147–48
Bayle, Pierre, 9–13, 17, 54; on Burnet, 142; on history, 64, 65n, 67, 71–72; on LaPeyrère, 73; on mathematics, 50; on nature, 138
Beaumont, John, 110, 124, 144
Beccari, Jacopo Bartolomeo, 156
Becher, Johann Joachim, 176
Behrens, Georg Henning, 119n
Bellini, Lorenzo, 14n
Bentley, Richard, 148
Beringer, J. B. A., 133n
Bernoulli, Jakob (I), 51
Bernoulli, Johann (I), 188n, 189
Bianchi, Giovanni Battista, 39, 51n, 91, 152n
Bianchini, Francesco, 69n
Bibliothèque raisonnée, 111–12, 224–25
Bignon, Jean-Paul, 11n, 24
Birch, Thomas, 164
Blair, Patrick, 114
Blount, Charles, 147
Boccone, Paolo, 27–29, 101–2, 125–29, 203, 218
Boerhaave, Hermann, 63, 185–86
Bolingbroke, Henry St. John, viscount, 198
Bonnet, Charles, 256, 258–60
Borel, Pierre, 53, 54
Borelli, Giovanni Alfonso, 37, 59, 181, 246
Borromeo, Clelia, 37
Bossuet, Jacques-Bénigne, 64, 74–75
Bouguer, Pierre, 254
Bourdelin, Claude, 61
Bourdelot, Pierre-Michon, 27, 29, 110, 113, 128–29
Bourguet, Louis: Alpine valleys, 243–45, 259; correspondence, 89–90, 133; geological debates, 117, 157–59, 161, 166–67, 171, 220–21; Italy, 37–38; Paris Academy, 32, 180, 218; timescale, 195–96
Bovet, Richard, 58
Boyle, Robert, 25, 37, 109; Bible, 79; deduction, 47–48; Descartes, 44; experiment, 60; mechanical philosophy, 62–63; repute, 21–22, 24, 33, 35, 52; sea floor, 174, 179, 219, 222–23, 227, 244; subterranean heat, 184–86, 189, 246; witnesses, 56–57

Breyne, J. P., 117
Browallius, Johan, 232–33
Browne, Thomas, 147
Buffon, Georges-Louis Leclerc, comte de: on cosmogony and geology, 225–26, 243–47; on history, 247–49; on method, 51, 239–42; reception of, 171–72, 232–33, 250–62
Buonamico, G. F., 125–27
Buonanni, Filippo, 55, 59, 91, 127–28
Buragna, Carlo, 45
Burigny, Jean Lévesque de, 18
Burnet, Thomas: his theory, 5, 79, 140–42, 145, 149, 193, 207, 238; reception of, 142–49, 153, 200, 226
Bussing, Caspar, 147

Calogerà, Angelo, 17
Camden, William, 53, 69, 85–86, 88, 261
Campailla, Tommaso, 46
Camusat, Denis-François, 30
Cartesianism, 50, 62–63; cosmogony, 196, 209; geology, 183–85, 187, 238; Burnet and, 141, 145; Steno and, 99–100, 202. *See also* Descartes, René
Cartier, Pierre, 158
Cary, Robert, 78
Cassini, Jean-Dominique (or Giandomenico), called Cassini I, 29, 36, 37, 78n, 91–92
Castel, Louis-Bertrand, 98, 170, 217, 261
Catastrophes, 5, 175, 190, 193, 223. *See also* Revolutions
Catesby, Mark, 85
Celsius, Anders, 95, 227–28, 232–33
Chambers, Ephraim, 224, 260
Charas, Moïse, 61, 184–85
Charron, Pierre, 41
Chemistry, 4, 110, 138, 176, 250, 260; crystallization, 178, 193
Childrey, Joshua, 54, 85, 110–11
China, antiquity of, 76–78, 87–88, 89, 91–92, 191
Chronology, 45n, 76–78, 191. *See also* Timescales
Clayton, John, 124–25
Clément, Pierre, 253–54
Clerk, John, 155–56
Coastlines, 127, 168–69, 219–20; Mediterranean, 98, 131, 159
Collections, collecting, 53–55, 85–87, 90–91, 251
Colonna, Fabio, 201

Conti, Antonio: ancient history, 98–99, 197–98, 220; the Bible, 169; correspondence, 38, 217; Neoplatonism, 132, 169; the sea level, 226, 233–34
Cornelio, Tommaso, 37
Craig, John, 45
Creation (biblical), 78, 80–81, 138; measuring the Six Days, 72, 141, 193; expanding the Six Days, 193–94, 197, 223–24
Croft, Herbert, 143, 148
Cuper, Gisbert, 69

DaCosta, E. Mendez, 135
Dale, Samuel, 162
Daubenton, Louis-Jean-Marie, 250, 260
Delisle, Joseph-Nicolas, 32, 91–92
Denis, Jean-Baptiste, 27–29, 34, 128
Denyse, Jean, 19
Dereham, Thomas, 25, 37, 88n
Derham, William, 111, 155, 170, 260
Descartes, René: cosmogony, 138; history, 64–65; method, 41, 43–44, 53n; critiques of, 21, 148, 200. *See also* Cartesianism
Des Maizeaux, Pierre, 8, 14
Desmarest, Nicolas, 199, 261
Deyron, Jacques, 88
Dezallier d'Argenville, Antoine-Joseph, 135, 171, 180, 191, 218, 225
Di Capua, Leonardo, 37
Diderot, Denis, 51, 260
Diluvial geology: defined, 136, 139; advocated, 118, 135, 140–42, 149–60; criticized, 140, 146, 155, 160–69, 212, 214–16, 219, 221, 240–41; debated, 169–72. *See also* Divine intervention; Elephant fossils; Flood
Divine freedom, 47–48, 115, 149
Divine intervention, 59, 71, 74, 82, 137–38. *See also* Flood
Divine providence, 64, 138n, 140–41, 149–50, 233
Divine purpose, 122, 127–28, 132, 148, 229; Buffon and, 242, 254–55
Donati, Vitaliano, 251, 258n
Dryden, John, 45–46
DuHamel, Jean-Baptiste, 29

Earthquakes, 71, 182–83, 204–8, 225–26. *See also* Hooke, Robert
Eckhart, Johann Georg, 211
Ehrhart, Balthasar, 23, 225

Elephant fossils, 112–18, 135, 208, 210
Epistemology. *See* Knowledge
Esper, Johann Friedrich, 117
Eternalism. *See* Timescales
Europe savante, 10, 14, 17–19
Evelyn, John, 142
Experiments, 59–63; in geology, 109–10, 138, 179, 182–85, 192–93, 210; value of, 95, 175, 176–78

Fairfax, Nathaniel, 54, 56
Fire. *See* Heat; Volcanoes
Flood (biblical), 72, 81, 172; local or universal, 72–73, 139–40, 163, 170; miraculous or natural, 140–41, 143–44, 149–50 (Burnet); 152–54, 156–57, 162–64, 167 (Woodward); 169–72 (the Jesuits); 219–21 (Vallisneri); 224 (Moro); 227 (Celsius); 240–42, 254 (Buffon). *See also* Diluvial geology
Floods, flooding, 98, 111, 126–28; John Ray on, 149, 161, 165
Foley, Samuel, 45
Fontenelle, Bernard le Bovier de: secretary of Paris Academy, 35, 211–12; on experiment, 62, 176; on geology, 32, 115–16, 129–30, 168, 180, 196, 211–18; on history, 59, 64, 92, 103–4; on mathematics, 44–45, 50–51; on myths, 70, 71, 80, 205; on nature (as historian), 95, 261; on timescales, 198–99, 239; his prose and readership, 18, 30–31, 38–39, 58, 131, 211, 218n. *See also* Academy of Sciences, Paris; Myths
Fossil populations, 122, 126, 166, 222, 236–37
Fossils: freshwater, 111n, 209, 213–14; marine, 105–8, 119–35; terrestrial, 96, 108–12, 112–19. *See also* Elephant fossils; *Glossopetrae*; Plastic powers; Seed theory
Fréret, Nicolas, 51, 82
Fréron, Elie, 253
Frisi, Paolo, 233n

Gale, Thomas, 144
Galilei, Galileo: mathematics, 43–44; the telescope, 41, 60; condemnation of, 220; repute of, 32–33, 35, 38
Gallois, Jean, 13, 28, 84n
Gassendi, Pierre, 26, 33, 35, 37, 41, 50
Gaubil, Antoine, 92
Gautier, Henri, 168
Generelli, G. C., 225

Geoffroy, Etienne-François, 15
Gibson, Edmund, 69
Gimma, Giacinto, 221n
Glanvill, Joseph, 58–59, 61
Gleditsch, Johann Gottlieb, 111
Glossopetrae, 99, 120, 126, 128–29, 203.
  See also Fossils: marine
Grandi, Jacopo, 107, 146, 203
Graverol, Jean, 147
Grew, Nehemiah, 110
Grimm, Friedrich Melchior, 253
Guettard, Jean-Etienne, 111
Guglielmini, Domenico, 36, 39, 48, 197n

Hale, Matthew, 79, 81, 148
Haller, Albrecht von, 251, 254, 256,
  258–59
Halley, Edmond, 29, 87, 188, 207–8;
  timescales, 197, 246
Harris, John, 152–53, 160, 163
Hartsoeker, Nicolaas, 165, 232
Hauksbee, Francis, 155, 162
Heat (subterranean), 180; non-volcanic,
  183–87, 189, 209, 242, 247; volcanic,
  181–83, 246
Henckel, Johann Friedrich, 156, 159,
  176n, 178, 193
Hevelius, Johannes, 21, 56–57
Hill, John, 84
History (of the earth): biblical
  interpretations, 136–38; syntheses,
  103–4, 233–39. *See also* Diluvial geology;
  Sea; Sedimentation; Timescales;
  Witnesses
History (human), 103–4; ancient times,
  70–71, 80–81; the Bible, 70–82;
  narratives, 66, 84, 172; scholarship,
  18–19, 66–70, 80, 82; scientists and,
  83–94. *See also* Knowledge
History of the Works of the Learned, 9, 10,
  14–15
Hobbes, Thomas, 27, 45, 46, 147
Holbach, Paul-Henri Thiry, baron d', 226,
  260
Holloway, Benjamin, 152, 237
Homberg, Wilhelm (or Guillaume),
  184–85
Honoré de Sainte-Marie, B.-V., 19, 76n
Hooke, Robert, 13, 24–25; ancient
  history, 87; collecting, 53; Design, 127,
  132; diluvialism, 155, 160; earthquake
  theory and its reception, 204–8, 233;
  experiments (geological), 176–77, 182,
  192; fossils, 106, 108–9, 117–18,
  119–21; knowledge, 46; microscopes,

60; myths, 71, 102–3; timescale, 192;
  witnesses (geological), 96–98, 239. *See
  also* Land, elevation of
Horn, Caspar, 29
Huet, Pierre-Daniel, 45, 50, 71, 76, 148
Hume, David, 79
Hutton, James, 136, 174, 180
Huygens, Christiaan, 27, 28, 48, 62–63,
  187

Jallabert, Jean, 166
Jansenism, 252, 254
Journal des savans, 7–19, 23, 25, 27–29;
  on Becher, 176; on Buffon, 252–53; on
  Burnet, 145–47; on diluvialism, 160,
  169; on fossils, 93, 111, 112, 130–31,
  210; on monuments, 69n; on Newton,
  47; on Spinoza, 74
Journals, 7–19. *See also* entries for specific
  titles, such as the *Acta eruditorum*, and
  for editors, such as Bacchini, Bayle,
  LeClerc, Nazari
Jüngken, Johann Helfrich, 45, 102n
Jussieu, Antoine de: on geological time,
  197–98, 217; on geology, 32, 98,
  117–18, 215–16, 239; and Buffon, 244;
  and Maillet, 229
Justel, Henri, 25, 27, 28, 87

Keill, John, 143, 144, 149
King, Charles (pseud. for Henry
  Rowlands, q.v.), 124–25, 165
Kircher, Athanasius, 22, 72; on China,
  77–78; his museum, 55, 90–91; on
  springs, 186–87
Knowledge: causal explanation, 47–49,
  52, 58–63; deduction in geology,
  99–102, 142, 239–42; deduction/
  mathematics, 42–52, 62–63; historical
  testimony, 56, 64–82; human testimony,
  56–60, 92; the senses, 21, 34, 51–63,
  150; verification, 22, 28, 34, 52,
  54–55, 59–60. *See also* Experiments;
  Witnesses

La Condamine, Charles-Marie de, 225
Lafitau, Joseph-François, 87n
La Hire, Gabriel-Philippe de (La Hire
  fils), 132, 212
La Hire, Philippe de (La Hire *père*), 109,
  132, 188, 212
Lamy, Bernard, 148n, 170
Land: elevation of, 175, 200, 227, 245;
  diluvialism and, 141, 151; earthquakes,
  volcanoes, and, 204, 207–8, 221–26;

Land (*cont.*)
  Steno on, 100, 121
Lang (Langius), Karl Nikolaus, 23, 130
LaPeyrère, Isaac de, 73, 104, 139, 231
LeCat, Claude-Nicholas, 20n, 115, 253n
LeClerc, Jean, 75–76, 78, 147; his journals, 10, 12, 14, 25
Leeuwenhoek, Antoni van, 23, 28, 111, 192
Lehmann, Johann Gottlob, 172
Leibniz, Gottfried Wilhelm: actualism, 238; experiment, 62, 95, 177–78; fossils, 114, 117, 237; history, 69–70, 89; journalism, 9; knowledge, 45–46, 51–52; timescale, 78, 195–96; his theory and its reception, 208–15, 219, 237, 241–42; witnesses, 96, 118
Leigh, Charles, 154
LeMascrier, Jean-Baptiste, 228–29
Lemery, Nicolas, 26–27, 182–83, 185, 246
Leopold, Johann Friedrich, 159
Lhwyd, Edward: ancient history, 85–87; diluvialism, 165; fossils, 106, 110, 121, 124, 131; timescale, 193–94; on Burnet, 143–44; on Woodward, 157, 161–62. *See also* Seed theory
Lignac, J.-A. Lelarge de, 254–55
Linnaeus, Carolus, 55, 94, 228
Lister, Martin, 55, 87, 106, 121–22, 236–37
Locke, John, 45, 67, 88, 177; on miracles, 82, 144
Logan, James, 194–95
Louville, Jacques-Eugène d'Allonville, chevalier de, 91
Lovell, Archibald, 143
Lusus naturae. See Fossils; Neoplatonism; Plastic powers
Lyell, Charles, 180

Mabillon, Jean, 16–17, 67–68, 79, 82, 139–40
Maffei, Scipione, 17, 91, 169, 225
Magalotti, Lorenzo, 34, 38, 58, 145, 190
Magliabechi, Antonio, 8, 23, 112
Maillet, Benoît de, 58, 201, 239n; his theory and its reception, 171, 228–32
Mairan, Jean-Jacques Dortous de, 185–86, 189, 246
Malebranche, Nicolas, 50, 65, 130, 132, 145
Malpighi, Marcello, 17, 23, 25, 34–36, 59; on Burnet, 142, 146; on fossils, 203; on spontaneous generation, 128

Mammoths. *See* Elephant fossils
Manfredi, Eustachio, 35, 232–33
Maraldi, Giacomo Filippo, 108, 129
Marchant, Nicolas, 52
Marine movements. *See* Sea
Mariotte, Edme, 29, 102, 187–88; on knowledge, 47, 48–50, 57
Marsili (or Marsigli), Anton Felice, 15, 35, 91
Marsili, Luigi Ferdinando: and academies, 31, 35, 36, 38, 91, 213; on the Danube Basin, 91; on diluvialism, 165–66, 170; on fossils, 116
Martini, Martino, 77–78
Mathematics. *See* Knowledge
Mather, Cotton, 115
Mather, Increase, 58
Mayow, John, 182, 187
Mechanical philosophy, 33, 35, 62, 63, 107, 135n. *See also* Cartesianism; Descartes; Gassendi
Medici, Leopoldo de', 44
Melle, Jacob von, 159
Mémoires de Trévoux, 10–15, 17–18; diluvialism, 169–72; geology, 39, 93, 138, 148, 210, 252
Mersenne, Marin, 7, 26, 43, 63
Misson, François-Maximilien, 108, 125
Molyneux, Thomas, 96, 115, 117, 118, 191
Monconys, Balthasar de, 21
Montaigne, Michel de, 41
Montesquieu, Charles-Louis de Secondat, baron de, 37, 43, 93
Monti, Giuseppe, 116, 135, 156, 167
Montmor, Henri-Louis Habert de, 27, 33, 50, 61
Monuments. *See* History (human): scholarship
More, Henry, 57, 107. *See also* Neoplatonism
Morgagni, Giovanni Battista, 35, 84
Moro, Anton Lazzaro: his theory, 5, 169, 195, 221–24, 237; its reception, 224–26, 233
Morton, John, 111–12, 133, 134, 154–55
Moscardo, Lodovico, 55
Mountains. *See* Land, elevation of
Muratori, Ludovico Antonio, 17, 38, 90, 220
Museums. *See* Collections
Myle, Abraham van der, 139, 170
Myths, 70–71, 81; evaluation of, 98–99, 212–13, 249; in geology, 97, 101–3, 205–6, 226, 231–32

Nazari, Francesco, 14, 25, 107, 113; on
  Steno, 101–2, 202–3
Neoplatonism, 107, 130, 132, 154, 169.
  See also More, Henry; Plastic powers
Newton, Isaac, and Newtonianism: biblical
  exegesis, 79, 138, 142–43; cosmogony,
  196, 238; divine intervention, 242;
  mathematics, 47, 50; optics, 46, 60;
  rules of reasoning, 62, 221–22
Nicholls, William, 148
Nouvelles ecclésiastiques. See Jansenism
Novelle letterarie, 224

Oldenburg, Henry, 8–9, 14, 21–25,
  27–28; on Plot, 85; on Spinoza, 74; on
  Steno, 101–2, 203; on verification, 56.
  See also Philosophical Transactions and
  Royal Society of London
Ovid. *See* Myths

P., L., 162–63, 238
Paragallo, Gaspare, 88, 181
Pascal, Blaise, 50, 54
Patin, Charles, 55, 68–69, 102
Pecquet, Jean, 27
Pepys, Samuel, 142
Perrault, Charles, 29, 34, 48, 60, 62, 92
Perrault, Pierre, 62, 184, 187
Petty, William, 51, 144
Pezron, Paul-Yves, 78, 87
Philosophical Transactions, 8, 13, 14, 21–25;
  on Burnet, 152; on China, 87–88; on
  fossils, 115. *See also* Oldenburg, Henry;
  Royal Society of London
Plastic powers, 106–7, 109–10, 119–22,
  125–28; rejected, 130–35, 219. *See also*
  More, Henry; Neoplatonism
Pliny, 53, 77n, 109
Plot, Robert, 85–87, 89; Design, 127;
  disciples, 125; flood, 139; fossils, 106,
  109, 113–14, 121–22, 131, 236;
  springs, 188; witnesses, 94, 206, 249
Pluche, Noël-Antoine, 138, 250–51
Porcia, Giovanni Artico, conte di, 42, 90,
  220
Port–Royal. *See* Arnauld, Antoine
Pouilly, Louis-Jean Lévesque de, 18, 19,
  80–81
Power, Henry, 61
Primitive rocks, 224–25, 237–38; as old as
  Creation, 165, 213; chemistry and, 250;
  igneous origins, 209, 242–43
Pryme, Abraham de la, 118, 207n

Quirini, Giovanni, 107

Ramazzini, Bernardino, 143, 145, 146,
  188
Ray, John, 36–37, 53, 86; on Burnet, 144;
  on Design, 132; earthquakes, 207, 226,
  246; the flood and flooding, 72, 124,
  139, 149–50, 161–62, 164–65; fossils,
  111, 121–24, 236–37; timescale, 194
Réaumur, René-Antoine Ferchault de, 32,
  92–93, 95, 135, 166, 180; Buffon,
  254–56; faluns of Touraine, 168,
  179–80, 215–17, 226, 239; timescale,
  193
Redi, Francesco, 34, 59, 61, 106–7, 128
Renaudot, Théophraste, 26
Revolutions (in geology), 214–16, 218n,
  258. *See also* Catastrophes
Rivière, Guillaume, 159
Roberval, Gilles Personne de, 48, 49–50,
  84
Robinson, Tancred, 153
Robinson, Thomas, 163
Rohault, Jacques, 26–28; geology, 102,
  184, 187; knowledge, 48, 63; sensory
  data, 52–54
Rowlands, Henry, 69n. *See also* King,
  Charles
Royal Society of London, 20–26, 33–35,
  37, 56, 61; on Burnet, 144; on
  diluvialism, 155; on fossils, 112, 114;
  on Hooke, 98; on Vallisneri, 221;
  publications of, 267
Rudbeck, Olaus (or Olof), 97n

Sack, A. F. W., 159–60
St. Clair, Robert, 143
Sallo, Denis de, 7–10, 13, 16
Saulmon, 213
Sauvages de la Croix, Pierre-Augustin de,
  218n
Scheuchzer, Johann (the younger),
  37–38, 156, 212
Scheuchzer, Johann Jakob (the elder), 14,
  37–38, 89–90, 166–67, 212–14; fossils,
  130–32, 151; timescale, 194, 196–97;
  Woodward, 25, 151, 156–57, 170
Scholasticism, 20, 25, 35, 42, 65. *See also*
  Aristotle, Aristotelianism
Scientific societies. *See* Academies
Scilla, Agostino, 91, 125–27, 140, 218
Scripture. *See* Chronology; Creation;
  Diluvial geology; Flood; History
Sea: bottom currents, 230, 244–45, 257;
  currents, 117–18, 166–67, 173–74;
  diminution, 200–1, 226–30, 257; level
  (variations of), 106, 215, 219–20,

Sea (*cont.*)
    222–23, 232–34. *See also* Boyle, Robert:
    sea floor
Sedimentation, 179–80, 192, 219–20,
    243–46. *See also* Diluvial geology:
    criticized
Seed theory, 106, 124, 125, 129, 212;
    rejected, 130–31
Shorelines. *See* Coastlines
Sibbald, Robert, 69
Simon, Richard, 11n, 65, 74–76, 79, 195
Skippon, Philip, 36–37
Sloane, Hans, 24, 86, 114–16, 217
Sluse (Slusius), René-François de, 107
Society of Jesus. *See* China; *Mémoires de
    Trévoux*
Somner, William, 96–97
Sorbière, Samuel, 27
Sorbonne, 252–53
Souciet, Etienne-Augustin, 171
Spada, Giacomo, 221
Spanheim, Ezechiel, 75, 102
Spener, C. M., 210
Spinoza, Baruch (or Benedict), and
    Spinozism, 59, 148; biblical scholarship,
    71, 73–76, 79; deduction, 45, 46;
    experiment, 62
Spontaneous generation, 34, 106–7, 128
Sprat, Thomas, 21, 87, 94
Springs, origin of, 11, 168, 186–89
Steno, Nicolaus: anatomy, 22, 27, 46;
    Design, 132; experiment (geological),
    176; the flood, 107, 140; fossils, 106,
    108–9, 113, 119–21; heat, 181;
    timescale, 191; his *Prodromus* and its
    reception, 99–102, 164, 201–4
Stillingfleet, Edward, 75–76, 80, 139
Stonehenge, 52, 97, 102, 120
Story, Thomas, 194–95
Strachey, John, 236–37
Stukeley, William, 159, 193
Swedenborg, Emanuel, 96, 192, 226–28

Targioni Tozzetti, Giovanni, 251, 256–58
Teleology. *See* Divine purpose
Tentzel, Wilhelm Ernst, 112–14, 116, 118.
    See also Elephant fossils
Thévenot, Melchisédech, 27, 33
Timescales, 175, 191; eternalism, 73, 148,
    190, 197, 217, 229–32; geology
    (lengthened time), 166–67, 194–99,
    217, 220, 224, 228, 246–47, 262;
    geology (short time), 191–94, 195, 202,
    254; human history, 73, 76–78, 191,

230–32; natural chronometers, 196–97
Torricelli, Evangelista, 37, 43–44, 52, 62
Tournefort, Joseph Pitton de: on fossils,
    106, 129–30, 131, 212; on myths, 103;
    on witnesses, 98, 249

Uniformity (laws), 95, 162, 164, 167;
    Buffon on, 238–42, 246; Hooke on,
    205–6. *See also* Actualism

Valletta, Giuseppe, 8, 15, 37
Valleys, 141, 174, 243–45, 256
Vallisneri, Antonio: antiquarianism, 90;
    correspondence, 37–38, 90, 98, 220;
    diluvialism, 166–69, 171; fossils, 157,
    236, 237; experiment, 95, 176;
    journalism, 11, 13, 14, 17;
    Neoplatonism, 132; springs, 188–89;
    timescale, 195–97; his treatise and its
    reception, 23, 90, 217–22, 224
Vallisneri, Antonio, Jr., 251
Varenius, Bernhard, 182, 244; his list of
    strata, 164–65, 173–74, 179, 194, 245
Vernon, Francis, 25
Vico, Giambattista, 36n, 42n, 51n
Viviani, Vincenzo, 84, 92
Volcanoes, 181–83, 221–26, 230, 246. *See
    also* Heat
Voltaire, François-Marie Arouet, called,
    121, 253
Vossius, Isaac, 22, 139

Wagner, Christian, 144–45
Wallis, John (1616–1703), 56–57, 97–98,
    122, 206, 261
Wallis, John (1714–93), 85n
Warren, Erasmus, 143
Webb, John, 77
Werner, Abraham Gottlob, 136
Whiston, William, 45, 78, 153–54, 207, 241
Wilkins, John, 45, 72, 79, 190
Witnesses: in geology, 95–103, 116–18,
    204–8, 209–10, 216–17, 223, 227,
    238–39, 248–49, 260–61; in history,
    64–65, 66–67; in law, 57; in the
    sciences, 56–58, 60
Wodrow, Robert, 115, 160–61, 162
Woodward, John, 24, 53, 86; on the
    Bible, 81, 150, 152; his theory and its
    reception, 136, 139, 140, 150–69, 171,
    173, 240–41; his influence, 200, 203–4,
    235–36. *See also* Diluvial geology

Zanotti, Francesco Maria, 31, 35–36, 58